Empirical Ecocriticism

Empirical Ecocriticism

Environmental Narratives for Social Change

Matthew Schneider-Mayerson, Alexa Weik von Mossner,
W. P. Malecki, and Frank Hakemulder, Editors

UNIVERSITY OF MINNESOTA PRESS
MINNEAPOLIS • LONDON

Portions of chapter 4 are adapted from "Environmental Literature as Persuasion: An Experimental Test of the Effects of Reading Climate Fiction" by Matthew Schneider-Mayerson, Abel Gustafson, Anthony Leiserowitz, Matthew H. Goldberg, Seth A. Rosenthal, and Matthew Ballew, *Environmental Communication* 17, no. 1: 35–50, https://doi.org/10.1080/17524032.2020.1814377.

Published by the University of Minnesota Press
111 Third Avenue South, Suite 290
Minneapolis, MN 55401-2520
http://www.upress.umn.edu

ISBN 978-1-5179-1534-6 (hc)
ISBN 978-1-5179-1535-3 (pb)

A Cataloging-in-Publication record for this book is available from the Library of Congress.

Printed in the United States of America on acid-free paper

32 31 30 29 28 27 26 25 24 23   10 9 8 7 6 5 4 3 2 1

Contents

Introduction

Toward an Integrated Approach to Environmental Narratives and Social Change

MATTHEW SCHNEIDER-MAYERSON, ALEXA WEIK VON MOSSNER, W. P. MALECKI, AND FRANK HAKEMULDER

Knowing that you need to tell a new story does not always mean you know what to say, or how to say it. This is, in some ways, the situation we find ourselves in today. Most environmentally engaged scholars, thinkers, and activists agree that to respond to the existential challenges we currently face, we need new narratives about who we are, how we're entangled with the rest of the natural world, and how we might think, feel, and act to preserve a stable biosphere and a livable future with as much justice as possible. But what kinds of stories should we tell? To which audiences? Through what media? Are some stories more impactful than others? Are some counterproductive? And how can scholars of literature, theater, art, digital media, film, television, and other cultural forms contribute to, expedite, or shape the historic socioecological transformation that is now underway?

In this introduction, we argue that to aid the planning, development, and execution of effective, justice-oriented strategies for cultural production and communication, environmentally engaged scholars ought to attempt to integrate insights, data, experiences, and hypotheses from both the humanities and the social sciences and ground their theories in available empirical scholarship. What is needed at this moment is a holistic, interdisciplinary, data-driven approach to environmental narrative, which might aid and inform cultural production and communication. What is needed, in short, is an empirical form of ecocriticism.

Empirical Ecocriticism is an invitation to this new area of research. It is at once a manifesto, a toolkit, a proof of concept, and a conversation. It familiarizes readers with some of the methods used by empirical ecocritics, demonstrates their application in concrete case studies, and provides critical reflections on the value, challenges, and potential of studying the reception of environmental storytelling. In a world that is experiencing regular, unprecedented, and escalating socioecological catastrophes, with the possibility of ecological and sociopolitical collapse on the horizon, the development of an empirical form of ecocriticism, synergistically combining the methods of and knowledge from the humanities and social sciences, is not just potentially fruitful. It is necessary.

Storytelling in an Age of Accelerating Crises

It now widely understood and acknowledged that we are living through an extraordinary time of accelerating socioecological crises. The primary and most ubiquitous manifestation is anthropogenic climate change. Readers of this book likely do not need a reminder of the scale of the climate crisis, but in 2020, California experienced its first gigafire, with over four million acres burning in two months; 2021 had major wildfires ravaging Southern Europe; and 2022 saw a recurring heat wave in Siberia, with largely uncontrolled blazes destroying remote forests. The Atlantic hurricane season gets more and more destructive, with economic costs of $60 billion per year. The consequences of climate change are now felt all over the world, with unprecedented heat waves from Australia to Europe making daily life unbearable and causing premature deaths. The chances of fatal floodings increase every year, from Nigeria to Pakistan. The psychological costs are incalculable. All of this death, violence, suffering, and trauma is distributed unequally—borne first and most by the people (and nonhuman animals) who bear the least responsibility.

While climate change is the most grave, urgent, and permanent of the challenges we face, it is accompanied by a broader environmental crisis, characterized by deforestation, ocean acidification, desertification, defaunation, species extinction, air pollution, and plastic pollution. Of course the last few years will be remembered as the time of the coronavirus pandemic. While it is not yet clear whether land use and climatic changes directly contributed to the spread of Covid-19 to humans, scientists note that the emergence of zoonotic viruses are linked to deforestation and other stressors on wildlife (Tollefson 2020), and they expect more viruses to emerge as the climate changes (Ryan et al. 2019). By the time you read these words, there will surely be more unprecedented symptoms of these overlapping and cascading emergencies. As their manifestations become more obvious and undeniable, awareness and concern are also growing. Finally, belatedly, we are witnessing a public recognition of the incredible gravity, urgency, and existential stakes of ongoing socioecological crises.

This recognition does not always include an acknowledgment that culture, values, and stories are at the heart of the problem (Hulme 2009). This is partially because of the modern tendency to view environmental issues as problems that are approached and ultimately resolved through science, technology, and policy. All three are important and necessary, but we often forget what they are, and what they are capable of doing. Science helps us understand the world and develop projections for different future trajectories, while technology and policy are tools to shape the world. But a tool can be used in many different ways. A hammer can be used to pound a nail into a board—the first step in constructing a sustainable modular house for climate refugees, let's say. But it can also be used to smash a window or bludgeon an animal to death. That's what our technology and policy have often been doing, except the windows we're breaking are part of our only home, the teeming biosphere of planet Earth, and the animal represents the growing number of birds, fish, reptiles, insects, and other creatures that are disappearing as a result of human activity.

What controls the planetary hammers we hold in our hands? The direction of the world we're shaping is being determined not by science, technology, and policy but by the desires, values, and priorities of those who wield these world-shaping tools, as well as the systems in which they're embedded. These desires, values, and priorities are in turn shaped by the environmental (or anti-environmental) attitudes, affects, beliefs, and behaviors of average people, as well as economic and political elites. And while economic and political systems operate according to their own logic, inertia, and path dependencies, they generally require the participation, consent, or, at the very least, quiescence of individuals and communities. As such, the attitudes, affects, beliefs, and values of average people—primarily those in high-consumption, rich, and geopolitically powerful countries—are a critical site of struggle in responding to the monumental socioecological challenges of the present moment.

Mediated Crises and the Importance of Narratives

One of the primary weapons in this struggle over desires, values, and priorities has been expository messages relying on statistical data and arguments. This strategy has engendered a mountain of admirably detailed and impeccably researched scientific reports, political manifestos, and educational pamphlets. It has also yielded stunningly meager results. A troubling number of people still deny that there is a global environmental crisis, while others admit its existence but do not believe it merits any significant changes in the global economy or their own lives. This is despite having been bombarded for quite a long time with expository messages explicitly stating that such beliefs are not only wrong but will lead to catastrophe. Why are such messages not more effective?

As with many seemingly irrational but common behaviors—such as people maintaining habits they know to be deadly, or failing to follow medical advice that is essential to their survival—the answer seems to lie in the nature of the human mind (Ariely 2008; Martin et al. 2018). The causes of our resistance to expository, fact-based environmental messages have been identified for some time now by scholars in fields such as environmental psychology and environmental communication. Among the primary culprits are disattention, incomprehension, and negative cognitive responding (Marshall 2014; Mercier 2016). Exploring these obstacles helps us understand why expository messages have disappointed, and why narratives—and empirical ecocriticism—hold such promise.

Let's take each of these in turn. First, because many people do not consider environmental issues to be relevant to their daily lives, they are simply not paying attention to environmental messages (Marshall 2014). Second, even if they are encouraged or forced to pay attention, they have a hard time translating those messages into immediately comprehensible terms because humans have a hard time understanding anything that is not directly accessible to everyday perception, including the macroscale processes of climate change

(Marshall 2014; Slovic 1998). Third, even if people comprehend the planetary processes to which environmental messages frequently refer, those who are more skeptical about the existence, anthropogenic nature, or gravity of these problems tend to engage in negative cognitive responding. They search for weaknesses in scientific reports and for corroboration in messages to the contrary; as a result, they give too much weight to the dubious claim that environmental issues are not major problems (Marshall 2014; Mercier 2016).

These tendencies have led many activists, science and environmental communicators, journalists, and strategists to turn to other forms of communication, and narratives have been identified as a promising possibility. Hyperbolic headlines such as "Stories to Save the World" (Armitstead 2021) and "Can Books Save the Planet?" (Ullrich 2015) epitomize the widespread hope in the environmental potential of stories, which is premised on their capacity to overcome the obstacles that limit the efficacy of expository messages. Narratives are well equipped to address the problem of disattention because a strong narrative frame can make any topic interesting and engaging, no matter how unimportant or dull we might otherwise find it (Malecki et al. 2019a). Stories can also help tackle the problem of incomprehension because humans, as "storytelling animals" (Gottschall 2013), inevitably use narratives to make sense of the world and their own lives (Green et al. 2002). This has been demonstrated by empirical research and has also become the common wisdom of the marketing, public relations, and political consulting industries, which typically rely on storytelling to move products and elect candidates (Salmon 2010). Through their emotional engagement and absorbing power—their transportation effects—narratives can make us less susceptible to internal doubts, and thereby make us more willing to seriously consider ideas and values that might have previously seemed dubious or objectionable (Green and Brock 2000; Nabi and Green 2015). There is a rising hope, then, that environmental narratives can be an important complement to statistics and factual arguments.

Beyond the theoretical, there is the basic fact that in many places today, environmental (or antienvironmental) attitudes, affects, beliefs, and values are inevitably influenced by culture and media because that is how we frequently interact with and learn about the world. They are inculcated first when we are young, by lullabies, cartoons, children's books, school textbooks, films, and advertisements. As we mature, they are inevitably shaped by the unique combination of media that each of us consumes and are exposed to throughout our lives, including but not limited to pop music, photographs, films, documentaries, shareable videos, GIFs, memes, TV and streaming shows, advertisements, radio programs, podcasts, plays, short stories, sermons, prayers, poems, novels, graphic novels, fine art, and video games, as well as Twitter and Instagram posts and TikTok videos. In the United States in 2020, the average person consumed 5.7 hours of traditional (linear) media every day. Another 7.5 hours was spent with digital media, some of which is likely to be shareable videos and other narrative content (Dolliver 2020). In Germany, the combined total was 10.3 hours (Enberg 2020); in the United Kingdom, 9.0 (Fisher 2020); in Japan, 7.5 (Cramer-Flood 2020b); in the world's most populous countries, China and India, the totals were 7.0 and 6.0 hours, and growing quickly (Cheung 2020; Cramer-Flood 2020a). In this context, it is undeniable that media constitute a key site of intervention.

Given the ubiquity of media and the proven potential of storytelling, there are good reasons to be hopeful about the persuasive power of environmental narratives. But for that hope to transform into something sufficiently solid to serve as a basis for effective communication strategies, we need to study the impact of such narratives directly and empirically. This is what empirical ecocriticism sets out to do.

Beyond producing new knowledge, our goal is to contribute to the social change that is needed to respond to the web of environmental, social, and political crises that the world finds itself enmeshed in today. At this point in time, nudges in consumer behavior and tweaks in policy will not address the scale of problems such as the climate crisis, widespread plastic pollution, mass extinction, a growing migration crisis, gaping inequality, and metastasizing fascism. While most of the case studies in this book focus on microlevel social change, or shifts in attitudes, beliefs, and behavior, these constitute a necessary foundation for macrolevel social change such as transformations of industrial, legal, and economic systems (Harper and Leicht 2018). Indeed, as numerous contributors note, small and temporary changes from a single cultural text mean we can expect more significant and lasting transformations from a torrent of texts, which is exactly what this moment demands. Given this desire for social change, we are interested not only in whether cultural texts foster more awareness of injustice and looming catastrophe, but whether they are more or less likely to lead to enduring changes in consciousness, behavior, and political engagement. In this way, we hope that research in empirical ecocriticism might be of some use to communicators and storytellers. We see ourselves working alongside tens of millions of authors, artists, filmmakers, and creative workers as well as activists, organizers, frontline responders, caregivers, teachers, innovators, engineers, policy makers, and others who are reorienting their work and their lives to respond to the challenges of this historical moment.

Empirical Ecocriticism and Interdisciplinary Synergy

Unfortunately, there is a gulf between common beliefs about the power of environmental storytelling—expressed by humanists, media and literary critics, authors, artists, and other cultural producers—and the state of research on this topic. This is partially because the scholarly literature on environmental narratives is so fragmented that it is difficult for researchers and practitioners to assemble a reliable picture of how compelling narratives really are, how they work, and how they affect actual audiences. While scholars in the environmental humanities provide an abundance of fascinating arguments about all kinds of formal, thematic, intertextual, cultural, and ideological factors that might contribute to the power of environmental stories, they typically do not test their claims with the aid of empirical methods. In fact, apart from a growing list of important exceptions (Caracciolo 2021; Easterlin 2012; Garrard et al. 2019; Slovic 1998; Weik von Mossner 2017), ecocritics rarely connect their arguments to the copious empirical data on the psychological and social work of stories. In contrast, while scholars in the environmental social sciences pay some attention to the psychological mechanisms of narrative impact in making claims about environmental stories and use empirical methods to test those claims, they typically neglect formal dimensions (such as voice, style, and narrative perspective) as well as intertextual aspects (such as genre and tradition).

This has led to a gap in the research on environmental storytelling that is an obstacle to the development of a holistic perspective on how environmental stories work and how they might do more. Few seem to notice this gap. Ecocritics tend not to see it because empirical methods and systematic data analysis are often considered foreign and even antithetical to contemporary literary and cultural criticism. Environmental social scientists tend not to see it because they often consider cultural production to be a subject of secondary or tertiary significance, and the kind of thematic, formal, and intertextual analysis that are central to ecocriticism are infrequently applied within the social sciences. Because these academic fields have been and remain siloed, few have remarked on this gap or seem interested in closing it.

This is the goal of empirical ecocriticism: to expand our understanding of the psychological, social, and political work of environmental narratives through synergy by integrating the environmental humanities and environmental social sciences. Empirical ecocriticism aims to take the most relevant claims about environmental storytelling made within the environmental humanities and contextualize them within the scholarship in the social sciences. Similarly, it connects relevant claims from the environmental social sciences to existing humanistic scholarship on environmental narratives and submits them to empirical tests, so that the resulting data can be analyzed under the brighter light of both bodies of knowledge. The aim is to obtain conclusions that will be valid according to the established conventions of the social sciences and appropriately sensitive to the aesthetic, ethical, psychological, cultural, historical, and political dimensions of narratives.

Empirical ecocriticism can draw from and contribute to a number of fields and subfields, including narratology, ecomedia studies, environmental aesthetics, ecomusicology, the philosophy of literature, media studies, environmental psychology, the sociology of literature, and the anthropology of popular culture. As it exists today, it is primarily indebted to three fields: ecocriticism, environmental communication, and the empirical study of literature. The established methods and insights of each of these fields inform and guide the interdisciplinary work of empirical ecocritical researchers. Instead of continuing to develop as isolated monocultures, empirical ecocriticism can be the soil in which more productive polycultures and interdisciplinary synergies might grow. Empirical ecocriticism can also offer something in return because the combination of ecocritical and empirical methods in the same study enables new insights. In the following sections, we briefly describe how empirical ecocriticism draws on the work that each of these fields is doing and how it might contribute to them.

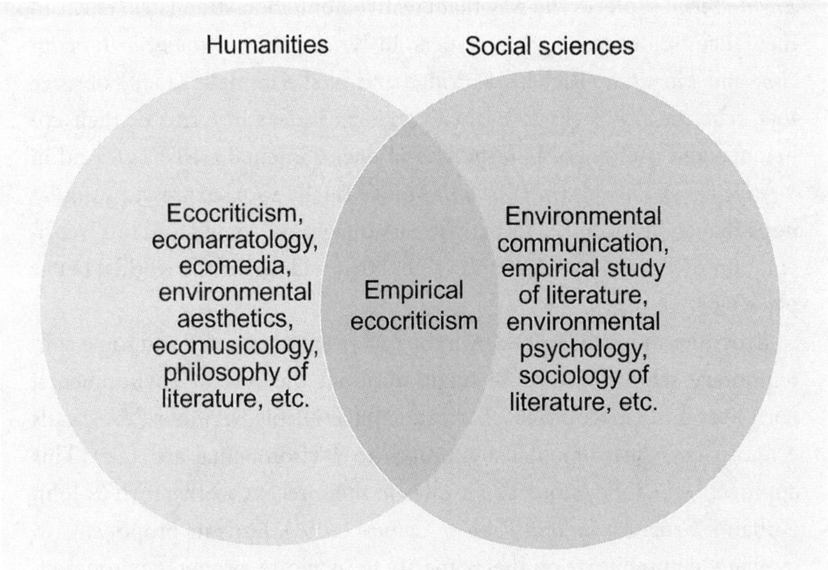

Figure I.1. Empirical ecocriticism's relationship to scholarship in the environmental humanities and social sciences.

Integrating Ecocriticism

As suggested by its name, empirical ecocriticism builds on ecocriticism's focus on the relationship between cultural texts and environmental issues. It is from ecocriticism, and the larger field of environmental humanities, that it takes its acute attention to the ecological dimensions of a broad range of narratives. Empirical ecocriticism has begun to contribute to this burgeoning field of research by investigating a core assumption of many ecocritics: that environmentally engaged narratives have a positive impact on readers' attitudes, feelings, and behaviors in relation to the more-than-human world.

This assumption is particularly pronounced in some of the early and pathbreaking works of ecocriticism. In her introduction to the foundational *Ecocriticism Reader,* Cheryl Glotfelty (1996) suggests that one of the typical questions posed by ecocritics is "How do our metaphors of the land influence the way we treat it?" (xix), highlighting the socioecological effects of narrativization. In *Seeking Awareness in American Nature Writing,* Scott Slovic (1998), the longtime editor of *ISLE: Interdisciplinary Studies in Literature and*

Environment, explores the way that creative nonfiction stimulates "environmental consciousness" (7) in readers. In *Writing the Environment: Ecocriticism and Literature,* Richard Kerridge and Neil Sammells (1998) observe that ecocriticism "seeks to evaluate texts and ideas in terms of their coherence and usefulness as responses to environmental crisis" (5). And in *Writing for an Endangered World,* Lawrence Buell (2001) expresses a similar hope that an "'ecocentric' form of literary imagining" would lead to a "reorientation of human attention and values" that will "make the world a better place" (6).

Ecocriticism has diversified over the past twenty-five years, and some contemporary scholars are more skeptical about the role of environmental narratives in socioecological change and what Nicole Seymour (2018) calls ecocriticism's "instrumentalist approach to environmental art" (26). This approach, which Seymour also locates in more recent works, such as John Parham's *Green Media and Popular Culture* (2016), leads its proponents to evaluate "cultural texts on their capacity to inculcate 'proper' environmentalist feelings," "educate the public, incite quantifiable environmental activism, or even solve environmental problems" (26). Seymour fears that this instrumentalism not only risks "confirm[ing] the negative reputation of environmentalism as didactic, prescriptive, and demanding" (27) but also overshadows environmental narratives' capacities, such as "bearing witness to crisis, enacting catharsis, serving as cultural diagnosis, and so on" (27). Studies that explore these capacities include not only Seymour's own work but also Mark Bould's trade book *The Anthropocene Unconscious* (2021), which claims that "all the stories we tell are stories about the Anthropocene" (18) and that what is urgently needed is a new way of reading them. Invested in a similar project, Min Hyoung Song's *Climate Lyricism* (2022) singles out the lyric "as a mode of literary attentiveness" (4) that reveals climate to be present in most literary texts, and calls on readers to engage with it. Such projects are valuable, timely, and important. Even so, we are still interested in learning how "bearing witness to crisis" or offering "a cultural diagnosis" affects the people exposed to those narratives. Even if it is true that all narratives are actually about the socioecological dynamics of the Anthropocene, this does not mean that they are likely to have the kinds of effects on audiences that we and some other ecocritics might wish.

Perhaps, in a moment of existential crisis, we ought to rethink "instrumentalism" as a pejorative term. In *Narrative in the Anthropocene,* Erin James (2022) divides environmental humanities scholars with an interest in narrative into two camps. There are those "who suggest that the top priority of such scholarship should be the pursuit of the right type of narrative," a group that includes Val Plumwood (2002), Ursula K. Heise (2016a), and Christophe Bonneuil and Jean-Baptiste Fressoz (2016), as well as Greg Garrard, Gary Handwerk, and Sabine Wilke (2014). Then there are those who are "critical of narrative's anthropocentrism" (James 2022, 9), including Claire Colebrook (2014), Timothy Morton (2013), and Timothy Clark (2015), who have suggested that rather than trusting in the power of narratives to change the world for the better, the environmental humanities should "focus on probing the limits of the human imagination" (James 2022, 9). We are skeptical about whether lines can be drawn so neatly; most ecocritics in the first camp are well aware of both the capacities and limitations of narratives, for example. Further, Seymour (2018) freely acknowledges that despite her critique of ecocritical instrumentalism, her own work engages "in some form of instrumentalism itself" in its attempt to "broaden the recognized repertoire of environmental affects" (28).

Empirical ecocriticism supplements the "instrumentalist" ecocritical approach by providing empirical data. In this way, it constitutes a tool that humanists can use to predict which texts are more likely to have desired psychological and political effects. It is interested not so much in the inevitable limitations of storytelling or in how one can or should read texts, but rather in how they are being read right now—not whether a text is "environmental" or not, but how it affects actual readers. What empirical ecocriticism can add to ecocriticism's theoretical claims, then, is empirical evidence that might support, problematize, or refine ecocritical hypotheses. Ideally, the field will develop in close conversation with other branches of ecocriticism, including feminist and environmental justice ecocriticism, environmental rhetoric, affect studies, Indigenous studies, and ecomedia, which will be mutually inspiring and fruitful, leading to new research questions, new research designs, and new insights about how environmental narratives engage and influence actual audiences (or fail to do so). Empirical ecocriticism intersects with some of these branches already, as demonstrated by the chapters in this book that touch on the emotions that are elicited by environmental documentaries and climate fiction, and the ways that narratives about environmental injustice can perpetuate or help mitigate the negative outcomes that affect some groups more than others. As the case studies in this volume also demonstrate, the answers provided by empirical research are not always straightforward, and they do not always confirm the hypotheses that informed them. They can be surprising; sometimes they challenge our intuitions and favored theories. But they add a richness to ecocritical investigations because they provide insight on what flesh-and-blood readers, viewers, and audiences do with the narratives they are exposed to.[1]

We expect that this volume will spark conversations about ecocriticism's relationship to the social sciences. Some ecocritics may worry that empirical methods are not fine-grained enough to capture the experience of encountering environmental narratives. Others might point out that such methods have a history of being used for morally, socially, and politically questionable purposes. Still others might feel such methods are incompatible with the cherished disciplinary identity and practices of the humanities—indeed, that they are part and parcel of a kind of disciplinary imperialism, a quantitative creep that threatens the very existence of the humanities. As trained humanists, we are sensitive to all of these concerns. We want to be clear about what empirical ecocriticism is and is not.

We note, first, that empirical ecocriticism includes qualitative, participatory, and action-oriented methods as well as quantitative instruments. As demonstrated by the case studies in this book, empirical instruments, when properly calibrated, can be sensitive to multiple dimensions of our interactions with texts. Second, empirical methods have a long history of being used for exemplary purposes, such as developing vaccines, advocating for pro-environmental policies, and supporting progressive movements. Ultimately, no method, including humanistic methods such as philosophical analysis and literary criticism, has an unblemished historical record. Third, empirical ecocriticism adopts a position of methodological pluralism and pragmatism, seeking merely to supplement the methods typically used within ecocriticism and the environmental humanities, and motivated primarily by research interests inherent to the field. We do not advocate for empirical research because we believe it is epistemologically superior but rather for pragmatic reasons: as the most appropriate tool for a specific job (Price 2010). We do not claim, for example, that empirical methods can determine the meaning or aesthetic value of a text, or provide historical or cultural context. And fourth, we note that ecocriticism as a field has always been open to and benefited from interdisciplinary methods and empirical research. Indeed, the flagship journal of ecocriticism is named *Interdisciplinary Studies in Literature and Environment,* and ecocritics such as Glen Love (2003), Nancy Easterlin (2012), and Ursula K. Heise (2016b) have drawn on the natural sciences, including ecology. While some ecocritics might be dubious about empirical methods, it is worth noting that these are often the same methods that scholars, citizens, and policy makers generally trust to provide us with information about climate change, declining biodiversity, and environmental injustice, among other things.[2]

Most of these points are elaborated on in the methodological chapters that follow this introduction, while other critiques and considerations are articulated by the senior scholars from ecocriticism, environmental communication, and the empirical study of literature we invited to write the short reflections that comprise the final section of this book.

Integrating Environmental Communication

Empirical ecocriticism brings ecocriticism and the environmental humanities into close contact with environmental communication. For the last two decades, researchers in this field have applied social scientific methodologies to understand different forms of environmental communication, with the (often unstated) goal of maximizing the efficacy of such communication to address urgent socioecological problems. This type of empirical research has been highly influential on contemporary climate communication. It has emphasized, for example, the need for messaging that addresses specific audiences, and that is authentic, bold, accurate, imaginative, and empowers people to take meaningful action (Boykoff 2019). While environmental communication researchers have been primarily interested in journalism and activist rhetoric (Comfort and Park 2018), a number of important studies have examined the influence of environmental narratives, such as film (Bilandzic and Sukalla 2019; Howell 2011; Leiserowitz 2004), on beliefs, attitudes, and behavior. The past few years have seen calls for a diversity of approaches to environmental communication (Moser 2016), along with an increased attention to the impact of environmental literature and art (Boykoff 2019).

However, since both environmental communication and ecocriticism assumed their current forms in the mid-1990s (Slovic, Rangarajan, and Sarveswaran 2019), the two fields have operated like trains running on parallel tracks. Both fields are heading in the same direction, powered by the same concerns, and now and then their passengers glance at their neighbors, make eye contact, and smile. Until recently, however, there has been a distinct lack of communication and collaboration, let alone integration. This is to the detriment of both fields as well as our collective ability to have a holistic understanding of the function, efficacy, and potential of environmental communication and environmental media at a moment in which environmental storytelling has assumed an absolutely critical significance.[3]

From environmental communication, empirical ecocriticism draws its application of social scientific methodologies, which help us establish whether the hypotheses generated by ecocritics, environmental humanists, and other scholars are verifiable according to the scientific method. In order to learn, for example, whether climate fiction influences readers' awareness of environmental injustice (Schneider-Mayerson 2020), or whether narrative empathy can make readers care about the plight of nonhuman species (Malecki et al. 2019b), an empirical ecocritic might choose to conduct interviews, a focus group study, a survey, or a controlled experiment. Empirical methods are not perfect—no method is—but they are the most reliable methods available to empirically examine the impact of any stimulus, and to predict the impact of similar stimuli in the future. When they are practiced well, they acknowledge and are transparent about their limitations.[4] Empirical ecocriticism has already led to groundbreaking collaborations between environmental communications scholars and ecocritics, such as the collaboration between one of the editors and five social scientists at the Yale Program on Climate Change Communication, which produced the first experimental test of the influence of climate fiction on readers (Schneider-Mayerson et al. 2020), a subject that hundreds of ecocritics, journalists, cultural critics, and authors have speculated about over the last decade. Such collaborations promise to enrich both fields while allowing them to jointly build on the knowledge developed within each of them.

Although empirical ecocriticism is primarily aimed at contributing to ecocriticism, we expect that it will also constitute a valuable contribution to environmental communication. First, empirical ecocriticism helps address a significant gap in environmental communication. While most research in environmental communication does not focus on narratives, empirical ecocriticism is squarely focused on the influence of novels, short stories, poetry, children's literature, film, television, video games, music, and theater, among other media—an orientation that might help empirical researchers of environmental media centralize their knowledge, learn from each other, and facilitate interdisciplinary collaboration. Second, environmental communication scholars have rarely been concerned with the formal and aesthetic features of the texts they have studied, whereas nuanced narrative elements such as narrative voice, perspective, genre, fictionality, and the construction of the protagonist are of crucial importance to empirical ecocritics. Third, environmental communication rarely includes textual analysis, whereas empirical ecocriticism can combine social scientific methodologies with the kind of textual analysis that has long been the métier of ecocriticism. Finally, and perhaps most important, empirical ecocriticism frequently draws on the theories and hypotheses that have emerged from decades of scholarship in ecocriticism and the environmental humanities, whereas environmental communication has almost entirely ignored this vast body of work.

Integrating the Empirical Study of Literature

Empirical ecocriticism also draws from and can contribute to the empirical study of literature, an approach to literary texts and their readers that also uses methods originating in the social sciences (Kuiken and Jacobs 2021). The empirical study of literature originated in the 1980s, when nascent collaborations between literary scholars, psychologists, and sociologists became more formalized. The alliances that were formed then were characterized by an interdisciplinary blend of literary theory and hypotheses that were based on close reading of texts and the use of rigorous methodologies, both qualitative and quantitative, to explore the processing and effects of literary reading.

Much of the work in the empirical study of literature concerns investigations into the processing of literary texts, and empirical ecocriticism can make use of such insights as a basis for its hypotheses. The differences between empirical ecocriticism and the empirical study of literature are in emphasis. Empirical ecocriticism is primarily interested in how its findings might be applied in society at large—in literature, theater, or education, for example. Second, research in the empirical study of literature often focuses on the role of specific textual features in the reception process; to date, empirical ecocriticism has taken a more global approach to texts. Finally, empirical ecocriticism is concentrated on a particular set of themes and issues, all revolving around the socioecological issues described earlier.

To some readers the empirical study of literature will be less familiar than ecocriticism or environmental communication. To explain its proximity to empirical ecocriticism, it may be helpful to describe two relevant areas of research in this field: studies that focus on the cognitive processing of literary texts and studies that examine the practical applications of the power of literature. The first group aims to reveal the more fundamental processes that underlie literary reading, which might explain the role that literary texts can have on, for instance, raising readers' awareness of environmental injustice. Results from research on the processing of metaphors (Bambini et al. 2019) and the emotions evoked by sound in poetry (Kraxenberger et al. 2018) could deepen our understanding of literary communication, helping us to more precisely locate those aspects of the text that are most impactful. Scholarship in narratology has generated numerous claims about how particular modes of narration affect readers; in the empirical study of literature, these hypotheses are tested experimentally (Bortolussi and Dixon 2013; Salem, Weskott, and Holler 2017). For example, the structure of stories is assumed to determine reader responses such as surprise, curiosity, and suspense (Brewer and Lichtenstein 1982), and these functions of narrative structure have been examined in controlled experiments through the manipulation of literary texts (Balint, Kuijpers, and Doicaru 2017). Likewise, assumptions about how readers' affiliations and sympathies can be influenced by narrative perspective can be tested empirically (Hakemulder and Van Peer 2015). All such narrative techniques that maintain or change readers' engagement with stories are of obvious relevance to researchers in ecocriticism and environmental communication. However, these connections have rarely been made.

Similarly, empirical scholarship on the impact of literary style has rarely been considered by ecocritics. Foregrounding (stylistic devices that deviate from "normal" forms of discourse) is often considered to be fundamental to literature, where it is responsible for deautomatizing readers' perceptions—a powerful aspect of fiction that can make readers experience, as if for the first time, the way that many humans treat nonhumans (for example). The response to foregrounding is one of the most systematically studied topics in the empirical study of literature (Van Peer et al. 2021), with research including the examination of the neurological pathways of processing literary versus nonliterary texts, and the role of foregrounding in readers' aesthetic appreciation (Jacobs 2015). It would be valuable for empirical ecocriticism to be informed by insights from such work.

A second group of studies pertains to the practical uses of literature in social contexts, such as literary education or therapeutic settings. While the first group of studies focuses on the reading process as a result of an interaction between specific readers and specific textual characteristics, the second considers reading strategies and didactic approaches. For example, Martijn Koek and colleagues (2016) examine whether a particular approach to literary education can enhance students' critical thinking, and Marloes Schrijvers and colleagues (2019) study whether literary education can stimulate reflection on oneself and others. Mark Bracher and colleagues (2019) investigate the potential of literature education to foster compassion. Generally, the results suggest that the impact of literary reading depends on individual variables (such as reading experience), textual characteristics (such as content and form), and how the texts are read. Such insights are crucial for empirical ecocriticism, suggesting it is probably not just simple exposure of any group of readers to environmental media that will have a desired impact. Research in the empirical study of literature can inform how impactful narrative encounters can be facilitated.

The empirical study of literature can profit from empirical ecocriticism in at least two ways. First, at a time when the humanities are increasingly defining themselves in terms of their relationship to ongoing environmental crises, interdisciplinary collaboration could contribute to the search for practical solutions. Thus, empirical ecocriticism amplifies the relevance and significance of the empirical study of literature. Second, empirical ecocriticism can be a source of new and important hypotheses for the empirical study of literature that the latter might not generate by itself.

Structure and Chapter Outlines

The purpose of this volume is to launch empirical ecocriticism and offer some critical reflections on its attributes and potential. As such, we have assembled a book that aims to be informative and instructive to readers who are new to this field while also providing valuable insights for those who are already familiar with it.

The volume is divided into three sections. The first section, Methods, presents a range of empirical methods borrowed from social science disciplines such as psychology, communication studies, and anthropology that can be used productively by empirical ecocritics, along with some pertinent examples. The first two chapters describe a variety of qualitative and quantitative methods, from phenomenological analysis to randomized controlled experiments. We highlight these methods because they are widely accepted in the social sciences and can yield generative answers to ecocritical questions. Because most humanists are not trained in these methods, we want to provide a sense of what they can achieve and the kinds of research questions for which they are best suited. Readers already familiar with these methods might nevertheless be interested in their specific application to environmentally oriented narratives. However, we do not assert that these are the only empirical methods that can be applied in empirical ecocritical research. The selection of methodology depends on the questions one wants to examine, and it is important to keep in mind that empirical methods are constantly evolving. What counts as a reliable and productive method can vary considerably from one discipline to another, and we have no desire to define or limit the range of appropriate methods for empirical ecocriticism, which is epistemologically flexible and open to new and exploratory methodologies. This is demonstrated by the third chapter in this section, which describes an innovative form of participatory action research.

First, in "Experimental Methods for the Environmental Humanities: Measuring Affects and Effects," W. P. Malecki discusses the experimental method, explaining why experiments can be useful for studying questions that are central to the environmental humanities and concern complex causal relations. Can narratives improve social attitudes toward nonhuman animals? Can they promote interspecies empathy? Could the dominant emotional tone of pro-environmental messages—dramatic, somber, and serious—be counterproductive? The chapter explains not only why experimental methods can be useful in answering such questions but also how to go about using them. It discusses a variety of experimental instruments and protocols, such as laboratory experiments, natural experiments, questionnaires, statistical analyses, brain scans, and implicit attitude tests where millisecond-long differences in how we respond to different stimuli reveal our unconscious biases. Malecki explains the advantages and limitations of these instruments and some of the epistemological and ethical challenges of experimental research.

In the second chapter, Paul Sopcak and Nicolette Sopcak discuss when and how researchers should use "Qualitative Approaches to Empirical Ecocriticism." These methods are most appropriate when we do not fully comprehend the concepts or processes we are dealing with and we want to deepen our insights while being attentive to the complexity of the phenomena we study. For example, what is the lived and felt experience of reading climate fiction? How can different qualitative methodologies get at this experience in different ways? Focusing on three major research traditions (phenomenology, ethnography, and grounded theory), Sopcak and Sopcak take readers on an armchair walk-through of research trajectories, helping them determine which kinds of research questions match which method; what kind of data should be collected; and what kind of results one can expect. As an introduction to qualitative methods, this chapter will help researchers select the most appropriate methodology and avoid the most common mistakes.

The method discussed in the third chapter, "Exploring the Environmental Humanities through Film Production," involves researchers becoming active participants in cultural production. Drawing on the established methods of active participant observation and participatory action research, its aim is not to generate knowledge *about* people but to generate knowledge *with* them through collective action. This novel method, known as field to media, is geared specifically to studying the process of creating music videos.

Researchers can cooperate on such videos with local communities, as performer, producer, editor, or director, and in this way contribute to a socially and culturally richer understanding of the process of making media and the perspectives and experiences of the communities involved. Rebecca Dirksen, Mark Pedelty, Yan Pang, and Elja Roy argue that this approach could be particularly useful for empirical ecocriticism because it can provide empirical insights into ecomedia that are experiential as well as culturally and socially nuanced. They illustrate this by describing studies they have conducted in four different locations—the Salish Sea region, Bangladesh, China, and Haiti—both participating in and gathering data on the creation of music videos that concern the environmental challenges that are particularly pressing in these places.

The second section, Case Studies, showcases six case studies, demonstrating a wide variety of possible methods, from controlled experiments to qualitative interviews to corpus linguistics. The texts at the center of these investigations are similarly diverse: from literature to film to theater, fiction to nonfiction. Some of the case studies present quantitative results that allow the authors to formulate general claims about impact, while others opt for a fine-grained analysis of how readers and viewers from specific cultural backgrounds respond to specific aspects of a narrative. We hope that by including such a wide variety of studies this section will not only be informative but will give readers a sense of what is possible, along with ideas for future research.

The first case study, "Does Climate Fiction Work? An Experimental Test of the Immediate and Delayed Effects of Reading Cli-Fi," is the result of a collaboration between an ecocritic, Matthew Schneider-Mayerson, and five environmental social scientists: Abel Gustafson, Anthony Leiserowitz, Matthew H. Goldberg, Seth A. Rosenthal, and Matthew Ballew. They conducted an experimental study to measure the immediate and delayed effects of climate fiction on readers via short stories by Paolo Bacigalupi and Helen Simpson. They found that whether the stimulus was a speculative dystopian story or a realist story exploring the psychological dynamics of climate change awareness and denial, reading climate fiction had small but significant positive effects on several important beliefs and attitudes about global warming, observed immediately after participants read the stories. Although these effects diminished to statistical nonsignificance after a one-month interval, the authors note that longer texts, such as novels, can be expected to have more significant and longer-lasting effects. Finally, they discuss the need for environmental media to not focus exclusively on threats but also promote self-efficacy and response efficacy—the sense that one can take action, and that such an action will be effective.

How do different narrative perspectives encourage different emotions? How do different emotions lead to different behaviors and levels of political support for environmental policies? In "The Roles of Exemplar Voice, Compassion, and Pity in Shaping Audience Responses to Environmental News Narratives," Jessica Gall Myrick and Mary Beth Oliver explore these critical questions through the lens of environmental news stories. Many environmental news stories are narratives, describing environmental problems from the perspective of those directly affected. News-based narratives about environmental injustice can encourage audiences to experience compassion for those directly affected and increase intentions to assist the victims of environmental injustice; they can also evoke pity, an ambivalent emotion that can promote negative stereotypes. Myrick and Oliver found that when the victims of environmental injustice are given more of a direct voice in a news narrative, readers experienced more compassion and less pity. Compassion was associated with greater political support for regulating water quality and heightened intentions to seek further information about the issue. In contrast, pity was unrelated to political support and was negatively related with the intention to seek more information. Their results demonstrate the significance of perspective in environmental justice narratives and highlight the connections between affect, behavior, and politics.

In "The Reception of Radical Texts: The Complicated Case of Alice Walker's 'Am I Blue?,'" a team of three ecocritics, Alexa Weik von Mossner, W. P. Malecki, and Matthew Schneider-Mayerson, and two psychologists, Marcus Mayorga and Paul Slovic, presents the results of an experiment conducted in the United States. The study is the latest in a series of attempts to tackle a conundrum posed by Alice Walker's notorious story, "Am I Blue?," which was banned in 1994 by the California State Board of Education for being anti–meat eating, and which has been hailed by activists and scholars alike as a text that effectively advocates for animal rights. The authors designed an experiment to study the narrative impact of the story on American readers, seeking to replicate the results of an earlier study, conducted in Poland, while also testing new hypotheses about the cultural situatedness of reception and the impact of two text-immanent features. Some of the results are counterintuitive, providing a reminder of the difficulty of predicting the reception of rich and complex literary texts. While partially confirming the researchers' hypotheses about the effects of human–animal comparisons and the depiction of emotional rather than physical violence against animals, this study suggests that sociopolitically "radical texts" (Ross 2011) may fail to have the desired effect on readers who do not already share their perspective.

Environmental humanists have frequently highlighted the multiple and conflicting temporalities of the Anthropocene, but pivotal concepts such as "slow violence" (Nixon 2011) have rarely been examined empirically. In "Screening Waste, Feeling Slow Violence: An Empirical Reception Study of the Environmental Documentary *Plastic China*," Nicolai Skiveren presents the results of a qualitative reception study of the 2016 documentary *Plastic China,* which portrays the social and environmental consequences of the international plastic recycling industry in China. The study investigates the experiences of a group of Danish viewers using qualitative interviewing to map their diverse affective reactions to the film as well as the active efforts they made to interpret it. In discussing their responses, Skiveren uses Stuart Hall's encoding/decoding model of communication, demonstrating one way that empirical ecocritics might utilize the framework of audience reception studies as a way to not only evaluate the capacity of environmental media to communicate or represent complex ecological issues but also to identify some of the obstacles.

A large number of people are or have been involved in environmental media not only as consumers but as producers. In "All the World's a Warming Stage: Applied Theater, Climate Change, and the Art of Community-Based Assessments," Sara Warner and Jeremy Jimenez discuss the influence of applied theater on the environmental beliefs and behaviors of participants—one of the first empirical studies of the impact of environmental theater. They describe their three-year project, conducted by a group of academics, artists, and local residents who created community-based plays about climate change's impacts on the Finger Lakes region of New York, and highlight the potential for applied theater to serve as both a method of science communication and mode of knowledge production. Drawing on participant surveys, interviews, and other assessment strategies embedded throughout the process, they draw tentative conclusions about the impact of their productions and discuss how applied theater can offer an alternative method for understanding what can count as valid data while simultaneously engaging participants in the work of creating new knowledge.

In the last chapter in this section, "Tracing the Language of Ecocriticism: Insights from an Automated Text Analysis of *ISLE: Interdisciplinary Studies in Literature and Environment,*" Scott Slovic and David M. Markowitz, an ecocritic and a corpus linguistics analyst, argue that a useful methodology for empirical ecocriticism is the automated analysis of texts. To demonstrate this potential, their chapter examines publications from ecocriticism's leading journal, *ISLE: Interdisciplinary Studies in Literature and Environment,* in search of patterns that reveal information about writing style and authorship over time. Slovic and Markowitz's data include fifteen years' worth of publications, and they used automated text analysis software to quantify language patterns across four key indicators: word count, rate of analytic thinking, jargon, and concreteness. The data suggest that articles published in *ISLE* have become longer, more specialized and theoretical, and more abstract over time. Through their analysis, they introduce a model for the automated analysis of environmental texts, which scholars might develop and apply in the future, while situating empirical ecocriticism within the history of ecocriticism.

Readers will note that most of the case studies in this book are coauthored. Indeed, collaboration is of vital importance for this type of research. Conducting social scientific studies often involves too many procedures, skills, and data for one researcher to handle. It requires knowledge of social scientific methodologies, careful planning of a research design, locating or recruiting an appropriate sample, collecting data, and submitting it to rigorous and unbiased analysis. Empirical ecocriticism is even more challenging, requiring an expertise in ecocriticism as well as social scientific methods. Few scholars have expertise in both areas, so empirical ecocriticism tends to be a team effort; the integration at the heart of empirical ecocriticism is reflected in the authors' home disciplines. We hope that this book will encourage readers to reach out to colleagues with complementary interests and skills.

The third section, Reflections, contains short essays on the value, limitations, and potential future directions of empirical ecocriticism. Ursula K. Heise, a leading ecocritic, situates empirical ecocriticism among approaches that examine the reception of literary and cultural texts and more recent quantitative and interdisciplinary approaches, including econarratology. Heise concludes that empirical ecocriticism "stands to play a crucial role in linking the study of environmental communication with the study of environmental literature" and to "enhance the many types of humanities research that understand academic work to be in dialogue with social activism and the collective search for justice among and beyond humans." Greg Garrard, also a leading ecocritic, explores some potential objections to empirical ecocriticism among humanists, such as "differences in epistemic culture . . . and scholarly discourse between the humanities and social sciences" and the potential for the reification of fluid subjectivities through quantification. Nonetheless, empirical ecocriticism "signals a welcome commitment to facts, procedural rigor, and productive interdisciplinarity," and as such, Garrard "applaud[s] its challenge to untested assumptions" and "its openness to interpretive as well as quantitative methods."

David I. Hanauer, a leading scholar of the empirical study of literature, cautions that while narrative persuasion is a critical topic, it is not a simple or straightforward applied research paradigm, and it does not always produce actionable results. This is especially so, he notes, with the most urgent issue of this historical moment: climate change. As such, Hanauer argues for an approach to environmental art, literature, and media that is similar to the theater productions developed by Warner and Jimenez for their case study, involving homegrown and immediate environmental issues of relevance for local publics, enacting collective action, and producing participatory artistic outcomes. Finally, Helena Bilandzic, a leading scholar of environmental communication, affirms the need for a more multifaceted and synergistic research approach to environmental narrative, including collaboration between researchers in ecocriticism and environmental communication. Bilandzic identifies and discusses three important areas of exploration for empirical ecocritics: the effects of different aesthetic and formal textual features on audiences, the risk and reward of persuasive intent in environmental media, and the multicausal and gradual process of changes in attitude and behavior.

Empirical Ecocriticism has been assembled in a spirit of interdisciplinarity and collaboration. Its primary goal is to demonstrate the work that empirical ecocriticism is capable of and invite scholars from various disciplines to join us and expand the range of narratives investigated, audiences examined, and methods used. We view this book not as a definitive statement on empirical ecocriticism but as an invitation to discussion, to further theorizing and integrated research, and ultimately to the development of more productive and justice-oriented forms of environmentally engaged literature, art, and media. Secondarily, we hope that the methodologies, case studies, and reflections in this volume precipitate a lively conversation about how ecocritics, environmental humanists, and other scholars of environmental media can contribute to ongoing efforts to address the existential socioecological crises of our time.

Notes

1. While this interest in the impact of literature on readers (and the world at large) is rare within traditional literary criticism, it is not unheard of. In its reader-focused orientation, empirical ecocriticism draws on the concerns of reader-response theory, a school of criticism that flourished from the 1970s to the 1990s. Associated with critics such as Stanley Fish and Wolfgang Iser, the reader-response school articulated an extreme antiformalist position, arguing that the meaning of a text is constructed (or co-constructed) by its readers. Though reader-response theory has fallen out of favor—in some guides to literary theory, the chapter on reader-response theory has been replaced with a chapter on ecocriticism—this interest in readers has been picked up by scholars in gender studies, queer studies, translation studies, historicism, and cognitive literary studies.

2. Many people also rely on other epistemic sources, such as personal experience and Indigenous knowledge, and we are not asserting that these sources are any more or less valuable. We are merely noting here that despite some humanists' concerns about empirical methodologies, we frequently rely on them in both our personal lives and our scholarship.

3. For a more extensive discussion about the relationship between ecocriticism and environmental communication, see Slovic, Rangarajan, and Sarveswaran (2019).

4. As editors, we tried to model the value of acknowledging limitations by inviting the authors of the essays in the Reflections section to write about the limitations of empirical ecocriticism, as well as its value and potential.

References

Ariely, Dan. 2008. *Predictably Irrational: The Hidden Forces That Shape Our Decisions.* New York: Harper.

Armitstead, Claire. 2021. "Stories to Save the World: The New Wave of Climate Fiction." *Guardian,* June 26, 2021. http://www.theguardian.com/books/2021/jun/26/stories-to-save-the-world-the-new-wave-of-climate-fiction.

Balint, Katalin, Moniek Kuijpers, and Miruna Doicaru. 2017. "The Effect of Suspense Structure on Felt Suspense and Narrative Absorption in Literature and Film." In *Narrative Absorption,* edited by Frank Hakemulder, Moniek M. Kuijpers, Ed S. Tan, Katalin Bálint, and Miruna M. Doicaru, 177–98. Amsterdam: John Benjamins.

Bambini, Valentina, Paolo Canal, Donatella Resta, and Mirko Grimaldi. 2019. "Time Course and Neurophysiological Underpinnings of Metaphor in Literary Context." *Discourse Processes* 56 (1): 77–97.

Bilandzic, Helena, and Freya Sukalla. 2019. "The Role of Fictional Film Exposure and Narrative Engagement for Personal Norms, Guilt and Intentions to Protect the Climate." *Environmental Communication* 13 (8): 1069–86.

Bonneuil, Christophe, and Jean-Baptiste Fressoz. 2016. *The Shock of the Anthropocene: The Earth, History, and Us.* Translated by David Fernbach. London: Verso.

Bortolussi, Marisa, and Peter Dixon. 2003. *Psychonarratology: Foundations for the Empirical Study of Literary Response.* Cambridge: Cambridge University Press.

Bould, Mark. 2021. *The Anthropocene Unconscious: Climate, Catastrophe, Culture.* New York: Verso.

Boykoff, Max. 2019. *Creative (Climate) Communications: Productive Pathways for Science, Policy, and Society.* Cambridge: Cambridge University Press.

Bracher, Mark, Deborah Barnbaum, Michael Byron, et al. 2019. "Compassion-Cultivating Pedagogy: Advancing Social Justice by Improving Social Cognition through Literary Study." *Scientific Study of Literature* 9 (2): 107–62.

Brewer, William F., and Edward H. Lichtenstein. 1982. "Stories Are to Entertain: A Structural-Affect Theory of Stories." *Journal of Pragmatics* 6 (5–6): 473–86.

Buell, Lawrence. 2001. *Writing for an Endangered World: Literature, Culture, and Environment in the U.S. and Beyond.* Cambridge, Mass.: Belknap Press of Harvard University Press.

Caracciolo, Marco. 2021. *Narrating the Mesh: Form and Story in the Anthropocene.* Charlottesville: University of Virginia Press.

Cheung, Man-Chung. 2020. "China Time Spent with Media, 2020." Insider Intelligence, May 1, 2020. https://www.insiderintelligence.com/content/china-time-spent-with-media-2020.

Clark, Timothy. 2015. *Ecocriticism on the Edge: The Anthropocene as a Threshold Concept.* London: Bloomsbury.

Colebrook, Claire. 2014. *Death of the Posthuman: Essays on Extinction, Vol. 1.* Open Humanities Press. https://archive.org/details/Colebrook-Death-Of-The-Posthuman.

Comfort, Suzannah Evans, and Youn Eun Park. 2018. "On the Field of Environmental Communication: A Systematic Review of the Peer-Reviewed Literature." *Environmental Communication* 12 (7): 862–75.

Cramer-Flood, Ethan. 2020a. "India Time Spent with Media, 2020." eMarketer, May 12, 2020. https://www.emarketer.com/content/india-time-spent-with-media-2020.

Cramer-Flood, Ethan. 2020b. "Japan Time Spent with Media, 2020." eMarketer, May 13, 2020. https://www.emarketer.com/content/japan-time-spent-with-media-2020.

Dolliver, Mark. 2020. "U.S. Time Spent with Media, 2020." eMarketer, April 29, 2020. https://www.insiderintelligence.com/content/us-time-spent-with-media-2020.

Easterlin, Nancy. 2012. *A Biocultural Approach to Literary Theory and Interpretation.* Baltimore: Johns Hopkins University Press.

Enberg, Jasmine. 2020. "France and Germany Time Spent with Media, 2020." eMarketer, May 5, 2020. https://www.insiderintelligence.com/content/france-and-germany-time-spent-with-media-2020.

Fisher, Bill. 2020. "U.K. Time Spent with Media, 2020." eMarketer, May 6, 2020. https://www.emarketer.com/content/uk-time-spent-with-media-2020.

Garrard, Greg, Axel Goodbody, George B. Handley, and Stephanie Posthumus. 2019. *Climate Change Skepticism: A Transnational Ecocritical Analysis.* London: Bloomsbury.

Garrard, Greg, Gary Handwerk, and Sabine Wilke. 2014. Introduction to "Imagining Anew: Challenges of Representing the Anthropocene," special section, *Environmental Humanities* 5:149–53.

Glotfelty, Cheryll. 1996. "Introduction: Literary Studies in an Age of Environmental Crisis." In *The Ecocriticism Reader: Landmarks in Literary Ecology,* edited by Cheryll Glotfelty and Harold Fromm, xv–xxxvii. Athens: University of Georgia Press.

Gottschall, Jonathan. 2013. *The Storytelling Animal: How Stories Make Us Human.* Boston: Mariner.

Green, Melanie C., and Timothy C. Brock. 2000. "The Role of Transportation in the Persuasiveness of Public Narratives." *Journal of Personality and Social Psychology* 79:701–21.

Green, Melanie C., Jeffrey J. Strange, and Timothy C. Brock, eds. 2002. *Narrative Impact: Social and Cognitive Foundations.* New York: Psychology Press.

Hakemulder, Frank, and Willie Van Peer. 2015. "Empirical Stylistics." In *Bloomsbury Companion to Stylistics,* edited by Violeta Sotirova, 189–207. Oxford: Bloomsbury.

Harper, Charles L., and Kevin T. Leicht. 2018. "By Way of Introduction." In *Exploring Social Change,* 7th ed., edited by Charles L. Harper and Kevin T. Leicht, 1–12. New York: Routledge.

Heise, Ursula K. 2016a. "The Environmental Humanities and the Futures of the Human." *New German Critique* 128 (43): 21–31.

Heise, Ursula K. 2016b. *Imagining Extinction: The Cultural Meanings of Endangered Species.* Chicago: University of Chicago Press.

Howell, Rachel A. 2011. "Lights, Camera . . . Action? Altered Attitudes and Behaviour in Response to the Climate Change Film *The Age of Stupid.*" *Global Environmental Change* 21 (1): 177–87.

Hulme, Mike. 2009. *Why We Disagree about Climate Change: Understanding Controversy, Inaction, and Opportunity.* Cambridge: Cambridge University Press.

Jacobs, Arthur M. 2015. "The Scientific Study of Literary Experience: Sampling the State of the Art." *Scientific Study of Literature* 5 (2): 139–70.

James, Erin. 2022. *Narrative in the Anthropocene.* Columbus: Ohio State University Press.

Kerridge, Richard, and Neil Sammells. 1998. *Writing the Environment: Ecocriticism and Literature.* London: Zed Books.

Koek, Martijn, Tanja Janssen, Frank Hakemulder, and Gert Rijlaarsdam. 2016. "Literary Reading and Critical Thinking: Measuring Students' Critical Literary Understanding in Secondary Education." *Scientific Study of Literature* 6 (2): 243–77.

Kraxenberger, Maria, Winfried Menninghaus, Anna Roth, and Mathias Scharinger. 2018. "Prosody-Based Sound–Emotion Associations in Poetry." *Frontiers in Psychology* 9:1284.

Kuiken, Donald, and Arthur M. Jacobs, eds. 2021. *Handbook of the Empirical Study of Literature.* Berlin: De Gruyter.

Leiserowitz, Anthony A. 2004. "Day after Tomorrow: Study of Climate Change Risk Perception." *Environment: Science and Policy for Sustainable Development* 46 (9): 22–39.

Love, Glen A. 2003. *Practical Ecocriticism: Literature, Biology, and the Environment.* Charlottesville: University of Virginia Press.

Malecki, Wojciech, Bogusław Pawłowski, Piotr Sorokowski, and Anna Oleszkiewicz. 2019a. "Feeling for Textual Animals: Narrative Empathy across Species Lines." *Poetics* 74:101334.

Malecki, Wojciech, Piotr Sorokowski, Bogusław Pawłowski, and Marcin Cieński. 2019b. *Human Minds and Animal Stories: How Narratives Make Us Care about Other Species.* New York: Routledge.

Marshall, George. 2014. *Don't Even Think about It: Why Our Brains Are Wired to Ignore Climate Change.* New York: Bloomsbury Publishing.

Martin, Leslie R., Cheyenne Feig, Chloe R. Maksoudian, Kenrick Wysong, and Kate Faasse. 2018. "A Perspective on Nonadherence to Drug Therapy: Psychological Barriers and Strategies to Overcome Nonadherence." *Patient Preference and Adherence* 12:1527–35.

Mercier, Hugo. 2016. "Confirmation Bias—Myside Bias." In *Cognitive Illusions,* edited by Rüdiger Pohl, 109–24. London: Psychology Press.
Morton, Timothy. 2013. *Hyperobjects: Philosophy and Ecology after the End of the World.* Minneapolis: University of Minnesota Press.
Moser, Susanne C. 2016. "Reflections on Climate Change Communication Research and Practice in the Second Decade of the 21st Century: What More Is There to Say?" *Wires Climate Change* 7 (3): 345–69.
Nabi, Robin, and Melanie C. Green. 2015. "The Role of a Narrative's Emotional Flow in Promoting Persuasive Outcomes." *Media Psychology* 18 (2): 137–62.
Nixon, Rob. 2011. *Slow Violence and the Environmentalism of the Poor.* Cambridge, Mass.: Harvard University Press.
Parham, John. 2016. *Green Media and Popular Culture: An Introduction.* Basingstoke, U.K.: Palgrave Macmillan.
Plumwood, Val. 2002. *Environmental Culture: The Ecological Crisis of Reason.* New York: Routledge.
Price, Huw. 2010. *Naturalism without Mirrors.* Oxford: Oxford University Press.
Ross, Sven. 2011. "The Encoding/Decoding Model Revisited." Paper presented at the annual meeting of the International Communication Association, Boston, Mass., May 25, 2011.
Ryan, Sadie J., Colin J. Carlson, Erin A. Mordecai, and Leah R. Johnson. 2019. "Global Expansion and Redistribution of *Aedes*-Borne Virus Transmission Risk with Climate Change." *PLoS Neglected Tropical Diseases* 13 (3): e0007213.
Salem, Susanna, Thomas Weskott, and Anke Holler. 2017. "Does Narrative Perspective Influence Readers' Perspective-Taking? An Empirical Study on Free Indirect Discourse, Psycho-narration and First-Person Narration." *Glossa* 2 (1): 61.
Salmon, Christian. 2010. *Storytelling: Bewitching the Modern Mind.* New York: Verso.
Schneider-Mayerson, Matthew. 2020. "'Just as in the Book'? The Influence of Literature on Readers' Awareness of Climate Injustice and Perception of Climate Migrants." *ISLE: Interdisciplinary Studies in Literature and Environment* 27 (2): 337–64.
Schneider-Mayerson, Matthew, Abel Gustafson, Anthony Leiserowitz, Matthew H. Goldberg, Seth A. Rosenthal, and Matthew Ballew. 2020. "Environmental Literature as Persuasion: An Experimental Test of the Effects of Reading Climate Fiction." *Environmental Communication,* 1–16.
Schrijvers, Marloes, Tanja Janssen, Olivia Fialho, and Gert Rijlaarsdam. 2019. "Gaining Insight into Human Nature: A Review of Literature Classroom Intervention Studies." *Review of Educational Research* 89 (1): 3–45.
Seymour, Nicole. 2018. *Bad Environmentalism: Irony and Irreverence in the Ecological Age.* Minneapolis: University of Minnesota Press.
Slovic, Scott. 1998. *Seeking Awareness in American Nature Writing: Henry Thoreau, Annie Dillard, Edward Abbey, Wendell Berry, Barry Lopez.* Salt Lake City: University of Utah Press.

Slovic, Scott, Swarnalatha Rangarajan, and Vidya Sarveswaran, eds. 2019. *The Routledge Handbook of Ecocriticism and Environmental Communication.* New York: Routledge.

Song, Min Hyoung. 2022. *Climate Lyricism.* Durham, N.C.: Duke University Press.

Tollefson, Jeff. 2020. "Why Deforestation and Extinctions Make Pandemics More Likely." *Nature* 584 (7820): 175–76.

Ullrich, J. K. "Climate Fiction: Can Books Save the Planet?" *Atlantic,* August 14, 2015. https://www.theatlantic.com/entertainment/archive/2015/08/climate-fiction-margaret-atwood-literature/400112/.

Van Peer, Willie, Paul Sopcak, Davide Castiglione, Olivia Fialho, Arthur M. Jacobs, and Frank Hakemulder. 2021. "Foregrounding." In *Handbook of Empirical Literary Studies,* edited by Donald Kuiken and Arthur M. Jacobs, 145–76. Berlin: De Gruyter.

Weik von Mossner, Alexa. 2017. *Affective Ecologies: Empathy, Emotion, and Environmental Narrative.* Columbus: Ohio State University Press.

PART I

Methods

Chapter 1

Experimental Methods for the Environmental Humanities

Measuring Affects and Effects

W. P. MALECKI

We often find it difficult to establish what causes a specific effect—or even to distinguish cause from effect in the first place. Consider, for example, the ongoing debate over the advantages of a plant-based diet. It is not uncommon to argue in its favor by pointing out various differences between vegans and omnivores, such as that vegans allegedly live longer and even make better lovers (Orlich et al. 2013; PETA UK 2020; Petre 2020). Perhaps they do, but how can we tell if this is due to their diet or to some other lifestyle element? Among other things, vegans tend to exercise more than the average person; they are also less likely to drink alcohol and smoke cigarettes—all of which influence both mortality rates and sexual health (Gajewska et al. 2020; Jiannine 2018; Penhollow and Young 2008; Penniecook-Sawyers et al. 2016; Tong et al. 2019). Or consider another example: the renewed controversy over the influence of violent video games (Farokhmanesh 2019; Francisco 2018). Even if it were true, as some claim, that people who play such games are more violent than the average person, how could we decide whether this is because violent games lead people to develop an inclination toward violence, or because that inclination leads them to develop an interest in those games?

Although such questions might seem to be challenges primarily for natural and social scientists, we environmental humanists face them too. This is because we also talk about causal regularities that pertain to complex phenomena. Many of us believe, for instance, that environmental literature can improve people's environmental attitudes, making them more concerned about climate change or the fate of endangered species (Armitstead 2021; Ammons 2010; Brady 2019; Buell 2001; Heller 2019; Malpas 2021; Rueckert 1978; Waldman 2018). Certainly it seems that people who are keen on such literature also tend to have above-average levels of environmental concern. But what if it is not that environmental literature caused them to develop such increased concerns, but rather that having such concerns caused them to develop an interest in the literature? Perhaps the belief in the power of environmental literature to change the minds of the public is a product of wishful thinking, a relic of a sentimentalist conception of art dating back to the nineteenth century (Solomon 2004). How could we tell?

Or consider the recent debates on so-called bad environmentalism, stimulated by the idea that the typical sensibilities of ecological campaigns such as "their sentimentality, their reverence, their serious fear-mongering" may contribute to "public negativity toward activism" (Seymour 2018, 5). Could changing those sensibilities to irony or satire, as in *Don't Look Up* (McKay 2021), help? Perhaps. But what if the "negativity towards activism" is due to entirely different factors? Might it be, for example, because people are generally aversive to changing their views and routines, which is precisely what environmental activists want them to do? What if it turns out that irony or satire is counterproductive?

These are only a few of the questions that interest environmental humanists and pertain to complex causal regularities. The number of such questions is only likely to grow as the field adopts a more activist attitude, becoming increasingly interested in how cultural phenomena contribute to the current environmental crises and how they can help to address them (Heise 2016; Streeby 2018). Experimental methods can be of great use in answering such questions, as they have been designed to disentangle complex causal matrices and counteract various cognitive biases that affect anyone inquiring into causal relations (Ruxton and Colegrave 2011; Webster and Sell 2007). Perhaps most important, human beings seem to be hardwired to look for causal relations everywhere, which makes it easy for us to neglect or downright reject the possibility that some regularities we observe might be due to chance (Kahneman 2013).[1] Experimental methods are designed to discern not only whether one factor causes another but also whether we are dealing with a causal relation at all.

My aim here is to illustrate the potential these methods hold for the environmental humanities. To this end, I will show how they have already been used to address a number of questions of key relevance to the field, in particular those that concern affects—emotions, attitudes, and other mental states that involve evaluative feeling (Frijda and Scherer 2009; Hogan 2016). I will discuss, among other things, whether narratives can promote interspecies empathy, whether they can influence attitudes toward nonhuman animals, and which emotions are more likely to lead to pro-environmental action. I will present a wide variety of experimental procedures and instruments, including laboratory experiments, natural experiments, brain scans, questionnaires, and implicit attitude tests, where millisecond differences in how we react to different stimuli can reveal our unconscious biases. These examples show that experimental methods can not only help answer questions we already have but also inspire new questions, some of which we might not ask otherwise.

Randomized Controlled Experiments

Of all experimental methods, randomized controlled studies are perhaps the most paradigmatic (Webster and Sell 2007). In turn, the most paradigmatic example of randomized controlled studies are clinical trials of drugs, where the advantages of the method are at their clearest and most socially useful. We all agree that scientists need to be as certain as possible about the therapeutic effects of a drug before it finds its way to the pharmacy, and we all understand that this requires eliminating the possibility that its perceived effects are due to confounding factors. This means, for instance, that in order to establish whether a drug helps people recover from Covid-19, it would not be enough to give it to a group of infected and symptomatic people and observe the effects. This is because we would not be able to tell whether those people improved because of the drug, because of the work of their immune system alone, or because of some other factor. It would not even be enough to merely compare their responses to the drug with the symptoms in a control group consisting of people not prescribed the drug. This is because we would then not be able to eliminate the possibility that any observed difference in symptoms is due to the composition of the groups. People in one group might have had fewer underlying conditions we were not aware of, or they might have been less sick to begin with. This is why, in conducting clinical trials of drugs, researchers not only have experimental and control groups, but also assign participants to the groups randomly (Cohen 2013; Shaughnessy, Zechmeister, and Zechmeister 2012).

In addition to their use in medical science, randomized controlled studies are also used in fields like psychology (Webster and Sell 2007), education (Ruxton and Colegrave 2011), philosophy (Knobe and Nichols 2008), and literary studies (Hakemulder 2000). They have been used to shed light on various phenomena environmental humanists are interested in too. Consider, for instance, that countless scholars, activists, and writers, including Leo Tolstoy and Thomas Hardy, have postulated that narratives can improve attitudes toward nonhuman animals (Keen 2011; West 2017). Consider also that countless organizations and media outlets have used animal stories to stoke support for pro-animal causes in the past, sometimes with great success. This was the case with Anna Sewell's *Black Beauty* ([1877] 2012), which inspired anticruelty legislation that significantly improved the lives of British horses in the last quarter of the nineteenth century (Chitty 1971; Johnson and Johnson 2002; Nyman 2016; Pearson 2011), as well as with Simon Wincer's 1993 family drama, *Free Willy,* which led to the real-life liberation of the orca that starred in the movie (Simon et al. 2009). But consider too that it is prima facie unclear whether such successful cases are examples of a larger tendency of animal narratives to impact attitudes and behavior, or whether they are mere anomalies, with narratives exhibiting such power only as a function of a rare combination of factors. In order to answer this question, one would need to conduct an experimental study not unlike the clinical trials described above. Indeed, this is precisely what my team and I did as part of a large research project on animal narratives that combined the methods of literary history, biological anthropology, and social psychology (Malecki et al. 2019).

As in clinical trials, we had an experimental group and a control group, and we assigned participants to both groups randomly. Our experimental group read an animal narrative; the control group did not. However, it was not just any animal story; it was one that shared its narrative structure with many classic tales that are said to have exerted a positive impact on attitudes toward animals in the past—a story of an animal who suffered at the hands of human beings, and suffered because it belonged to a particular nonhuman species. If the plight of the animal protagonist of *Black Beauty* was representative of the exploitative practices toward horses that were common in nineteenth-century England, then the plight of the protagonist of our experimental story is representative of various exploitative practices toward monkeys that are still common in various parts of the world. In our story, an animal, Clotho, is first taken from the jungle in which she was born, then separated from her family, sold to animal traders, and sent to Europe, where she experiences a series of atrocities, including being abused by a circus trainer and being painfully experimented on. The story was an excerpt from a crime novel entitled *The Master of Numbers* by Polish best-selling author Marek Krajewski (2014). Beyond sharing its general narrative structure with the historically impactful animal stories mentioned above, it was similarly gripping and visceral. If attitudinal impact was something one could count on from such stories, then we could count on it from this particular story.

This was also because, as with most clinical trials, we designed our study to avoid triggering the placebo effect. There is ample evidence that the mere conviction that one has taken a drug can have a therapeutic effect, so in clinical trials, people in the control group are typically provided a placebo—that is, a substance that has the same appearance as the substance provided to the experimental group but without the active ingredient (Cohen 2013; Harrington 1997). Similarly, in our study, we wanted to be certain that any observed attitudinal effects of our experimental narrative were the result of our participants having read an animal story of a particular kind, not the result of their simply having read a narrative. We therefore asked people in the control group to read a narrative placebo. This narrative did not concern any topic relevant to our study (nonhuman animals, animal welfare, and so on), but it was as similar to the experimental story as possible. It was written in the same style and genre, and it described events taking place in the same historical period and cultural context. This was achieved by picking an excerpt from the same novel as the experimental story.

With this design in place, we could then compare the attitudinal impact the stories had on both groups. We did that with a scale, a set of questionnaire items that were designed to measure changes across a given psychological dimension and that possessed the necessary psychometric features, such as validity and consistency (Maio and Haddock 2012). Scales are frequently used in experimental social science, including to measure phenomena that are of great interest to environmental humanists. There is, for instance, the New Ecological Paradigm scale (Dunlap et al. 2000), which consists of statements such as, "When humans interfere with nature, it often produces disastrous consequences." This scale has been used to measure pro-environmental orientation in thousands of studies. Our scale measures attitudes toward animals and consists of seven statements, including "Human needs should always come before the needs of animals," "The low costs of food production do not justify maintaining animals under poor conditions," and "Apes should be granted rights similar to human rights." We called it the Attitudes Toward Animal Welfare scale, or ATAW (Malecki, Pawłowski, and Sorokowski 2016).

The ATAW scale uses a Likert scale; that is, the only kind of response it allows consists of indicating one of a sequence of numbers that represent the level of agreement or disagreement with a given statement (Harris 2003). In our case, these were the numbers 1 to 7, representing answers ranging from "I completely disagree" to "I completely agree." Taken together, the responses to all seven items at the ATAW scale yielded a score that numerically represented participants' attitudes toward animals, which allowed us to compare their attitudes with the attitudes of other participants, and eventually to compare the average results obtained in one group to the average results obtained in the other. In order to be able to use that comparison to establish whether the animal narrative had an effect on our participants, we submitted our results to null hypothesis significance testing, as is standard in experimental social sciences (Cohen 2013). This kind of testing takes into account the distribution of one's data in order to assess the probability that the result the data points to could have been obtained as a result of chance or error even if the null hypothesis (in our case, that animal narratives do not have an attitudinal impact) were true (Shaughnessy, Zechmeister, and Zechmeister 2012). It is conventionally agreed that the result is valid, or statistically significant, if that probability is lower than 5 percent, commonly expressed as "$p < .05$," where p stands for statistical significance (Lindgren 1993).

Our results turned out to be statistically significant, and we replicated them in a string of experiments involving thousands of participants and a number of different animal narratives representing various genres and national literatures, including the famous horse abuse scene from Dostoyevsky's *Crime and Punishment* ([1866] 1993) and Oriana Fallaci's journalistic story "The Dead Body and the Living Brain" ([1967] 2010), on the infamous head transplant experiments conducted by Dr. Robert White (Malecki et al. 2019). This means that we now have solid experimental evidence that the common belief in the attitudinal power of animal narratives is not mere wishful thinking but reflects an actual social phenomenon. Most important, this means it is not a waste of time, effort, and funds for animal welfare organizations, advocates, and activists to try to encourage support for their goals by using narratives, including fiction.

Randomized controlled experiments are useful not only for testing the validity of commonly held beliefs but also for illuminating issues that are the subject of controversy. Consider, for instance, the debate over the moral impact of fiction versus nonfiction, specifically as it pertains to the impact of narratives depicting the plight of a specific social group or species (Djikic, Oatley, and Moldoveanu 2013; Mar et al. 2006). The concrete animal narratives I have discussed so far—*Black Beauty, Free Willy,* and the story of Clotho—are fictional, but there are also plenty of nonfiction narratives of animal abuse and exploitation, be they TV documentaries, newspaper stories, or personal testimonies. Should research on the persuasive power of animal stories be concentrating on those instead? After all, it seems only reasonable to expect that fictional depictions of animal suffering would have a smaller impact than nonfictional ones because real suffering tends to matter more than imagined suffering. Fictional depictions of animal suffering might also be expected to have a smaller impact because our reactions to fictional depictions of suffering are typically in an aesthetic rather than ethical mode, in that we focus more on the pleasant feelings of a touching story, as opposed to its moral implications (Solomon 2004).

Yet according to some theories, it is precisely this aesthetic frame that gives fiction an advantage over nonfiction because it allows us to yield to our compassionate feelings more freely than otherwise, which might translate into a significant moral change (Oatley 2002; Shusterman 2001). Consider also that some people might expect nonfictional representations of animal suffering to be too gruesome or too morally demanding to watch, inducing a sense of guilt or implying an obligation to take action (Eitzen 2005). A fictional frame might neutralize such expectations. After all, one might be less likely to expect similar guilt or obligations from a story about the suffering of characters who are not real, and many people seem to accept or even enjoy representations of intense suffering provided that these are framed as fictional, as evidenced by the fact that so many global blockbusters are incredibly violent (Oatley 1999). Perhaps, then, fiction is the best path. Maybe animal rights activists should do more to reach out to the wider public via fictional works such as Donna Leon's 2012 bestseller *Beastly Things,* which uses the frame of a detective story to educate the reader about the atrocities of industrial meat production.

In order to investigate these questions, my team and I conducted a randomized controlled study. This time we had three groups: two experimental groups and a control group. People in the control group read a narrative placebo, and people in the experimental groups read an animal story. The animal story was the same for both experimental groups, but in one of them, the participants were led to believe that the text was fiction, an excerpt from a detective novel, and in the other, they were led to believe it was nonfiction, an excerpt from a work of journalism. This was possible because the text, which came from Gail Eisnitz's 2007 journalistic book *Slaughterhouse: The Shocking Story of Greed, Neglect, and Inhumane Treatment inside the U.S. Meat Industry,* could be interpreted either way. While it reported a conversation the author had with a federal prison inmate about his involvement in the illegal slaughter of horses, it read like something out of a detective novel. After asking participants in all three groups to respond to the ATAW scale after reading their respective narratives, and after having submitted the resulting data to statistical analysis, we were able to assess which of the versions of the text had the bigger impact, thus permitting us to shed experimental light on the fiction versus nonfiction debate.

The results were surprising to us, but not because we considered one side in this debate to be right and it was proven to be wrong. They revealed something we did not consider at all: that both versions of the text had a significant impact on our participants, and to the same degree. This was baffling, as one would expect there to be at least some difference in impact between what is seen as fiction and what is seen as truth. Why, then, were there no differences?

Normally, when one obtains a counterintuitive result such as ours, one should consider the possibility that this was due to chance or error. But it turned out that there are other experimental studies showing no difference between the impact of texts perceived as fictional and those perceived as nonfictional on specific attitudes, including attitudes toward those who are mentally ill or grieving (Green and Brock 2000; Koopman 2015). It therefore seemed that we were dealing with a genuine psychological phenomenon, and the only mystery was the reason behind it. We eventually concluded that it must be related to the emotionally charged content of the stories used in the aforementioned studies—animal suffering, grief, and depression. Our idea was that the readers' response to such content in a fictional story may simply overshadow their cognitive awareness that the portrayed events are not real. What makes this likely is that the part of the brain responsible for perceiving emotions (the limbic system) is much older in evolutionary terms than the part of the brain responsible for generating propositional knowledge, including whether a message is fictional or not (the prefrontal cortex), and can easily override its influence (Panksepp 1998). In other words, the mechanism behind our results would be similar to what happens when we get scared of zombies in a horror movie or enthuse over the fortunes of our favorite character in a TV drama, apparently forgetting that the zombies and our favorite character do not exist. It would therefore seem that if an animal story is sufficiently powerful in affective terms, it may not matter to readers whether it is fictional or not. Because the same may be true for climate fiction, or other kinds of environmental narratives, this chimes with recent calls for the environmental humanities to focus not only on how emotions are represented in environmental texts, but also on their role in the reception of such works (Weik von Mossner 2017).

Understanding the emotional impact of narratives requires empirical research. Although emotions are notoriously complex and elusive, there are tried and tested instruments for measuring all sorts of emotions experimentally, including those focusing on physiological changes, such as electroencephalography, and self-reports, such as questionnaires (Mauss and Robinson

2009). Questionnaire items were used, for instance, in a study by Claudia Schneider et al. (2017) that contributes to the aforementioned debate on the sensibilities of contemporary environmentalism by showing anticipated pride from engaging in a pro-environmental action to be more motivating than anticipated guilt resulting from not engaging in it. They were also used in a study testing the power of stories to induce empathic concern for nonhuman protagonists, which has been often hypothesized by environmental humanists, writers, and activists (Angantyr, Eklund, and Hansen 2011; Cosslett 2006; DeMello 2013; Elick 2015; Gruen 2015; Herman 2018; Pearson 2011; Weik von Mossner 2018; West 2017). Its results showed that feeling with an animal protagonist may eventually lead not only to feeling for that protagonist, but also to improved attitudes toward its entire species, and toward nonhuman animals in general (Malecki et al. 2018).

Similar experimental methods could be used to test other conjectures about the emotional impact of environmental stories that have been made by environmental humanists. For instance, on the basis of the results of his exploratory correlational study on the reception of Paolo Bacigalupi's *The Water Knife* (2015), Schneider-Mayerson (2020) suggests that novels portraying "desperate climate migrants engaged in a self-interested and often violent struggle for survival can backfire, since [their] readers . . . might not empathize with climate migrants but fear them" (356). Indeed, as admitted by one of the participants in that study, who identified as "very liberal," Bacigalupi's novel made him think about what climate migrants "would do to me to have access to the water I have access to for just an hour. That thought frightens me" (356). Although it is frightening to think that climate fiction might induce such sentiments on a larger scale, only experimental research could tell us whether that is indeed the case, or whether examples such as this one are mere outliers, with the impact of *The Water Knife* and similar novels on perceptions of climate migrants being positive overall.

But experiments can reach beyond how emotions and other mental phenomena manifest themselves in surveys. In fact, they can penetrate deep below the level of conscious experience itself, down to its neurological basis. There are neurological studies, for instance, showing that reading a work of fiction increases brain connectivity in unexpected places, including in the bilateral somatosensory cortex, which is responsible for the feeling of our own bodies (Berns et al. 2013). It is not unlikely that this is due to our brain's simulating the emotions and bodily sensations of the characters in the book (Freedberg and Gallese 2007), which raises the question of whether the same mechanism works for stories with nonhuman protagonists, and whether repeated exposure to such narratives could serve to retrain our neural pathways in a significant way (Weik von Mossner 2016).

One can also measure behavioral impact experimentally. In a 2018 study, for instance, we tried to establish whether animal stories can have a significant impact on pro-animal behavior by testing whether people who have been exposed to such a narrative would allot more money to a pro-animal charity than those who have not. The results were again surprising. Even though the story used in the study had been previously shown to positively impact attitudes toward animals, it turned out to have no impact whatsoever on pro-animal behavior (Malecki et al. 2018). But there was a catch. We tried to test people's behavioral responses by giving them a list of charities, only one of which concerned animals, and then asking to distribute money between them. Although this seemed reasonable when we designed the experiment, we eventually realized that in real life, people rarely face such decisions, so the effect we observed might have been the result of an artificial situation we created. This brings us to another major challenge that every experimental researcher must face: how to make sure the results of one's experiment are applicable to real life.

Ecological Validity

The ecological validity (or external validity) of a study is the extent to which its results represent the general conditions outside of the study's controlled settings (Kellogg 2002). That researchers engaging in randomized controlled studies have to pay special attention to ecological validity stems from the very feature that makes such studies a more convenient method of establishing causal regularities than simply observing the world around us. Many phenomena occur in contexts so complex that one cannot be sure what causes what, while randomized controlled studies recreate such phenomena in contexts where potentially confounding factors are largely eliminated. But in this way, the experimental conditions may become so artificial that the results they yield cannot be reliably extrapolated to phenomena taking place outside of the experimental setting.

Consider, for instance, that in order to be able to control as many potential confounders as possible, randomized controlled studies are often conducted in laboratory settings (Webster and Sell 2007). Participants are invited to a laboratory and told what to do by a researcher, who then carefully observes their actions. Such a procedure has many advantages. First, the environment is known to researchers and can be strictly controlled so as to avoid confounding factors. Second, laboratory settings allow for the close observation of participants so researchers can tell whether participants actually follow the instructions and do not engage in actions that might skew the results, such as discussing their experiences with each other. But these advantages come at a price because most of the phenomena laboratory studies concern do not typically occur in labs. For example, people do not normally read or watch environmental narratives in laboratory spaces while being carefully watched by a researcher. Indeed, to many, such conditions may be the very antithesis of the conditions in which they watch or read such content—a comfortable environment of their choice, without a stranger observing them (Burke 2011). If there is a discrepancy between one's research conditions and the real-life conditions one aims to study, it is only natural to expect a discrepancy between one's results and how things are outside the lab.

Fortunately, there are ways of neutralizing such limitations, one of which is to look for situations where real life resembles randomized controlled conditions. Studies that take advantage of such situations are called natural experiments (Dunning 2012), and their paradigmatic example once again comes from the field of medicine. In the 1850s, England was plagued by cholera, and efforts to contain it yielded disappointing results. This was, as we now know, because they wrongly supposed the etiology of the disease to be miasmatic vapors, as opposed to drinking water contaminated with waste. Among the few who suspected contaminated water was Dr. John Snow. At first he had only intuition and anecdotal information to support his view (Vinten-Johansen 2003); he was also unable to conduct a regular laboratory study to confirm it. Given the lethality of the disease, this would have been unethical. However, it happened that events in London coincided to create laboratory-like conditions at the metropolitan level, allowing Snow to conduct an experiment without intentionally exposing anyone to risk.

There were two companies supplying Londoners with water at the time, and both provided it to people of all demographic backgrounds living in all of the city's districts. It would often happen that some tenants living in a given building would obtain their water from one company, while the remaining tenants obtained theirs from the other. In scientific jargon, Londoners were assigned to these companies randomly. Add to this that at some point one of those companies began to draw its water from a source that lay below sewage outlets on the Thames, while the other kept its source above those outlets, and the potential for a randomized controlled study on the capacity of contaminated water to contribute to cholera outbreaks becomes clear. Seizing the opportunity, Snow gathered data on cholera infections from clients of both companies, which allowed him to confirm his hypothesis and make a contribution to both the science of medicine and scientific methodology (Dunning 2012; Snow 1857, 1965).

Real-life situations that resemble experimental conditions are obviously rare, but one can arrange them. This was the case with the first of the studies on animal narratives that I describe here, which compared the impact of an animal narrative with the impact of a narrative placebo to establish whether animal narratives can improve people's attitudes toward animal welfare (Malecki, Pawłowski, and Sorokowski 2016). In conducting that study, we wanted our participants to not only read a certain kind of animal narrative, but to read it in an environment of their choice, to read it out of their own free will, and to express their attitudinal responses in a way that would allow us to measure them. We wanted to compare those responses with the responses of a control group, whose members would read a control narrative and respond to it in conditions that met the same requirements. And we needed to randomly assign the participants to these groups.

To do so we enlisted help from Marek Krajewski, who allowed us to take advantage of the naturally occurring event that was the publication of his then-forthcoming book, *The Master of Numbers* (2014). First, he agreed to include in his novel an animal narrative written according to our suggestions (the story of Clotho). Then, before the book's publication, he announced on Facebook that he and his publisher had teamed up with researchers studying how reading is related to personality and political and moral beliefs. He also asked his followers to help with that study. As compensation, Krajewski said, each participant would be able to read an advance excerpt from *The Master of Numbers* and have the opportunity to win a copy of the book. In fact, there were two excerpts from the book on offer: the story of Clotho, and an excerpt that did not concern animals at all, which served as our control narrative. The participants who registered to take part in the study were directed to a special website where they were randomly assigned to either the experimental or the control condition. Then, having read their respective excerpts, participants filled out a questionnaire that ostensibly measured personality traits and attitudes, but that included, camouflaged among other items, the ATAW scale.

In this way, we were able to conduct a randomized controlled study on animal narratives that not only did not take place in a laboratory, but also allowed the participants to read the texts in an environment of their choice. Those participants, moreover, did not choose to read those texts merely because they agreed to participate in a study, as typically happens in experimental research on reading, but because they really wanted to read it, and they would have chosen to read it outside of the experimental conditions. Finally, the whole idea of reading an excerpt from a forthcoming book must have been something they were familiar with, because writers make excerpts available all the time. In other words, despite our active coordination, the result was a natural experiment with a high level of ecological validity.

This study also succeeded in addressing two other limitations that often lower the external validity of experimental studies. One of those is related to the demographic composition of participant samples. It is an open secret that many social scientific experiments are conducted on students—students in social science courses in particular—and that their results might only be applicable to that particular population. As noted in an influential article, if psychology journals were to bear names that reflect the demographic scope of the studies they tend to publish, then they would have titles such as *The Journal of Personality and Social Psychology of American Undergraduate*

Psychology Students (Henrich, Heine, and Norenzayan 2010, 63). Thanks to the cooperation of a popular author, our study avoided this problem. Its participants were nearly two thousand Polish people between the ages of fourteen and eighty-one who lived in both small villages and big cities, and who were of different social and educational backgrounds. They were still quite specific in the sense of being avid readers of books, and being Polish, but most readers of animal narratives are avid readers of books, so this is precisely the kind of population one wants to have data about if one wants to know how animal narratives work in the real world.

The second reason why the external validity of experiments is sometimes quite low is that their artificial settings may make it easy for the participants to deduce what the hypothesis is, which might make their responses reflect how they want to be perceived in light of that hypothesis (for example, that they care about animals), rather than whether the hypothesis reflects actual social facts (Tedeschi 2013). These are known as impression management effects, which are the scourge of experimenters as well as pollsters, who routinely fail at predicting political outcomes because some people prefer not to reveal their political preferences (Marder et al. 2016).

However, there are strategies that allow for mitigating such effects. One such strategy involves the use of implicit association tests, or IATs, which take advantage of a phenomenon called priming. Priming occurs whenever our brain prepares the ground for certain sets of ideas, associations, and affects, making them more easily available to our consciousness (Molden 2014). It has been repeatedly shown, for instance, that a single stimulus related to a given phenomenon can make us recall words related to that phenomenon faster than other words. The word "doctor" or the picture of a doctor can make it temporarily easier for us to recall words such as "nurse," "injection," "surgery," and so on. On the basis of such findings, one reliable way of establishing whether our attitude toward something is negative or positive is to measure whether exposure to a stimulus related to it has an effect on the recall time of negative and positive terms. This is precisely what IATs do (Greenwald, McGhee, and Schwartz 1998). Indeed, they have been used to study all sorts of attitudes, including racial and ethnic biases and biases against people with disabilities (Pruett and Chan 2006). Needless to say, they could be used just as well to study various attitudes of relevance to the environmental humanities. For instance, just as is the case with many people who harbor sexist attitudes despite their professed feminist beliefs, many self-identifying environmentalists may harbor negative attitudes toward climate refugees. IATs would be a good way to reveal that.

Another way of avoiding impression management effects is to try to be as vague as possible about the actual purpose of the study and to design one's instruments (questionnaires and the like) so as to make their actual purpose difficult to discern (Shaughnessy, Zechmeister, and Zechmeister 2012). This is a widely accepted practice in social science, and we used it in our natural experiment as well. Our participants were initially told that the study concerned how reading literature is related to one's personality and political beliefs, and we hid the seven ATAW statements among dozens of statements measuring unrelated attitudes and beliefs. We thought that such statements would look perfectly normal in a questionnaire probing one's personality and political beliefs, and this assumption proved to be right. Our participants did not express any awareness of the actual purpose of the study; they learned about it only afterward, as part of a procedure known as debriefing. We could therefore assume that the reactions we observed were accurate expressions of the subjects' attitudes rather than artifacts of the experimental setting. Although this was a clear advantage, that aspect of our study may raise some doubts as it involved misleading our participants. Could such a procedure be acceptable from an ethical point of view?

Ethical and Epistemological Considerations

One way of approaching the deception question is to argue that deception can be justified by the purpose of the study, provided that its benefits—the invention of a new drug or a better understanding of socially significant facts—outweigh the moral cost of misleading the participants. This point implies that one should always think twice about whether one's study is worth carrying out in the first place. Because it is easy to be wrong about that, one should always seek approval from an appropriate research ethics committee. Seeking approval from a research ethics committee also helps to minimize the possibility of any other ethical shortcoming—and history demonstrates that every experimental researcher should acknowledge such a possibility.

This brings us to another ethical concern about experimental studies, which is that they can be conducted in an unethical way, for unethical purposes, causing direct or indirect harm to human or nonhuman subjects, as has happened in the past. Some of the worst examples include Nazi experiments on concentration camp prisoners and the Tuskegee syphilis study, which, from 1932 to 1972, involved denying proper treatment to a group of African American men with the disease (Farrimond 2013; Reverby 2000). These are terrible historical facts that we should never forget. However, it is also true that there are no methods whose historical ethical record would be significantly cleaner, even when it comes to such seemingly benign instruments as literary interpretation and theoretical speculation, which both have been used in the past for unethical purposes, including to promote racism, sexism, and speciesism (Harrowitz 1994; Shapiro 2004; Valls 2005). More important, the ethical failures of experimental studies should not be attributed to the method itself. This is demonstrated by the many experiments that not only did not cause any harm to their participants but also served commendable purposes, such as curing diseases, developing vaccines, and exposing all sorts of implicit biases (Caviola and Capraro 2020; Eberhardt et al. 2004; Moss-Racusin et al. 2012).

Experimental research has also been criticized because many experiments do not involve marginalized communities, primarily focusing instead on privileged subjects and their perspectives. Much experimental work in social science, for instance, has been critiqued as concerning people that are WEIRD (as in Western, Educated, Industrialized, Rich, and Democratic), yet generalizing those results to human beings in general, which is both an ethical and an epistemological shortcoming (Henrich, Heine, and Norenzayan 2010). This again is not something that is inherent to experimental social scientific research, but pertains to its application and can be remedied by encouraging studies that reflect nonwhite and nonwestern perspectives, by promoting comparative approaches, and by acknowledging limitations.

Another worry that is of both ethical and epistemological character concerns the quantitative nature of experimental studies. True, it might be said, quantitative methods allow for statistically significant results and measuring differences on a given dimension that are too minuscule to be expressed in words. But the price for this is limiting the participants' responses to a range of choices that are predetermined by the researchers and that therefore reflect a cultural and social perspective that the participants may not share. The concepts we used in our ATAW scale, for instance, may not be shared by all our participants, and therefore may not represent their true attitudes toward animals. We may thus be losing something important by not allowing our participants to freely express their minds. However, by using experimental methods, we are also gaining something important we would not be able to gain otherwise, and that empirical research is not a zero-sum game. There are kinds of empirical studies that do allow participants to express themselves relatively freely, such as surveys (De Vaus 2014; Schneider-Mayerson 2018). We can combine such methods with quantitative approaches by including open-ended questions in our questionnaires or designing quantitative studies that seek to confirm or reject hypotheses that result from qualitative research.

The final epistemological worry concerns the replication crisis in the experimental sciences. There is a growing recognition that a significant number of experimental studies across all kinds of fields, most notably medicine and psychology, cannot be replicated (Ioannidis 2005; Maxwell, Lau, and Howard 2015; Pashler and Harris 2012; Schooler 2014). Again, this does not mean that there is something inherently wrong with the method itself; rather, it indicates that some researchers have been insufficiently careful in gathering and analyzing their data, and that some have even falsified it (Head et al. 2015). Fortunately there are strategies that help to ensure that the general conclusions derived from experimental research about causal regularities are more reliable. One such strategy is testing one's hypotheses in as many studies as possible, and with as many different groups of participants representing as many different demographics as possible. This has been increasingly used across the experimental sciences, even within the experimental environmental humanities. For example, the finding on the impact of narrative on attitudes toward animals reported above has been confirmed in fifteen experiments involving more than four thousand participants from a variety of demographic backgrounds (Malecki et al. 2019).

Conclusion

Experiments are a valuable and underutilized instrument for environmental humanists to study causal phenomena, including the influence of emotions on pro-environmental behavior, the impact of stories on our attitudes toward animals, and the influence of fiction on our concern for climate change (Schneider-Mayerson et al. 2020). They can provide solid evidence for our theories on these important questions, put our theories into doubt, and shed light on competing theories where other methods would not be helpful.

That said, there is one practical problem here. While the experimental instruments that I describe here—questionnaires, IATs, and brain scans—do indeed open an array of new possibilities for the environmental humanities, environmental humanists are not typically trained to use them. Would it, then, not be too cumbersome and challenging for environmental humanists to actually do so? Aren't these instruments, including those that use standard questionnaires, more of a problem than a promise?

They do not have to be. There are experimental scientists in psychology, neurology, sociology, communication studies, and related fields who would be excited to cooperate with environmental humanists, lending their skills at designing and conducting studies and processing data. Such collaborations can be beneficial to all the fields involved because they allow environmental humanities to borrow methods and ideas from the experimental sciences and then transfer these ideas and approaches to other ends. But more important, they can be beneficial to the world outside academia. As I tried to show here, combining environmental humanist and experimental approaches allows us to better understand not only how cultural texts shape attitudes toward the environment and nonhuman others, but also how they can help us tackle environmental challenges—which is precisely why we tend to be involved in this work in the first place.

Note

1. One remarkable example of that bias is that "during the intensive rocket bombing of London in World War II, it was generally believed that the bombing could not be random because a map of the hits revealed conspicuous gaps. Some suspected that German spies were located in the unharmed areas. A careful statistical analysis revealed that the distribution of hits was typical of a random process—and typical as well in evoking a strong impression that it was not random. 'To the untrained eye,' . . . remarks [mathematician William Feller], 'randomness appears as regularity or tendency to cluster'" (Kahneman 2013, 215).

References

Ammons, Elizabeth. 2010. *Brave New Words: How Literature Will Save the Planet.* Iowa City: University of Iowa Press.

Angantyr, Malin, Jakob Eklund, and Eric M. Hansen. 2011. "A Comparison of Empathy for Humans and Empathy for Animals." *Anthrozoös* 24 (4): 369–77.

Armitstead, Claire. 2021. "Stories to Save the World: The New Wave of Climate Fiction." *Guardian*, June 26, 2021.

Bacigalupi, Paolo. 2015. *The Water Knife*. New York: Knopf.

Berns, Gregory S., Kristina Blaine, Michael J. Prietula, and Brandon E. Pye. 2013. "Short- and Long-Term Effects of a Novel on Connectivity in the Brain." *Brain Connectivity* 3 (6): 590–600.

Brady, Amy. 2019. "Climate Fiction: A Special Issue." *Guernica*, March 4, 2019. https://www.guernicamag.com/climate-fiction/.

Buell, Lawrence. 2001. *Writing for an Endangered World: Literature, Culture, and Environment in the U.S. and Beyond*. Cambridge, Mass.: Belknap Press of Harvard University Press.

Burke, Michael. 2011. *Literary Reading, Cognition and Emotion: An Exploration of the Oceanic Mind*. New York: Routledge.

Caviola, Lucius, and Valerio Capraro. 2020. "Liking but Devaluing Animals: Emotional and Deliberative Paths to Speciesism." *Social Psychological and Personality Science* 11 (8): 1080–88.

Chitty, Susan. 1971. *The Woman Who Wrote "Black Beauty": A Life of Anna Sewell*. London: Hodder and Stoughton.

Cohen, Barry H. 2013. *Explaining Psychological Statistics*. 4th ed. Hoboken, N.J.: Wiley.

Cosslett, Tess. 2006. *Talking Animals in British Children's Fiction, 1786–1914*. Burlington, Vt.: Ashgate.

De Vaus, D. A. 2014. *Surveys in Social Research*. 6th ed. London: Routledge.

DeMello, Margo, ed. 2013. *Speaking for Animals: Animal Autobiographical Writing*. New York: Routledge.

Djikic, Maja, Keith Oatley, and Mihnea C. Moldoveanu. 2013. "Reading Other Minds: Effects of Literature on Empathy." *Scientific Study of Literature* 3 (1): 28–47.

Dostoyevsky, Fyodor. (1866) 1993. *Crime and Punishment*. Translated by Richard Pevear and Larissa Volokhonsky. New York: Knopf.

Dunlap, Riley E., Kent D. Van Liere, Angela G. Mertig, and Robert Emmet Jones. 2000. "Measuring Endorsement of the New Ecological Paradigm: A Revised NEP Scale." *Journal of Social Issues* 56 (3): 425–42.

Dunning, Thad. 2012. *Natural Experiments in the Social Sciences: A Design-Based Approach*. Cambridge: Cambridge University Press.

Eberhardt, Jennifer L., Phillip Atiba Goff, Valerie J. Purdie, and Paul G. Davies. 2004. "Seeing Black: Race, Crime, and Visual Processing." *Journal of Personality and Social Psychology* 87 (6): 876–93.

Eisnitz, Gail A. 2007. *Slaughterhouse: The Shocking Story of Greed, Neglect, and Inhumane Treatment inside the U.S. Meat Industry*. Amherst, N.Y.: Prometheus Books.

Eitzen, Dirk. 2005. "Documentary's Peculiar Appeals." In *Moving Image Theory,* edited by Joseph D. Anderson and Barbara Fisher Anderson, 183–99. Carbondale: Southern Illinois University Press.

Elick, Catherine L. 2015. *Talking Animals in Children's Fiction: A Critical Study.* Jefferson, N.C.: McFarland.

Fallaci, Oriana. (1967) 2010. "The Dead Body and the Living Brain." In *Other Nations: Animals in Modern Literature,* edited by Tom Regan and Andrew Linzey, 117–24. Waco, Tex.: Baylor University Press.

Farokhmanesh, Megan. 2019. "Trump and Republicans Continue to Blame Video Games for Their Failures on Gun Control." *Verge,* August 5, 2019. https://www.theverge.com/2019/8/5/20754793/trump-gun-control-video-games-violence-republicans-no-evidence-dayton-el-paso-texas-ohio.

Farrimond, Hannah. 2013. *Doing Ethical Research.* New York: Palgrave Macmillan.

Freedberg, David, and Vittorio Gallese. 2007. "Motion, Emotion and Empathy in Esthetic Experience." *Trends in Cognitive Sciences* 11 (5): 197–203.

Frijda, Nico, and Klaus Scherer. 2009. "Affect (Psychological Perspectives)." In *The Oxford Companion to Emotion and the Affective Sciences,* edited by David Sander and Klaus R. Scherer, 183–84. Oxford: Oxford University Press.

Gajewska, Danuta, Paulina Katarzyna Kęszycka, Martyna Sandzewicz, Paweł Kozłowski, and Joanna Myszkowska-Ryciak. 2020. "Intake of Dietary Salicylates from Herbs and Spices among Adult Polish Omnivores and Vegans." *Nutrients* 12 (9): 2727.

Green, Melanie C., and Timothy C. Brock. 2000. "The Role of Transportation in the Persuasiveness of Public Narratives." *Journal of Personality and Social Psychology* 79 (5): 701–21.

Greenwald, Anthony G., Debbie E. McGhee, and Jordan L. K. Schwartz. 1998. "Measuring Individual Differences in Implicit Cognition: The Implicit Association Test." *Journal of Personality and Social Psychology* 74 (6): 1464–80.

Gruen, Lori. 2015. *Entangled Empathy: An Alternative Ethic for Our Relationships with Animals.* New York: Lantern.

Hakemulder, Jèmeljan. 2000. *The Moral Laboratory: Experiments Examining the Effects of Reading Literature on Social Perception and Moral Self-Concept.* Philadelphia: J. Benjamins.

Harrington, Anne, ed. 1997. *The Placebo Effect: An Interdisciplinary Exploration.* Cambridge, Mass.: Harvard University Press.

Harris, Richard J. 2003. "Traditional Nomothetic Approaches." In *Handbook of Research Methods in Experimental Psychology,* 39–65. Hoboken, N.J.: Wiley.

Harrowitz, Nancy A., ed. 1994. *Tainted Greatness: Antisemitism and Cultural Heroes.* Philadelphia: Temple University Press.

Head, Megan L., Luke Holman, Rob Lanfear, Andrew T. Kahn, and Michael D. Jennions. 2015. "The Extent and Consequences of P-Hacking in Science." *PLoS Biology* 13 (3): e1002106.

Heise, Ursula K. 2016. *Imagining Extinction: The Cultural Meanings of Endangered Species.* Chicago: University of Chicago Press.

Heller, Jason. 2019. "These Cli-Fi Classics Are Cautionary Tales for Today." NPR, July 26, 2019. https://www.npr.org/2019/07/26/745379270/these-cli-fi-classics-are-cautionary-tales-for-today.

Henrich, Joseph, Steven J. Heine, and Ara Norenzayan. 2010. "The Weirdest People in the World?" *Behavioral and Brain Sciences* 33 (2–3): 61–83.

Herman, David. 2018. *Narratology beyond the Human: Storytelling and Animal Life.* Oxford: Oxford University Press.

Hogan, Patrick Colm. 2016. "Affect Studies." *Oxford Research Encyclopedia of Literature,* August 31, 2016. https://doi.org/10.1093/acrefore/9780190201098.013.105.

Ioannidis, John P. A. 2005. "Contradicted and Initially Stronger Effects in Highly Cited Clinical Research." *JAMA* 294 (2): 218–28.

Jiannine, Lia M. 2018. "An Investigation of the Relationship between Physical Fitness, Self-Concept, and Sexual Functioning." *Journal of Education and Health Promotion* 7:57. https://doi.org/10.4103/jehp.jehp_157_17.

Johnson, Claudia Durst, and Vernon E. Johnson. 2002. *The Social Impact of the Novel: A Reference Guide.* Westport, Conn.: Greenwood Press.

Kahneman, Daniel. 2013. *Thinking, Fast and Slow.* New York: Farrar, Straus and Giroux.

Keen, Suzanne. 2011. "Empathetic Hardy: Bounded, Ambassadorial, and Broadcast Strategies of Narrative Empathy." *Poetics Today* 32 (2): 349–89.

Kellogg, Ronald T. 2002. *Cognitive Psychology.* London: Sage.

Knobe, Joshua Michael, and Shaun Nichols, eds. 2008. *Experimental Philosophy.* Oxford: Oxford University Press.

Koopman, Eva Maria (Emy). 2015. "Empathic Reactions after Reading: The Role of Genre, Personal Factors and Affective Responses." *Poetics* 50:62–79.

Krajewski, Marek. 2014. *Władca liczb* (The master of numbers). Kraków: Znak.

Leon, Donna. 2012. *Beastly Things.* New York: Atlantic Monthly Press.

Lindgren, B. W. 1993. *Statistical Theory.* 4th ed. New York: Chapman and Hall.

Maio, Gregory R., and Geoffrey Haddock. 2012. *The Psychology of Attitudes and Attitude Change.* Los Angeles: Sage.

Malecki, Wojciech, Bogusław Pawłowski, Marcin Cieński, and Piotr Sorokowski. 2018. "Can Fiction Make Us Kinder to Other Species? The Impact of Fiction on Pro-animal Attitudes and Behavior." *Poetics* 66:54–63.

Malecki, Wojciech, Bogusław Pawłowski, and Piotr Sorokowski. 2016. "Literary Fiction Influences Attitudes toward Animal Welfare." *PLoS One* 11 (12): e0168695.

Malecki, Wojciech, Piotr Sorokowski, Bogusław Pawłowski, and Marcin Cieński. 2019. *Human Minds and Animal Stories: How Narratives Make Us Care About Other Species.* New York: Routledge.

Malpas, Imogen. 2021. "Climate Fiction Is a Vital Tool for Producing Better Planetary Futures." *Lancet Planetary Health* 5 (1): e12–13.

Mar, Raymond A., Keith Oatley, Jacob Hirsh, Jennifer dela Paz, and Jordan B. Peterson. 2006. "Bookworms versus Nerds: Exposure to Fiction versus Non-fiction, Divergent Associations with Social Ability, and the Simulation of Fictional Social Worlds." *Journal of Research in Personality* 40 (5): 694–712.

Marder, Ben, Emma Slade, David Houghton, and Chris Archer-Brown. 2016. "'I Like Them, but Won't "Like" Them': An Examination of Impression Management Associated with Visible Political Party Affiliation on Facebook." *Computers in Human Behavior* 61:280–87.

Mauss, Iris B., and Michael D. Robinson. 2009. "Measures of Emotion: A Review." *Cognition and Emotion* 23 (2): 209–37.

Maxwell, Scott E., Michael Y. Lau, and George S. Howard. 2015. "Is Psychology Suffering from a Replication Crisis? What Does 'Failure to Replicate' Really Mean?" *American Psychologist* 70 (6): 487–98.

McKay, Adam, dir. 2021. *Don't Look Up.* Film starring Leonardo DiCaprio, Jennifer Lawrence, and Meryl Streep. Los Angeles: Netflix.

Molden, Daniel C., ed. 2014. *Understanding Priming Effects in Social Psychology.* New York and London: Guilford.

Moss-Racusin, Corinne A., John F. Dovidio, Victoria L. Brescoll, Mark J. Graham, and Jo Handelsman. 2012. "Science Faculty's Subtle Gender Biases Favor Male Students." *Proceedings of the National Academy of Sciences of the United States of America* 109 (41): 16474–79.

Nyman, Jopi. 2016. "Re-reading Sentimentalism in Anna Sewell's *Black Beauty:* Affect, Performativity, and Hybrid Spaces." In *Affect, Space, and Animals,* edited by Jopi Nyman and Nora Schuurman, 65–79. New York: Routledge.

Oatley, Keith. 1999. "Meetings of Minds: Dialogue, Sympathy, and Identification, in Reading Fiction." *Poetics* 26 (5–6): 439–54.

Oatley, Keith. 2002. "Emotions and the Story Worlds of Fiction." In *Narrative Impact: Social and Cognitive Foundations,* edited by Melanie C. Green, Jeffrey J. Strange, and Timothy C. Brock, 39–69. Mahwah, N.J.: Erlbaum.

Orlich, Michael J., Pramil N. Singh, Joan Sabaté, et al. 2013. "Vegetarian Dietary Patterns and Mortality in Adventist Health Study 2." *JAMA Internal Medicine* 173 (13): 1230–38.

Panksepp, Jaak. 1998. *Affective Neuroscience: The Foundations of Human and Animal Emotions.* Oxford: Oxford University Press.

Pashler, Harold, and Christine R. Harris. 2012. "Is the Replicability Crisis Overblown? Three Arguments Examined." *Perspectives on Psychological Science* 7 (6): 531–36.

Pearson, Susan J. 2011. *The Rights of the Defenseless: Protecting Animals and Children in Gilded Age America.* Chicago: University of Chicago Press.

Penhollow, Tina M., and Michael Young. 2008. "Predictors of Sexual Satisfaction: The Role of Body Image and Fitness." University of Texas–Arlington Libraries Research Commons, October 15, 2008. https://rc.library.uta.edu/uta-ir/handle/10106/24331.

Penniecook-Sawyers, Jason A., Karen Jaceldo-Siegl, Jing Fan, et al. 2016. "Vegetarian Dietary Patterns and the Risk of Breast Cancer in a Low-Risk Population." *British Journal of Nutrition* 115 (10): 1790–97.

PETA UK. 2020. "Vegans Make Better Lovers—Here's Why." PETA UK (blog), February 11, 2020. https://www.peta.org.uk/blog/vegan-lovers/.

Petre, Alina. 2020. "Do Vegans Live Longer than Non-vegans?" Healthline, May 7, 2020. https://www.healthline.com/nutrition/do-vegans-live-longer.

Pruett, Steven R., and Fong Chan. 2006. "The Development and Psychometric Validation of the Disability Attitude Implicit Association Test." *Rehabilitation Psychology* 51 (3): 202–13.

Reverby, Susan, ed. 2000. *Tuskegee's Truths: Rethinking the Tuskegee Syphilis Study.* Chapel Hill: University of North Carolina Press.

Rueckert, William. 1978. "Nature and Ecology: An Experiment in Ecocriticism." *Iowa Review* 9 (1): 71–86.

Ruxton, Graeme, and Nick Colegrave. 2011. *Experimental Design for the Life Sciences.* Oxford: Oxford University Press.

Schneider, Claudia R., Lisa Zaval, Elke U. Weber, and Ezra M. Markowitz. 2017. "The Influence of Anticipated Pride and Guilt on Pro-environmental Decision Making." *PLoS One* 12 (11): e0188781.

Schneider-Mayerson, Matthew. 2018. "The Influence of Climate Fiction: An Empirical Survey of Readers." *Environmental Humanities* 10 (2): 473–500.

Schneider-Mayerson, Matthew. 2020. "'Just as in the Book'? The Influence of Literature on Readers' Awareness of Climate Injustice and Perception of Climate Migrants." *ISLE: Interdisciplinary Studies in Literature and Environment* 27 (2): 337–64.

Schneider-Mayerson, Matthew, Abel Gustafson, Anthony Leiserowitz, Matthew H. Goldberg, Seth A. Rosenthal, and Matthew Ballew. 2020. "Environmental Literature as Persuasion: An Experimental Test of the Effects of Reading Climate Fiction." *Environmental Communication,* 1–16.

Schooler, Jonathan W. 2014. "Metascience Could Rescue the 'Replication Crisis.'" *Nature News* 515 (7525): 9.

Sewell, Anna. (1877) 2012. *Black Beauty.* Oxford: Oxford University Press.

Seymour, Nicole. 2018. *Bad Environmentalism: Irony and Irreverence in the Ecological Age.* Minneapolis: University of Minnesota Press.

Shapiro, Lisa. 2004. "Some Thoughts on the Place of Women in Early Modern Philosophy." In *Feminist Reflections on the History of Philosophy,* edited by Lilli Alanen and Charlotte Witt, 219–50. Dordrecht: Springer.

Shaughnessy, John J., Eugene B. Zechmeister, and Jeanne S. Zechmeister. 2012. *Research Methods in Psychology.* New York: McGraw-Hill.

Shusterman, Richard. 2001. "Art as Dramatization." *Journal of Aesthetics and Art Criticism* 59 (4): 363–72.

Siddiqui, Sabrina, and Olivia Solon. 2018. "Trump Meeting with Video Game Bosses Revives Tenuous Link to Gun Violence." *Guardian,* March 8, 2018.

Simon, M., M. B. Hanson, L. Murrey, J. Tougaard, and F. Ugarte. 2009. "From Captivity to the Wild and Back: An Attempt to Release Keiko the Killer Whale." *Marine Mammal Science* 25 (3): 693–705.

Snow, John. 1857. "Cholera, and the Water Supply in the South Districts of London." *British Medical Journal* 1 (42): 864–65.

Snow, John. 1965. *Snow on Cholera.* New York: Commonwealth Fund, Oxford University Press.

Solomon, Robert C. 2004. *In Defense of Sentimentality.* Oxford: Oxford University Press.

Streeby, Shelley. 2018. *Imagining the Future of Climate Change: World-making through Science Fiction and Activism.* Berkeley: University of California Press.

Tedeschi, James T. 2013. *Impression Management Theory and Social Psychological Research.* New York: Academic Press.

Tong, Tammy Y. N., Paul N. Appleby, Kathryn E. Bradbury, et al. 2019. "Risks of Ischaemic Heart Disease and Stroke in Meat Eaters, Fish Eaters, and Vegetarians over 18 Years of Follow-up: Results from the Prospective EPIC-Oxford Study." *British Medical Journal* 366:l4897.

Valls, Andrew, ed. 2005. *Race and Racism in Modern Philosophy.* Ithaca, N.Y.: Cornell University Press.

Vinten-Johansen, Peter, ed. 2003. *Cholera, Chloroform, and the Science of Medicine: A Life of John Snow.* Oxford: Oxford University Press.

Waldman, Katy. 2018. "How Climate-Change Fiction, or 'Cli-Fi,' Forces Us to Confront the Incipient Death of the Planet." *New Yorker,* November 9, 2018. https://www.newyorker.com/books/page-turner/how-climate-change-fiction-or-cli-fi-forces-us-to-confront-the-incipient-death-of-the-planet.

Webster, Murray, and Jane Sell, eds. 2007. *Laboratory Experiments in the Social Sciences.* Amsterdam: Academic Press/Elsevier.

Weik von Mossner, Alexa. 2016. "Environmental Narrative, Embodiment, and Emotion." In *Handbook of Ecocriticism and Cultural Ecology,* edited by Hubert Zapf, 534–50. Berlin: De Gruyter.

Weik von Mossner, Alexa. 2017. *Affective Ecologies: Empathy, Emotion, and Environmental Narrative.* Columbus: Ohio State University Press.

Weik von Mossner, Alexa. 2018. "Engaging Animals in Wildlife Documentaries: From Anthropomorphism to Trans-species Empathy." In *Cognitive Theory and Documentary Film,* edited by Catalin Brylla and Mette Kramer, 163–79. Cham, Switzerland: Springer International.

West, Anna. 2017. *Thomas Hardy and Animals.* Cambridge: Cambridge University Press.

Wincer, Simon, dir. 1993. *Free Willy.* Burbank, Calif.: Warner Bros. Pictures.

Chapter 2

Qualitative Approaches to Empirical Ecocriticism

Understanding Multidimensional Concepts, Experiences, and Processes

PAUL SOPCAK AND NICOLETTE SOPCAK

Over the past decade, the interdisciplinary field of environmental humanities has been successfully establishing itself in different areas of the humanities, such as history, anthropology, and philosophy, as well as in interdisciplinary departments and programs of study. Under this umbrella, scholars in the emerging field of empirical ecocriticism have begun applying empirical methods to examine whether and how environmental art and media transform public values surrounding the environment and potentially inspire action. Studies have looked at, for instance, whether reading fiction might positively affect our attitudes and behavior toward nonhuman animals (Malecki, Pawłowski, and Sorokowski 2016; Malecki et al. 2018); whether literary texts dealing with climate issues change readers' attitudes toward and practices affecting the environment (Schneider-Mayerson 2018); and whether differences in activist art elicit different emotional and cognitive responses (Sommer and Klöckner 2019). Examples from media studies include investigating whether interaction with a website designed to elicit a sense of emotional attachment to a certain location, while also providing information on climate change, affects users' climate change concerns (Monani et al. 2018); and how viewing climate change documentary films (Bieniek-Tobasco et al. 2019) or narrative engagement with fictional film (Bilandzic and Sukalla 2019; Howell 2011) transform public values surrounding the environment and impact viewers' efficacy beliefs, outcome expectations, emotions, and action orientation toward addressing climate change.

Most studies in empirical ecocriticism have been quantitative; some have collected and analyzed qualitative data in the form of open-ended survey questions (Monani et al. 2018; Schneider-Mayerson 2018, 2020). To date, however, few have used qualitative methods or methodologies. This may contribute to the fact that empirical ecocritical efforts encounter similar skepticism and distrust by some in the humanities (Seymour 2018) to that which empirical studies of literature have faced. The clash is often epistemological, where the aim for hard, quantifiable scientific evidence is perceived to be reductionistic and theoretically impoverished by humanists working in hermeneutic traditions (Gadamer 1982), who, following Wilhelm Dilthey (1992), believe that meaningful human actions and products cannot be fruitfully studied with methodologies developed to explain natural phenomena; they require a methodology of interpretation (i.e., hermeneutics) aimed at understanding. A view of knowledge that aims to avoid ambiguity and vagueness collides with one that values the rich polyvalence and epistemic import of the poetic and aesthetic. On one end of this spectrum there is wariness of the hypertheorized and jargon-laden hermeneutic enterprises that are perceived to produce unverifiable claims about the meaning or reception of a given work of art. On the other end is worry that the work of art is reduced to a psychological stimulus, the features and intricacies of which fail to be captured by the overly general and abstract categories of a neo-positivist enterprise. Although this may be somewhat overdrawing the contrast to make a point, in our experience it is not far off. For an emerging empirical field, the potential marginalization of empirical ecocriticism within the humanities presents a challenge. In addition to this internal challenge, empirical ecocritics must be wary of sacrificing conceptual and theoretical depth that draws on their expertise in specifically ecocritical phenomena for the allure of the perceived rigor and precision of quantitative approaches. Qualitative approaches may be uniquely capable not only of bridging this gap but also of providing a solid conceptual and theoretical footing for an emerging empirical ecocriticism.

Quantitative research is most frequently and most fruitfully used when theoretically embedded hypotheses are tested empirically and the constructs it works with are solidly anchored in theory and rigorously validated. Qualitative methods, in contrast, are most appropriate when concepts and their subjective meanings are yet to be clarified; existing theories are underdetermined or reductive; when the purpose is to gain a detailed understanding of multidimensional and contextualized phenomena without sacrificing complexity; and when the research aims at adding depth to existing, or discovering new, insights, as well as developing new theoretical frameworks (Richards and Morse 2007). Especially considering the field's nascent state, empirical ecocriticism has much to gain from qualitative approaches to its most pressing questions that are solidly grounded in qualitative methods and methodologies.

Overall, beyond the contributions that qualitative approaches can make in their own right, they also provide a rich conceptual and theoretical grounding for quantitative approaches. Our aim with this chapter is to provide some insight into qualitative principles and options for qualitative methods in the context of empirical ecocriticism.

We begin with a discussion of some foundational terms, concepts, and principles of qualitative inquiry and describe how these differ from quantitative approaches. Using phenomenology, ethnography, and grounded theory as examples, we then turn to the application of qualitative methods and provide an overview of different types of qualitative research questions and how to match these to the appropriate method, research strategies and analysis techniques, as well as what findings can be expected for each of these three methods. Then we describe two popular qualitative research methods, phenomenology and grounded theory, in greater detail by looking at philosophical and theoretical underpinnings as well as data collection, data analysis, and outcomes. For those looking for less resource-intensive forms of qualitative inquiry, we present two pragmatic qualitative approaches, qualitative and interpretive description, before ending the chapter with ethical considerations and an overview of the common mistakes made by novice qualitative researchers.

Our objective in this chapter is to provide a summary of the qualitative methods and methodologies we see as most promising in the context of empirical ecocriticism; it is not to provide exhaustive or detailed instructions for their application. Readers will find information that will facilitate their choice of an appropriate method/methodology and help avoid common mistakes, but they will need to consult further sources (suggestions provided) before designing and launching their study.

What Is Qualitative Inquiry?

Broadly, qualitative inquiry proceeds inductively and abductively[1] toward understandings and theories, as opposed to the deductive hypothesis testing of quantitative approaches. What can be challenging for newcomers to qualitative inquiry, however, is the sheer volume of methods and methodologies, as well as the disagreement among qualitative researchers (and different schools of qualitative research) regarding their form and value (Mayan 2009). This is partly because qualitative inquiry encompasses a variety of different philosophical and historical traditions; its methods are not unified but rather are multiple and rooted in different disciplines such as anthropology, sociology, and philosophy. What qualitative approaches have in common is their inductive and interpretivist nature (Mason 1996). Rather than deductively testing a theory or a specific hypothesis, qualitative researchers explore a research topic through induction and abduction. They depart from the data already available on a given phenomenon and investigate it as it appears embedded in its context (Flick 2002).

Beyond the methodological plurality and debate, a further challenge for the newcomer to qualitative approaches is the inconsistent and sometimes imprecise use of terms like *method, methodology,* and *research strategy.* In this chapter, we adopt Lyn Richards and Janice Morse's (2007) terminology and distinguish these foundational terms as follows. While the term *method* refers to "a more or less consistent and coherent way of thinking about and making data, interpreting and analyzing data, and way of judging the resulting theoretical outcome" (10), the term *methodology* "will link these strategies together" (11) by using a set of inherited practices, principles, and beliefs that are guided by theoretical frameworks. Lastly, the term *research strategy* means "a way of approaching data with a combination of techniques that are ideally consistent with the method" (11). Concretely, grounded theory and phenomenology are considered methodologies because their methods are anchored in and inseparable from their respective philosophical and theoretical underpinnings. Methods refer to the principles organizing data collection, interpretation, and analysis, whether tied to a methodology or not. Qualitative description and interpretive description are examples of methods independent from any specific philosophical and theoretical underpinning. Finally, research strategies are the specific techniques used and can include focus groups, interviews, and observation.

Some of the most common qualitative methodologies and respective methods in the social sciences are ethnography, grounded theory, phenomenology, discourse analysis, participatory action research, narrative inquiry, and case studies. Because our goal in this chapter is to provide a beginner's guide for an emergent interdisciplinary field, we will limit our discussion to three major methods that are well established in the social sciences.

Although phenomenology, ethnography, and grounded theory are often presented and used as methods, they also refer to the methodologies in which these methods are rooted and which in turn are solidly grounded in their respective philosophical and theoretical traditions and disciplines. Phenomenological qualitative inquiry has its roots in the philosophical discipline of phenomenology, established by Edmund Husserl early in the twentieth century, as well as in some of its subsequent developments initiated by his students and followers, such as Martin Heidegger, Jean-Paul Sartre, and Maurice Merleau-Ponty (existential), Hans-Georg Gadamer and Paul Ricoeur (hermeneutic), Alfred Schutz (social construction), and Simone De Beauvoir and Emmanuel Levinas (ethical). Ethnographical qualitative inquiry is rooted in the discipline of ethnography, but it evolved from two different perspectives—social anthropology from Europe and cultural anthropology from the United States—and over time its scope shifted from anthropology to sociology (Daly 2007). Ethnographies are used to describe a social group or specific culture (and its social patterns, values, and behaviors) different from a researcher's own (Fettermann 2010). Likewise, grounded theory, rooted in symbolic interactionism, developed different systems of data analysis (Anselm Strauss and Juliet Corbin's versus Barney Glaser's approach) and has been differently framed, such as constructivist grounded theory by Kathy Charmaz (2006), or feminist grounded theory by Judith Wuest and Marilyn Merritt-Gray (2001).

Although qualitative inquiry has elements that make it more intuitive and less prescriptive in its application than quantitative research—and thus more familiar and suited to researchers trained in the humanities—good qualitative research is rigorous and principled. Ecocriticism, as a new and emerging field, can benefit from qualitative inquiry, as it allows us to explore complex, multilayered phenomena. Because we live in a world heavily biased toward quantification (grades in school, IQ, cost of living, etc.), we are used to specific, definite, and quantifiable answers; qualitative research, by contrast, is interested in the story behind the numbers and in (re)presenting a phenomenon in its complexity and messiness, which can be challenging for researchers (Mayan 2009), but which is also rewarding for humanist ecocritics who might resist oversimplification and who are skeptical of easy answers when it comes to studying humans' relationship to the natural environment.

Two principles that are particularly useful for qualitative researchers are *methodological purposiveness* and *methodological congruence.* Methodological purposiveness emphasizes the need for a researcher to identify a research purpose that includes reviewing the literature and reflecting and focusing on a specific research area or a specific problem. It is key that the purpose of the research question determines the method (and the research strategies) the researcher will choose, not vice versa. Methodological congruence means that the purpose of the research and the research questions need to be congruent with the underlying theory, method, and research strategies. Methodological congruence does not imply that the design and execution of the research is predetermined and inflexible; instead, it emphasizes that methods and methodologies are connected to philosophical and theoretical assumptions and are guided by specific principles that should not be ignored. This is illustrated in Figure 2.1, the armchair walk-through, which can be read horizontally as well as vertically. Read horizontally, this figure can help to identify or refine the research question (methodological purposiveness). Researchers interested in climate activism, for instance, might consider whether they are interested in the phenomenon of climate activism, the process (how one becomes a climate activist), or the culture of climate activists. Once a decision on the research question is made in this manner, the figure can be consulted vertically. For instance, researchers interested in the research question "How do readers of literary fiction become engaged in climate activism?" will select grounded theory as a possible method, as this is a process-oriented question. Continuing down the grounded theory column, empirical ecocritics will see that it is suggested that they get familiar with symbolic interactionism as theoretical background. In a next step, they will be prompted to consider whether they have the means at their disposal to implement the congruent research strategies (in-depth interviews, memos) and whether the outcome, a theoretical model, is the desired result of their research efforts (methodological congruence). This brief description of the importance of methodological purposiveness and methodological congruence will help empirical ecocritics avoid confusing the flexibility of qualitative research with a lack of rigor.

Applying Qualitative Methods

An armchair walk-through (as coined by Morse 1992) is helpful to identify which kind of research question will fit with which method, what kind of data to collect, and what kind of findings to expect (Mayan 2009). Figure 2.1 illustrates the armchair walk-through for a possible ecocriticism question with main differences between phenomenology, grounded theory, and ethnography.

Phenomenology: A Closer Look

As is often the case with complex, rich, and dynamic intellectual disciplines and movements, there is debate within phenomenology regarding its scope, methods, and identity. Although we will not get into this debate or discuss the intellectual history of phenomenology in much depth, we will briefly touch on those philosophical underpinnings of a phenomenological qualitative methodology that we deem indispensable to working in this tradition. As mentioned earlier, phenomenology, like grounded theory, is not simply a qualitative method but rather is a methodology, which means its methods are inextricably linked to its philosophical and theoretical foundation. Studies claiming to apply phenomenological methods without this rootedness are not phenomenological in any strict sense.

The key to phenomenological inquiry is to understand that it involves a unique style of thinking and doing on the part of the researcher, rather than a method that can simply be followed like a recipe (Adams and Van Manen 2008). Although commitment will vary from one school or style of phenomenology to another (existential, hermeneutic, social construction, narrative,

Method/Methodology	*Phenomenology*	*Grounded Theory*	*Ethnography*
Theoretical/philosophical background	Phenomenological philosophy	Symbolic interactionism	Anthropology
Type of research question	What is the meaning/lived experience of reading dystopian climate fiction?	What is the process of readers of literary fiction engaging in climate activism?	What is the culture of environmental activists?
Research strategies	In-depth conversations or semistructured interviews, phenomenological writing	In-depth interviews, participant observation, memos	Observations, interviews, field notes
Analysis technique	Phenomenological attitude, eidetic variation, and explication; phenomenological writing, rewriting	Constant comparison, open and theoretical coding, memoing	Coding, comparing different types of data, developing a puzzle, "rich points"
Expected findings	In-depth, reflective description (of categories/structure) of the essential structures of lived experience (of a phenomenon)	Theoretical model, basic social psychological problem	Descriptions of day-to-day events of a specific group or culture in great detail (thick description)

Figure 2.1. Armchair walk-through. Adapted and modified from Morse (1992) and Mayan (2009).

ethical; Wertz 2011), the following are arguably necessary actions a researcher must perform for an inquiry to be called phenomenological: assumption of the phenomenological attitude (epoché and phenomenological reduction), performance of the eidetic reduction, and explication (of the lived experience of the object of study).[2]

Phenomenological attitude. Although closely related, the epoché and phenomenological reduction form distinct parts of a functional unity called the phenomenological attitude. The epoché interferes with what Husserl (2001) terms the natural attitude, which is our everyday mode of making sense of objects of experience. To effectively and efficiently navigate the everyday world, our perception takes shortcuts, so to speak, by taking objects and experiences for granted and assimilating them into preexisting categories. This is efficient because it renders perception mostly automatic, prereflective, and passive. In the epoché, the phenomenologist suspends (neutralizes, puts into brackets) this habitual, dogmatic, and passive attitude and creates space for an open attentiveness in which the phenomenological reduction can take hold. The phenomenological reduction then involves a focus on the experience of a given object at that moment, as well as a (meta)reflection on this experience itself. What is reflectively recovered in the phenomenological reduction is an active "ego as the peculiar center of the lived-experiencing" (Husserl 2001, 17) and the active co-relation between one's subjectivity and the world. Thus, the phenomenological attitude suspends our tendency to quickly and passively assimilate an object of experience into preexisting categories, then demands active, effortful sense making that involves (self-) reflection on our lived experience.

Eidetic variation. Once researchers have suspended the natural attitude through the epoché and applied the phenomenological reduction, the eidetic reduction can take hold and reveal essential structures of an intentional object (phenomenon).[3] Whereas the epoché suspends the natural attitude, eidetic variation now brackets the contingent and accidental objects and acts of consciousness, and focuses on the essential features of these (Husserl 1983). By varying the example or the features of an intentional object, we "eventually come up against something that cannot be varied without destroying that [intentional] object as an instance of its kind. It will be inconceivable that an object of that kind might lack a given feature" (Smith 2005, 564). In a phenomenological qualitative study, these variations often involve comparative analysis of commentaries and interview segments.

Explication. The concept of reflective writing, also referred to as explication, is closely tied to the phenomenological attitude. The form of reflection involved in explication pays attention to what Eugene T. Gendlin (1978–79) characterizes as a "felt sense" (50). This felt sense is implicitly meaningful; it becomes explicitly so by attending to the experiencing and what comes to presence at that very moment. Explication, when done successfully, involves making the studied phenomenon tangibly meaningful. A hallmark of a well-written phenomenological study is that it evokes the described lived experience in the reader. Merleau-Ponty (1962) describes this as follows: "The process of expression, when it is successful, does not merely leave for the reader and the writer himself a kind of reminder, it brings the meaning into existence as a thing at the very heart of the text" (212). Although the phenomenological attitude, eidetic variation, and explication are essential features of any phenomenological qualitative study, there are numerous variations in phenomenological methods. We will mention some of these, but we will focus on a variant of phenomenological inquiry that was developed in the 1990s by Max Van Manen (1997), the phenomenology of practice, since it is particularly accessible to newcomers.

Because these phenomenological concepts may be foreign and therefore not readily accessible to some readers, we will provide a brief hypothetical example of them in action. Let us assume ecocritics are interested in studying the lived experience of reading dystopian climate fiction. One of the texts (or excerpts thereof) chosen as response material is Omar El Akkad's 2017 novel *American War,* which opens as follows:

> I was happy then.
>
> The sun broke through a pilgrimage of clouds and cast its unblinking eye upon the Mississippi Sea.
>
> The coastal waters were brown and still. The sea's mouth opened wide over ruined marshland, and every year grew wider, the water picking away at the silt and sand and clay, until the old riverside plantations and plastics factories and marine railways became unstable. Before the buildings slid into

> the water for good, they were stripped of their usable parts by the delta's last holdout residents. The water swallowed the land. To the southeast, the once glorious city of New Orleans became a well within the walls of its levees. The baptismal rites of a new America. (9)

The data collected in interviews or lived-experience descriptions revolve around reflections on four existential themes of lived experiences: lived space (spatiality), lived body (corporeality), lived time (temporality), and lived human relation (relationality). Adopting a phenomenological attitude in the interview, and later in data analysis, first involves suspending the natural attitude, which would lead researchers to quickly identify and categorize themes according to established beliefs, feelings, and theories. The task is to bracket or neutralize "expectations that would prevent one from coming to terms with a phenomenon or experience as it is lived through" (Van Manen 1997, 185). In the second step, researchers will rigorously direct their attention to the lived experience being described by the interviewee. In order to do this, they suspend judgment and direct their attention to what comes to the fore and how it resonates with their own experience of the passage, particularly as it involves their sense of spatiality, corporeality, temporality, and relationality. In the comparative analysis, they are attuned to thematic similarities in the lived experience descriptions, then vary the features of the described experiences of reading dystopian climate fiction to determine the essential features of the phenomenon. Finally, they write an explication of the target phenomenon, which captures its essential features in a form that evokes (makes tangible) the experience as described by the participants.

The opening passage of *American War* might, for instance, evoke a sense of nature reclaiming its territory from humans and a sense of living in a hostile, ruined environment. Perhaps readers will report a sense of guilt, nostalgia, and fear along with a sense of the decay and fragility of human organization and achievement. Their responses might include a sense of bodily threat, feelings of existential anguish, and an acute sense of their vulnerability and finitude when reading about the workings of "deep time" (Ginn et al. 2018). They might also sense the alienation and disruption of community that this postapocalyptic scene engenders. Finally, they might attempt to reconcile the sense of loss and decay with the opening statement, "I was happy then," in an exploration of their paradoxical experience of a sense of hope and happiness amid this chaos and obvious destruction.

These are, of course, merely possibilities, and only a careful analysis of the empirical data would bring researchers closer to answering the question of how readers experience reading dystopian climate fiction. The focus on the lived experiences and avoiding reductionistic abstractions are important. Phenomenological data often include paradoxes and rich descriptions that defy simple categorization.[4]

Grounded Theory in a Nutshell

Grounded theory was introduced as a qualitative method in 1967 by sociologists Barney G. Glaser and Anselm L. Strauss. Grounded theory is based on the assumption that people actively shape and negotiate their world (Richards and Morse 2007). Although the term *grounded theory* is also sometimes used to refer to coding processes grounded in data, it is a "highly developed, rigorous set of procedures for producing formal, substantive theory of social phenomena" (Schwandt 2001, 110). Whereas a phenomenologist might be interested in the prereflectively experienced lifeworld (Adams and Van Manen 2008), a grounded theorist is interested in the concepts of what is known as a basic social psychological problem (BSPP). Grounded theory is not only a method but also a product of inquiry (a theory) that is often presented in the form of a theoretical model grounded in data (Charmaz 2006). Using the systematic set of procedures coherent with grounded theory, a model or theory grounded in the accounts and experiences of participants will be developed (Glaser 1978).

The Armchair Walk-through

To demonstrate how applying these qualitative methods differs, we restrict this armchair walk-through to phenomenology and grounded theory. How would research questions differ for a phenomenological as opposed to grounded theory approach? What would participant selection look like? How would we collect and analyze the data, and what would the expected outcome be?

Research questions. Phenomenology does not attempt to answer specific causal or process questions but is interested in a phenomenon itself (e.g., What is the lived experience of X? or What is the essential structure of experience X?). Asking specific questions may risk simplifying the problem or missing the essence of a phenomenon (Van Manen 1997). Questions are often asked implicitly or broadly. Phenomenological inquiry attempts to get at the meaning of a phenomenon as it is lived.

Research questions for ecocritical phenomenological inquiries might draw directly from current topics, discussions, and expertise in the field. For instance, John Charles Ryan (2018) points to theoretical work on the affects, embodiments, and "conceptual perplexities and contradictions of the Anthropocene," such as the paradoxical coinciding of a "profound abstraction and radical depersonalization," "temporal alienation," violence, and haunting, with a sense of wonder and enchantment in the experience of "deep time" (102–5). Phenomenology offers an exceedingly rich discussion of time consciousness, and its methodologies are aimed at explicating readers', viewers', and consumers' experiences of these temporalities in their interactions with environmental art.

Further research questions for phenomenological inquiries in the field of ecocriticism might look as follows: What is it like to read literary fiction that touches on climate change? What is the temporal, spatial, bodily affective, and relational experience of viewing films depicting suffering caused by environmental catastrophes? What is the structure of the experience of viewing environmental art?[5] That is, does it include paradoxical simultaneities? Does it express an ambiguous relationship with nature or the environment? Does it reconcile the existential drive to dominate and create a semblance of order and security with the sense of being at the mercy of an indifferent universe? These are questions that ask about lived experiences and are thus particularly well suited to be explored through a phenomenological approach.

The central question of grounded theory is, what is happening here? (Glaser 1978). The key element of a grounded theory is the focus on the process, steps, and stages of change. If the purpose of the study is to explore the process of reading literary fiction, viewing film, or experiencing art in the context of climate change, then grounded theory is the right method. Here too researchers might contextualize their research question directly within ongoing ecocritical discussions. For instance, Ryan (2018), who calls for a qualitative study of ecopoetry, mentions work in ecopoetics which discusses "experiments in community making, from poetry and visual art to foraging and cooking" (112). This is a topic for which grounded theory (and ethnographic) approaches can provide detailed, in-depth understandings. Further examples of ecocritical questions suited for a grounded theory approach are: How do readers of literary fiction become engaged in climate activism? What processes take place when experiencing art in the context of climate change?

Participant selection. Most qualitative methods follow the principle of purposive sampling. Purposive sampling consists of selecting participants that hold insider knowledge, have gone through the experience the researcher is interested in, or have other relevant expertise in the area of interest. Snowball sampling is a sampling technique in which participants in a study will be invited to assist in finding further participants from their circle of acquaintances. It is typically used in groups where participants are hard to find. So, for instance, when conducting a grounded theory study on how readers of literary fiction become engaged in climate activism, researchers might initially only have a few participants who can speak to this process. However, given their interests and activities, these participants can likely refer further participants with relevant experiences to the researchers.

In phenomenological inquiry, researchers are not particularly interested in the personal experience or story; rather, they use people's accounts to "become more experienced" themselves (Van Manen 1997, 62). For phenomenology, the goal is to get to the nature of a phenomenon as it is lived, whereas for grounded theory, participants' accounts are conceptualized, abstracted, and integrated in an emerging theory. The purpose of a hermeneutic phenomenological interview is to gather experiential material as a resource for deeper understanding and to develop a conversation about the meaning of a specific experience (Van Manen 1997). Therefore, participants need to have enough experience in the phenomenon, and they must be willing to share their experiences and engage in this rather intense form of interviewing. Participants are usually interviewed once, and the sample size for phenomenological studies is often quite small (six to eight interviewees for a Van Manen–style study) because the objective of a phenomenological study is the in-depth, reflective description (explication) of the essential structures of a lived experience (of a phenomenon) and not a focus on its idiosyncrasies. Experience has shown that the point of saturation—the concept of finding repetition in categories and themes, and the absence of emergence of new codes or themes—in Van Manen–style studies is reached at around six to eight participants. This relatively low number is partly attributable to the fact that participants are asked to write lived-experience descriptions, or they are prompted in interviews to describe the experience or phenomenon being studied, with attention to the four existential themes of lived experiences. Other forms of phenomenological methods that work with fewer data points per participant or have more noise in their data require a larger participant pool (Creswell 1998; Kuiken and Miall 2001).

In contrast, for grounded theory, a broad variation of participants will be selected; this variation is likely to result in articulation of different aspects and experiences related to a phenomenon or process. There is no determined sample size, but thirty to fifty interviews are suggested as a guideline for a grounded theory (Richards and Morse 2007). The larger sample size is necessary to gather variability (including opposing views), to reach saturation, and to adhere to the iterative process (collecting more data after a first round of coding). For a grounded theory on reading literary fiction in the context of climate change, it would be important to seek out readers of literary fiction (those who read climate fiction and those who do not), readers from different backgrounds, and readers at different stages of their lives. Further, the selection of participants will change throughout the process of theoretical sampling (Glaser 1978). This means that, for example, a second round of participants will be selected on the basis of the concepts that emerge in the analysis of the first few interviews. The procedure includes an inductive search for specific negative cases—for instance, when one reader of literary fiction contradicts all the other accounts of readers of literary fiction that support the emerging theory (Glaser 1978). The procedure also has a deductive component in that readers of literary fiction would be asked specifically about emerging concepts to saturate categories. This can also involve going back to the participants to clarify a concept or gather new information that did not emerge during the first interview (Richards and Morse 2007).

Data Collection and Analysis in Phenomenology

The outcome of a phenomenology of reading climate fiction would be a description of the lived experiences and their essential structures. Semistructured in-depth interviews are particularly useful "to gather descriptions of the life-world of the interviewee with respect to interpretation of the meaning of the described phenomena" (Kvale 1983, 174). In-depth interviews have the advantage of focusing on the interviewee's experience while remaining open to ambiguities and changes in interpretations (Kvale 1983). Not only during data analysis but also before and during the interviews, the researcher must assume the phenomenological attitude in order to get closer to the essence of the investigated phenomenon (Schwandt 2001; Van Manen 1997).[6] It facilitates an enlivened, open, and receptive mode of consciousness, one aimed at our experience and constitution of a lifeworld. This way of seeing the world involves profound wonder that allows us "to return to the world as lived in an enriched and deepened fashion" (Van Manen 1997, 185).

Examples of interview questions are, "How do you talk about literary fiction on climate change with others?" and "How do you feel before/after you view a film related to environmental catastrophes?" The important task for interviewers is to keep the discourse as close as possible to the concrete experience and to avoid a general, more abstract conversation about the topic (Van Manen 1997). Because this form of interviewing is quite in depth and involved, participants need to be even more comfortable with the interview process than in other forms of qualitative inquiry.

Another essential part of data collection and analysis in phenomenological research is phenomenological writing (explication), which involves an ongoing reflective process. Text is not only a product of a phenomenological study but is also crucial in the ongoing process of thinking about the phenomenon. Such writing is seen as an exercise, which sharpens the ability to see, reflect on, and bring to expression the essential structure of the target phenomenon. According to Husserl, language and reflection are intertwined. Phenomenological research consequently involves writing and rewriting to bring experience and concepts to expression.

Data Collection and Analysis in Grounded Theory

Interviews are also the main strategy for data collection in grounded theory, but other data sources such as focus groups, surveys, and observations are used as well. Grounded theorists are interested in questions regarding interactions and process (Richards and Morse 2007).

The process of data collection and analysis in grounded theory can be visualized as an expedition through a funnel that is broad at its top (the first interview) and gets more and more narrow at its end (eventually a theoretical model). Such a journey will not be linear; rather, it will be a continuous back-and-forth in which "the analyst jointly collects, codes, and analyzes his data and decides what data to collect next and where to find them, in order to develop his theory as it emerges" (Glaser and Strauss 1967, 45). The first interviews are kept as broad as possible—for instance, with the prompt, "Please tell me a little bit about how you became interested in climate change/climate fiction/environmental art." The grounded theorist lets participants take the lead during interviews, with minimal prompts for clarification (Morse and Field 1995). This type of unstructured interview gives participants liberty and flexibility in expressing and describing the way they experience and perceive the world (Duffy, Ferguson, and Watson 2004). The rationale behind this procedure in grounded theory is that categories and concepts will be built and abstracted from the data (Glaser 1978). Particularly in the early stage of open coding, during which researchers get a first sense of the data, it is considered crucial to follow participants' streams of thought (Glaser and Strauss 1967). Interview questions will become more specific and narrow as concepts and ideas emerge during the constant comparison being made between data collection and data analysis. The focus of the analysis is not the essential features of the lived experience but rather the theoretical abstraction of key contents into categories and concepts. Therefore, the interviews for a given study are intended to reach increasingly deeper levels in interviewees' accounts over the research process (Glaser 2002).

Another type of data that needs to be collected and analyzed is memos. Memos can be words, paragraphs, or pages of notes and sketches, which are important tools for the further research process, specifically for coding and conceptualizing (Glaser 1978). Glaser (1978) emphasizes the significance of theoretical memos for the analysis: "Memos are the theorizing write-up of ideas about codes and their relationships as they strike the analyst while coding" (83). The documentation of the memoing process is not only important to support the process of abstraction, conceptualization, and analysis but is also necessary in the process of developing concepts, categories, and theories (Wuest 2007).

One of the key elements of grounded theory is the back-and-forth between data collection, building concepts and categories, further data collection, adjusting concepts and categories, and so forth (Glaser and Strauss 1967). The coding process in grounded theory is seen as a way to open up the data in order to bring the analysis to a more conceptual and abstract level (Richards and Morse 2007). The coding process presented in this chapter is based on Glaser's (1978) approach, which distinguishes between *substantive* (first open, then selective coding) and *theoretical* coding. If we use the possible research question "How do readers of literary fiction become engaged in climate activism?" as an example, the first step of *open coding* is reading the transcripts line by line using a broad lens in order to get a general sense of the data. In this phase, initial codes are also called in vivo codes because the words and phrases of the participants are used whenever possible. These codes can be words such as "anxious" or phrases like "climate change bothers me," or "our environment has changed drastically over my lifetime." Glaser (1978) calls this phase "running the data open" (56) because analysts are supposed to come up with as many different codes as possible. This is necessary to avoid limiting the analysis to preconceived codes, and it forces analysts to reflect on the different (often contradictory) meanings that emerge from the data. Eventually, analysts will define a basic social psychological problem that is assumed to be at the core of the investigated process. Glaser (1978) suggests that the researcher should think about the core category in order to facilitate the analytical process and keep "the analyst from getting lost in the re-experiencing of his data" (57).

At the next level, *selective coding* helps identify and permit grouping of significant passages into substantive categories using constant comparison between codes and concepts to start developing patterns and themes. At this point, it is important to focus on the one BSPP or core that is closest or most relevant to the research question. Going back to our example of how readers of literary fiction become engaged in climate activism, we might find that other aspects emerge as important for climate activism, such as cultural factors, level of education, or witnessing impacts in the nearby environment. Nevertheless, in order to develop a meaningful model, only codes related to the BSPP and the research question will be selected. At this point in the ongoing data collection process, guided by themes emerging from the data, the interview questions might change in order to increase the level of abstraction and analysis (Richards and Morse 2007). Glaser (1978) suggests using the "6 Cs" (74) to describe whether a code or concept has the nature of cause, consequence, condition, context, covariance, and contingency, as well as the collected memos to facilitate the process of identifying concepts on the basis of empirical indicators. Because in academia we are accustomed to thorough analysis and fast abstractions, it can seem difficult or counterintuitive to take a step back in the middle of analysis. With our sample research question, one emerging concept could be "political identity." We might observe that there is a strong division between readers' political background as to how much they become engaged in climate activism or not. However, it would be necessary to take a step back and see if this concept holds if we let go of it in order to differentiate between *received distinctions* and *earned distinctions* of concepts and to avoid "one-upping," which means jumping too quickly to abstraction and conceptualization (Glaser 1978, 59). Earned distinctions of concepts can be recognized by their persistence; they will reappear even if they were dismissed at an earlier point.

As Glaser (1978) notes, taking a step back during the analysis is necessary to see patterns emerge. During the theoretical coding process, categories are formed and linked together by exploring possible relationships to a core category. At this point, "theoretical codes conceptualize how the substantive codes may relate to each other as hypotheses to be integrated into a theory" (72). Theoretical codes also need to "earn" their relevance, as they are only meaningful if they can be related back to the open and selective codes. At this juncture, when returning to data collection, participants will be theoretically sampled in order to confirm or challenge the constructed model and the assumed relationships between the labeled categories until saturation is reached (Wuest 2007). In the last phase of iterative coding, linkages and relationships between concepts and a core concept or BSPP will be identified and summarized in a theoretical model.

Outcomes. A phenomenological study of the kind used as an exemplar here results in an in-depth description in form of a written text on, for instance, what the experience (of time, space, body, affects, relationality) is when art evokes reflections on the environment. Such phenomenological texts are reflective and prereflective descriptions that encompass cognitive and emotional understanding. In opposition to grounded theory, such phenomenological texts are not theoretical. Meaning and lived experiences are not expressed through abstract reasoning, causal linkages, inferences, or opinions (Adams and Van Manen 2008). Phenomenological texts may include anecdotal narratives, reflections, and descriptions of lived-through moments and are characterized by their oriented, deep, strong, and rich quality (Van Manen 1997). This form of phenomenology is interested in the here and now, the moment as it is lived (Van Manen, n.d.). Although language is limited to fully describe a phenomenon, thorough phenomenological writing that follows Van Manen's (1997) method of inquiry aims at enabling the reader to vicariously live through the experience of a phenomenon. Thus, the outcome of an ecocritical phenomenological study would be a phenomenological text that would evoke deep and rich understanding of the art-induced experience of, for instance, time, space, the body, or affects in the context of the climate crisis.

Conversely, the outcome of a grounded theory study is a theoretical model of the relationships between categories (Creswell et al. 2007). Such a model is based on a research process that is in opposition to phenomenology because grounded theorists abstract, conceptualize, reconceptualize, and theorize from the beginning to the end of a grounded theory study. However, because the theory emerges directly from the data, it also fits the real world (Glaser and Strauss 1967). The generated theory emphasizes process and change over time. With the research question "How do readers of literary fiction become engaged in climate activism?" the theory would outline what processes take place when reading literary fiction, viewing film, or consuming art in the context of climate change. A completed grounded theory has the capacity to interpret what happens during these activities, potentially with explanatory power to identify concepts that may foster or hinder engagement in climate change activism (Glaser 1978).

Pragmatic Approaches to Qualitative Inquiry

These two armchair walk-through examples of what to consider when conducting a phenomenological or grounded theory study demonstrate some of the complexities of such undertakings, such as the necessity of knowing the method used and its related literature well, engaging in continuous reflection on process, and understanding the amount of time and work that qualitative researchers need to invest. Although these qualitative methods are well established, some qualitative researchers may not have the resources or the disciplinary background necessary to conduct a rigorous phenomenological or grounded theory study. A problem to avoid is labeling a study as phenomenological or grounded theory without the methodological purposiveness or methodological congruence that each of these methodologies entails. On this point, Roy Suddaby (2006), a reviewer of qualitative research, has expressed his surprise "by the profound misunderstanding of what constitutes qualitative research" (633). An explanation for this phenomenon might be that authors are being pushed (by a journal, funder, or supervisor) to label their qualitative research with a specific method or methodology (Sandelowski 2000).

As scholars in an evolving discipline, empirical ecocritics will explore which methods best fit their specific research questions. Two qualitative approaches developed in the health sciences, but applicable more widely, that have allowed researchers to answer qualitative research questions in a pragmatic, coherent, and practice-relevant way are qualitative description (Sandelowski 2000, 2010) and interpretative description (Thorne 2008). Both approaches have the advantage that they can be used in any discipline or field without extensive study and consideration of philosophical or disciplinary underpinnings.

Qualitative description. Although qualitative description is less popular than some of the other methodologies described above, it is a valuable option when endeavoring to answer qualitative research questions with fewer theoretical and philosophical underpinnings (Lambert and Lambert 2012). Qualitative description is "the method of choice when straight descriptions of phenomena are desired" (Sandelowski 2000, 334). Qualitative description is helpful to answer questions such as, "What are people's concerns about climate change?" "What are people's responses (e.g., thoughts, feelings, attitudes) toward climate change?" and "What reasons do people have for engaging or not engaging in climate activism?" (Sandelowski 2000). Although researchers who use qualitative description do interpret and analyze data using content analysis, the outcome, a qualitative description, remains close to the data (Sandelowski 2010). Although qualitative description involves an element of interpretation, the interpretation is low inference, meaning that different researchers will agree on the accuracy of the description of the "facts" (Sandelowski 2000). Qualitative description is less interpretive and theoretical than grounded theory or phenomenology but more interpretive than quantitative descriptions such as surveys (Sandelowski 2000). Using qualitative description can also solve the common problem of mislabeling or misrepresenting qualitative findings as grounded theory, phenomenology, or ethnography, when such methods were not used: "researchers can unashamedly name their method as qualitative description" (Sandelowski 2000, 339).[7]

Interpretive description. Interpretive description, in contrast, moves beyond mere description to conceptualization in the analysis, with the aim of interpreting and explaining a phenomenon in its context (Thorne 2008). Interpretive description has become a well-established qualitative research approach that can be tailored to any discipline, allowing incorporation of its ideas and underlying assumptions. It is increasingly used in applied fields because of "its potential to address practice problems across applied disciplines" (Oliver 2012, 409), and its contextual nature and pragmatic orientation make it well suited to an inexperienced qualitative researcher.[8] Rooted in the nursing discipline (Thorne, Reimer Kirkham, and O'Flynn-Magee 2004), interpretive description is a valuable approach when the objective is to identify themes associated with a specific research question in applied practice settings (Thorne 2008). Because interpretive description builds on and incorporates existing knowledge of a topic, an empirical ecocritical research question would be anchored in and informed by the researcher's disciplinary, philosophical, or theoretical orientations and knowledge. Because interpretive description does not follow a single procedure for how to design and conduct a study, there is some flexibility to develop a theoretical framework that follows its own logic. However, there are some foundational underpinnings pertaining to the research design of interpretive description, which are based on the naturalistic inquiry tradition by Yvonna S. Lincoln and Egon G. Guba (1985). These include the relationship between human experience, ethical consideration, reality as "multiple constructed realities," and the acknowledgment of the "inseparable relationship between the knower and the known, such as the inquirer and the 'object' of that inquiry interact to influence one another" (Thorne 2008, 74). As with other qualitative approaches described in this chapter, research questions are built on previous knowledge and often begin with "how," "what," and "when" questions. Furthermore, careful attention should be given to participant selection, which data collection strategies to use, and what contribution to the field answering a specific research question would make.

Both methods, qualitative description and interpretive description, use purposeful sampling and data collection strategies such as focus groups and semistructured or open-ended interviews. Data analysis, as with other qualitative approaches, is iterative and is based on the data (instead of preconceived categories). The data analysis process is "reflexive and interactive as researchers continuously modify their treatment of data to accommodate new data and new insights about those data" (Sandelowski 2000, 338). It is important to emphasize that with interpretative description, as with qualitative description, following and transparently documenting the research process remain key. Interpretive description uses the technique of constant comparative analysis, which has its origins in grounded theory and describes a technique that compares parts of the data (interviews, movie segments, artwork, and emerging themes) with other parts of data and the entire data set to conceptualize themes and theorize about their relationships (Glaser and Strauss 1967; Thorne 2008). During the first stage of data analysis in interpretive description, the main question for a researcher is, what is going on here? This allows the researcher to get familiar with the data on a macro level and to code segments and paragraphs with generic terms or labels. In a second step, the specific and previously developed research question is used to identify and create preliminary themes. The iterative process of constant comparison and reflection on how the data connect (or do not connect) with other parts of the data help researchers "move beyond the self-evident and superficial in linking the groupings and patterns" (Thorne 2008, 149). This includes remaining open to unexpected or surprising information or letting go of ideas that are not sufficiently supported. The outcome of a study using interpretive description is "a coherent conceptual description" (Thorne, Reimer Kirkham, and O'Flynn-Magee 2004, 7) "that is generated on the basis of informed questioning, using techniques of reflective, critical examination, and which will ultimately guide and inform disciplinary thought in some manner" (6).

Ethical Considerations in Qualitative Research

"Do no harm" is the most often cited and arguably the most important principle of research ethics to be considered by researchers (Hammersley and Traianou 2012). Because qualitative research typically involves human participation, it demands the ethical consideration of potential harm as well as moral and legal obligations (Saldana, Leavy, and Beretvas 2011). Qualitative research involves openness on the part of researchers, as well as immersion in a field that is not always well known and one that offers unique, often complex, relationships with participants. The onus of careful planning and acting in a professional and ethical fashion is on the qualitative researchers (Reid et al. 2018). For instance, certain questions or topics can cause participants emotional or psychological distress, or can be culturally inappropriate (Liamputtong 2010). Research on climate activism in particular could make participants feel hopeless or pressured to continue as study participants. Qualitative researchers also need to be reflective of power dynamics between researcher and participants and protect participants' identities (Reid et al. 2018). Researchers often have multiple roles including instructor, manager, or mentor. Researchers investigating climate activism may also hold another position that could influence the relationship with participants, particularly if the participant is a student or employee, for instance. Furthermore, in a smaller community or setting, the topic of climate activism could bring differing opinions to the surface that then in turn create or intensify tensions between individuals or groups.

Ethical considerations should also include cultural sensitivity, which is most important in cross-cultural research because participants may have experienced harm, particularly in the context of colonialization and marginalization of vulnerable populations (Liamputtong 2010). Researchers should consider how to (re)present the findings and how to exit a research study ethically. Focus groups and interviews, for example, are based on rapport between researcher and participants, and they often involve participants' shared lived experiences or stories over a period of time. Researchers need to think about the emotional and psychological impact of the involvement in research and how the relationship is ended, ideally with a closure that is known and acceptable to the participants (Morrison, Gregory, and Thibodeau 2012). Last, while ethical considerations are often focused on the participants engaged in data collection, there are other groups that can be harmed, such as funding organizations, institutions where the research is taking place, colleagues, communities, and researchers themselves (Hammersley and Traianou 2012).

Common Mistakes

Now that some of the basic considerations and methods of qualitative inquiry have been discussed, we will briefly turn to some of the most common mistakes that beginner qualitative researchers make, so that empirical ecocritics embarking on their first qualitative journey can avoid them.

Figure 2.2 addresses examples of common mistakes made by beginner qualitative researchers. A concise introductory resource is Deborah Cohen and Benjamin Crabtree's (2006) online "Qualitative Research Guidelines Project," in which the authors review the reasons reviewers and editors of a scholarly journal provide for rejecting manuscripts using qualitative approaches, subsuming them under the following categories: "lacks focus, too jargony, sample insufficient, analysis lacks depth, methods lack adequate description, and data quality concerns." For each of these common pitfalls, Cohen and Crabtree provide concrete examples from reviewers' comments—and, more important, links to concise online resources that explain how to avoid or address them.

Overall, we recommend that learning about qualitative inquiry be approached with similar expectations regarding the necessary discipline and time investment to those related to learning about quantitative research methods, like attending qualitative methods courses, workshops, or tutorials.

Study design	• Failure to design the study according to a method or methodology before collecting data. • Research question does not align with methods or methodology, data collection strategy, or desired outcome. • Collecting and analyzing qualitative data, and then trying to package it in a method after the fact and insufficient knowledge of methods in general. • Using deduction versus induction/abduction.
Research question	• Too specific. (Reflecting on and reformulating research questions are important for iterative process and decision-making points. Too specific a research question can lead the researcher down a rabbit hole.) • Too vague or unspecified. (Thorough review of literature and refining the question can help.) • Too many research questions at once. • Does not fit method selected.
Data collection	• Lack of purposive sampling process. • Underestimating the volume of data that the research might generate. • Linear instead of iterative. • Lack of participant diversity.
Data analysis	• Missing complexity, contradictions in data (superficial analysis of data). • Analyzing data with preconceived notions (unconsciously looking for bias confirmation instead of looking for contradictions and complexity). • Confusing organizing data with analyzing data. (Software programs help organize data but cannot replace a researcher's thinking, conceptualizing, and interpretation of the data.) • Lack of self-awareness and reflexivity.
(Re)presentation	• Failure to communicate why the method or methodology of data was chosen out of the many available. • Inadequate description of how participants were selected and/or how data were collected and analyzed. • Misuse of terminology, or use of jargon without precise definitions.

Figure 2.2. Common mistakes that beginner qualitative researchers should avoid.

Conducting and reflecting on the armchair walk-through for methodological purposiveness and congruence, and remaining open and sensitive to emerging data are just a few examples of how to ensure a qualitative study is rigorous. Another effective way to proceed is to collaborate with an experienced qualitative researcher.

As daunting as embarking on the first qualitative journey may seem, it is well worth the effort. If empirical ecocriticism is to further our understanding of the powerful relationship between art and action, it will need to avoid scientism, reductionism, and positivism and find ways to do justice to the complexity of human experience. Qualitative research is uniquely situated to do so. It takes seriously the epistemic import of the poetic and the aesthetic, and avoids treating works of art as psychological stimuli; it allows for ambiguity, messiness, and rich polyvalence without sacrificing rigor. Qualitative research promotes a kind of "burrowing down into the depths of the particular" and a "hovering in thought and imagination around the enigmatic complexities of the seen particular," which evokes in researchers, as well as those who read the research, "a process of reflection and (self)-discovery" (Nussbaum 2001, 69). In its effort to engender the change of mind and heart that humanity is in dire need of to secure its survival, empirical ecocriticism should turn to qualitative research for support.

Notes

1. *Abduction* refers to the reasoning involved in generating and justifying hypotheses and is often referred to in modern literature as "inference to best explanation." In contrast to deduction, abduction is the logical form responsible for generating new ideas. For a thorough discussion of abduction, see Eco (1988).

2. In support of this view, see Giorgi (2012) and Van Manen (1997, 2017). For a discussion critical of this view, see Zahavi (2019, 2020).

3. At this point a quick note is in order to counter charges of naive essentialism. Husserl (1983) distinguishes exact essences from morphological essences, holding that the former can only be intuited for intentional objects pertaining to the exact sciences, such as geometry. For the descriptive sciences, in contrast, which do not operate with ideal concepts such as triangles and squares, it is an epistemological fallacy to presuppose "exactness in the essences themselves which are seized upon" (165). Rather than with exact essences, the descriptive sciences deal with morphological essences, which are vague and "fluid," and which are "directly seized upon on the basis of sensuous intuition" (166).

4. Some examples of phenomenological qualitative research include Wertz (2011) and Sopcak (2010).

5. Scholars looking for examples of research questions applied in phenomenological studies of literary reading may want to consult Collier and Kuiken (1977); Kuiken and Miall (2001); Sikora, Kuiken, and Miall (2011); Sopcak (2011, 2013); and Sopcak and Kuiken (2012).

6. Petitmengin's microphenomenology is a new and increasingly popular form of phenomenological inquiry in which not only researchers but also participants untrained in phenomenology perform the phenomenological reduction (Petitmengin 2006; Petitmengin, Remillieux, and Valenzuela-Moguillansky 2019).

7. While Margaret Sandelowski (2000, 2010) is not the originator of qualitative description, her articles provide useful information on topics such as sampling, data collection, and analysis and representation of data.

8. Thorne (2008) provides a detailed and comprehensive guide for researchers interested in using interpretive description.

References

Adams, Catherine, and Max Van Manen. 2008. "Phenomenology." In *The Sage Encyclopedia of Qualitative Research Methods,* edited by Lisa Given, 615–19. Thousand Oaks, Calif.: Sage.

Bieniek-Tobasco, Ashley, Sabrina McCormick, Rajiv N. Rimal, Cherise B. Harrington, Madelyn Shafer, and Hina Shaikh. 2019. "Communicating Climate Change through Documentary Film: Imagery, Emotion, and Efficacy." *Climatic Change* 154 (1–2): 1–18.

Bilandzic, Helena, and Freya Sukalla. 2019. "The Role of Fictional Film Exposure and Narrative Engagement for Personal Norms, Guilt and Intentions to Protect the Climate." *Environmental Communication* 13 (8): 1069–86.

Charmaz, Kathy. 2006. *Constructing Grounded Theory: A Practical Guide through Qualitative Analysis.* Thousand Oaks, Calif.: Sage.

Cohen, Deborah J., and Benjamin F. Crabtree. 2006. "Qualitative Research Guidelines Project." Robert Wood Johnson Foundation, July 2006. http://www.qualres.org/HomeComm-3869.html.

Collier, Gary, and Don Kuiken. 1977. "A Phenomenological Study of the Experience of Poetry." *Journal of Phenomenological Psychology* 7:250–73.

Creswell, John W. 1998. *Qualitative Inquiry and Research Design: Choosing among Five Traditions.* Thousand Oaks, Calif.: Sage.

Creswell, John W., and William E. Hanson, Vicki L. Plano Clark, and Alejandro Morales. 2007. "Qualitative Research Designs: Selection and Implementation." *Counseling Psychologist* 35:236–64.

Daly, Kerry J. 2007. *Qualitative Methods for Family Studies and Human Development.* Thousand Oaks, Calif.: Sage.

Dilthey, Wilhem. 1962. "Einleitung in die Geisteswissenschaften. Versuch einer Grundlegung fuer das Studium der Gesellschaft und der Geschichte." In *Gesammelte Schriften,* edited by Bernhard Groethuysen. Goettingen, Ger.: Vandenhoeck and Ruprecht.

Duffy, Kathleen, Colette Ferguson, and Hazel Watson. 2004. "Data Collecting in Grounded Theory: Some Practical Issues." *Nurse Researcher* 67: 67–78.

Eco, Umberto. 1988. *The Sign of Three: Dupin, Holmes, Peirce.* Bloomington: Indiana University Press.

El Akkad, Omar. 2017. *American War.* Toronto: McClelland & Stewart.

Fetterman, David M. 2010. *Ethnography: Step-by-Step.* 3rd ed. Los Angeles: Sage.

Flick, Uwe. 2002. *An Introduction to Qualitative Research.* Thousand Oaks, Calif.: Sage.

Gadamer, Hans-Georg. 1982. *Truth and Method.* New York: Crossroad.

Gendlin, Eugene T. 1978–79. "Befindlichkeit: Heidegger and the Philosophy of Psychology." *Review of Existential Psychology and Psychiatry* 16 (1–3): 43–71.

Ginn, Franklin, Michelle Bastian, David Farrier, and Jeremy Kidwell. 2018. "Introduction." *Environmental Humanities* 10 (1): 213–25.

Giorgi, Amedeo. 2012. "The Descriptive Phenomenological Psychological Method." *Journal of Phenomenological Psychology* 43 (1): 3–12.

Glaser, Barney G. 1978. *Theoretical Sensitivity: Advances in the Methodology of Grounded Theory.* Mill Valley, Calif.: Sociology Press.

Glaser, Barney G. 2002. "Constructivist Grounded Theory?" *Forum Qualitative Sozialforschung / Forum: Qualitative Social Research* 3 (3): article 12.

Glaser, Barney G., and Anselm L. Strauss. 1967. *The Discovery of Grounded Theory: Strategies for Qualitative Research.* Chicago: Aldine.

Hammersley, Martyn, and Anna Traianou. 2012. *Ethics in Qualitative Research: Controversies and Contexts.* London: Sage.

Howell, Rachel A. 2011. "Lights, Camera . . . Action? Altered Attitudes and Behaviour in Response to the Climate Change Film *The Age of Stupid.*" *Global Environmental Change* 21 (1): 177–87.

Husserl, Edmund. 1983. *Ideas Pertaining to a Pure Phenomenology and to a Phenomenological Philosophy: First Book.* Translated by F. Kersten. The Hague: M. Nijhoff.

Husserl, Edmund. 2001. *Analyses Concerning Passive and Active Synthesis: Lectures on Transcendental Logic.* Dordrecht: Springer.

Kuiken, Don, and David S. Miall. 2001. "Numerically Aided Phenomenology: Procedures for Investigating Categories of Experience." *Forum Qualitative Sozialforschung / Forum: Qualitative Social Research* 2 (1): article 15. https://doi.org/10.17169/fqs-2.1.976.

Kvale, Steinar. 1983. "The Qualitative Research Interview: A Phenomenological and a Hermeneutical Mode of Understanding." *Journal of Phenomenological Psychology* 14 (1–2): 171–96.

Lambert, Vickie A., and Clinton E. Lambert. 2012. "Editorial: Qualitative Descriptive Research: An Acceptable Design." *Pacific Rim International Journal of Nursing Research* 16 (4): 255–56.

Liamputtong, Pranee. 2010. *Performing Qualitative Cross-Cultural Research.* Cambridge: Cambridge University Press.

Lincoln, Yvonna S., and Egon G. Guba. 1985. *Naturalistic Inquiry.* Beverly Hills, Calif.: Sage.

Malecki, Wojciech, Bogusław Pawłowski, and Piotr Sorokowski. 2016. "Literary Fiction Influences Attitudes toward Animal Welfare." *PLoS One* 11 (12): e0168695.

Malecki, Wojciech, Bogusław Pawłowski, Marcin Cieński, and Piotr Sorokowski. 2018. "Can Fiction Make Us Kinder to Other Species? The Impact of Fiction on Pro-animal Attitudes and Behavior." *Poetics* 66:54–63.

Mason, Jennifer. 1996. *Qualitative Researching.* Thousand Oaks, Calif.: Sage.

Mayan, Maria J. 2009. *Essentials of Qualitative Inquiry.* Walnut Creek, Calif.: Left Coast.

Merleau-Ponty, Maurice. 1962. *Phenomenology of Perception.* Translated by Donald Landes. London: Routledge & Kegan Paul.

Monani, Salma, Sarah Principato, Dori Gorczyca, and Elizabeth Cooper. 2018. "Loving Glacier National Park Online: Climate Change Communication and Virtual Place Attachment." In *Handbook of Climate Change Communication: Vol. 3,* edited by Walter Leal Filho, Evangelos Manolas, Anabela Marisa Azul, Ulisses M. Azeiteiro, and Henry McGhie, 63–83. Cham, Switzerland: Springer International.

Morrison, Zachary J., David Gregory, and Steven Thibodeau. 2012. "Thanks for Using Me: An Exploration of Exit Strategy in Qualitative Research." *International Journal of Qualitative Methods* 11:416–27.

Morse, Janice M. 1992. *Qualitative Health Research.* Newbury Park, Calif.: Sage.

Morse, Janice M., and Peggy Anne Field. 1995. *Qualitative Research Methods for Health Professionals.* Thousand Oaks, Calif.: Sage.

Nussbaum, Martha C. 2001. *The Fragility of Goodness: Luck and Ethics in Greek Tragedy and Philosophy.* Cambridge: Cambridge University Press.

Oliver, Carolyn. 2012. "The Relationship between Symbolic Interactionism and Interpretive Description." *Qualitative Health Research* 22:409–15.

Petitmengin, Claire. 2006. "Describing One's Subjective Experience in the Second Person: An Interview Method for the Science of Consciousness." *Phenomenology and the Cognitive Sciences* 5 (3–4): 229–69.

Petitmengin, Claire, Anne Remillieux, and Camila Valenzuela-Moguillansky. 2019. "Discovering the Structures of Lived Experience: Towards a Micro-phenomenological Analysis Method." *Phenomenology and the Cognitive Sciences* 18 (4): 691–730.

Reid, Anne-Marie, Jeremy M. Brown, Julie M. Smith, Alexandra C. Cope, and Susan Jamieson. 2018. "Ethical Dilemmas and Reflexivity in Qualitative Research." *Perspectives on Medical Education* 7 (2): 69–75.

Richards, Lyn, and Janice Morse. 2007. *Readme First for a User's Guide to Qualitative Methods.* Thousand Oaks, Calif.: Sage.

Ryan, John Charles. 2018. "Ecocriticism." In *The Year's Work in Critical and Cultural Theory* 27 (1): 100–22.

Saldana, Johnny, Patricia Leavy, and Natasha Beretvas. 2011. *Fundamentals of Qualitative Research.* Oxford: Oxford University Press.

Sandelowki, Margarete. 2000. "Focus on Research Methods: Whatever Happened to Qualitative Description?" *Research in Nursing and Health* 23: 334–40.

Sandelowski, Margarete. 2010. "What's in a Name? Qualitative Description Revisited." *Research in Nursing and Health* 33 (1): 77–84.

Schneider-Mayerson, Matthew. 2018. "The Influence of Climate Fiction: An Empirical Survey of Readers." *Environmental Humanities* 10 (2): 473–500.

Schneider-Mayerson, Matthew. 2020. "Just as in the Book? The Influence of Literature on Readers' Awareness of Climate Injustice and Perception of Climate Migrants." *ISLE: Interdisciplinary Studies in Literature and Environment* 27 (2): 337–64.

Schwandt, Thomas A. 2001. *Dictionary of Qualitative Inquiry.* Thousand Oaks, Calif.: Sage.

Seymour, Nicole. 2018. *Bad Environmentalism: Irony and Irreverence in the Ecological Age.* Minneapolis: University of Minnesota Press.

Sikora, Shelley, Don Kuiken, and David S. Miall. 2011. "Expressive Reading: A Phenomenological Study of Readers' Experience of Coleridge's *The Rime of the Ancient Mariner.*" *Psychology of Aesthetics, Creativity, and the Arts* 5 (3): 258–68.

Smith, Joel. 2005. "Merleau-Ponty and the Phenomenological Reduction." *Inquiry* 48 (6): 553–71.

Sommer, Laura Kim, and Christian Andreas Klöckner. 2019. "Does Activist Art Have the Capacity to Raise Awareness in Audiences? A Study on Climate Change Art at the ArtCOP21 Event in Paris." *Psychology of Aesthetics, Creativity, and the Arts* 15 (1): 60–75.

Sopcak, Paul. 2010. "In Memoriam: The Experience of Eulogizing a Loved One." *Phenomenology and Practice* 4 (1): 88–96.

Sopcak, Paul. 2011. "A Numerically Aided Phenomenological Study of Existential Reading." In *De Stralende Lezer: Wetenschappelijk Onderzoek Naar De Invloed Van Het Lezen,* edited by Frank Hakemulder, 123–52. Den Haag, The Netherlands: Stichting Lezen.

Sopcak, Paul. 2013. "'He Made Her Feel the Beauty': Readers' Responses to Maurice Blanchot and Virginia Woolf's Treatments of Finitude." *Scientific Study of Literature* 3 (2): 209–39.

Sopcak, Paul, and Don Kuiken. 2012. "Readers' Engagement with Virginia Woolf's *Mrs. Dalloway:* From Knowing about Death to the Experience of Finitude." *Mémoires du livre* 3 (2). https://doi.org/10.7202/1009348ar.

Suddaby, Roy. 2006. "From the Editors: What Grounded Theory Is Not." *Academy of Management Journal* 49 (4): 633–42.

Thorne, Sally. 2008. *Interpretive Description.* Walnut Creek, Calif.: Left Coast.

Thorne, Sally, Sheryl Reimer Kirkham, and Katherine O'Flynn-Magee. 2004. "The Analytic Challenge in Interpretive Description." *International Journal of Qualitative Methods* 3 (1): 1–11.

Van Manen, Max. 1997. *Researching Lived Experience: Human Science for an Action Sensitive Pedagogy.* London, Ontario, Canada: Althouse Press.

Van Manen, Max. 2017. "Phenomenology in Its Original Sense." *Qualitative Health Research* 27 (6): 810–25.

Van Manen, Max. n.d. "Inquiry." Phenomenology Online, accessed May 27, 2022. http://www.phenomenologyonline.com/inquiry/.

Wertz, Frederick J. 2011. *Five Ways of Doing Qualitative Analysis: Phenomenological Psychology, Grounded Theory, Discourse Analysis, Narrative Research, and Intuitive Inquiry.* New York: Guilford.

Wuest, Judith. 2007. "Grounded Theory: The Method." In *Nursing Research: A Qualitative Perspective,* edited by Patricia L. Munhall, 239–71. Sudbury, Mass.: Jones and Bartlett.

Wuest, Judith, and Marilyn Merritt-Gray. 2001. "Feminist Grounded Theory Revisited: Practical Issues and New Understandings." In *Using Grounded Theory in Nursing,* edited by Rita S. Schreiber and Phyllis N. Stern, 159–76. New York: Springer.

Zahavi, Dan. 2019. "Applied Phenomenology: Why It Is Safe to Ignore the Epoché." *Continental Philosophy Review* 54: 259–73.

Zahavi, Dan. 2020. "The Practice of Phenomenology: The Case of Max van Manen." *Nursing Philosophy* 21 (2): e12276.

Chapter 3

Exploring the Environmental Humanities through Film Production

REBECCA DIRKSEN, MARK PEDELTY, YAN PANG, AND ELJA ROY

If we accept as our starting point—as this volume's editors do in their Introduction—that environmentally engaged texts matter for social, cultural, and political reasons, then surely we must hold the conviction that the ways in which these texts are created and produced matter as well. That is, as environmentally concerned scholars invested in ecocritical debate, we ought to take concerted interest in the methodologies behind first the making of these texts, whether literary, sonic, filmic, or of some other media format, and second the studying of these texts. This chapter reflects on the exploratory process of defining one such text-making methodology by an interdisciplinary group of scholars and their colleagues at five separate research sites around the world. Initiated by a Humanities without Walls (HWW) consortium challenge to examine "The Work of the Humanities in a Changing Climate," we have completed a series of music videos that speak to urgent environmental issues in Bangladesh, China, Haiti, Tanzania, and the U.S./Canadian border region of the Salish Sea.[1] The thematic issues portrayed in these videos range from noise and water pollution to deforestation and declining biodiversity, and the musical treatments—from choice of musical genre to perspectives on musical activism—vary greatly. With environmental justice at the heart of our collaboration, each video issues a call to action directed at local and global audiences.

We came to these field site conversations with the concept of a methodology that we call field to media, a term intended to describe the pragmatic use of video production during research to study and amplify ecomusical responses to environmental challenges (Pedelty et al. 2020). Generally speaking, the goal is to marry the classic ethnographic practices of active participant observation and participatory action research with an applied or activist bent that focuses on film as the medium for communication. Participant observation is a standard methodology in disciplines concerned with ethnographic research as a means of data collection (anthropology, ethnomusicology, communication studies, sociology) that emphasizes immersive observation and participation in the contexts in which the researcher is living and studying. Active participant observation places greater emphasis on the researcher's direct and active participation than on a more passive form of interaction through observation; it assesses the researcher's carefully considered acts of participation themselves as advantageous and essential to the process of gathering quality data (Johnson, Avenarius, and Weatherford 2006). Further emphasizing the importance of active involvement, participatory action research (PAR) is a key methodology in ethnography-based disciplines that emphasizes democratization of knowledge and deep collective inquiry (based on collaboration between researcher and the community/group participating in that research) in order to effect change for the better. The emphasis is on research and action taken with people, not for or on behalf of people. Practitioners of this methodology often cite influences from critical pedagogy stemming from Freirian educational philosophy (Freire [1970] 2000; MacDonald 2020b; Steinberg and Down 2020) and relate their PAR practice to social justice movements. PAR has been used especially in regions classified as the Global South—for example, in Latin America and South Asia (Reason and Bradbury 2007).

With field to media, our goal is to combine the strengths of these core methodologies, then extend it a step further by positioning filmmaking as the key mode of action and the focal point for generating discussion. We aim to think of each step—from conceptualization and planning to production and distribution—as integral to our efforts at cocreation and negotiation around which we debate ideas vastly larger than the music videos themselves—in this case, related to the environment. As film work becomes increasingly prioritized as a mode of music-related scholarship (D'Amico 2020; Gubner 2018; Harbert 2018; MacDonald 2020a; Ndaliko 2016; Ranocchiari and Giorgianni

2020), these efforts to expand the methodological tool kit of ethnographic and experiential research are timely and necessary.

In further defining and detailing this methodology, it is instructive to draw out the commonalities between the five case studies.[2] As we reflect on the entire experience to date, it is first apparent that by actively participating in cocreating music videos, we have emphasized the sensorial and experiential learning grounded in the numerous exchanges and negotiations it took to make these performances happen. This experiential learning also extended to the intimate involvement in the artistic and technical aspects of producing, filming, and editing.[3] Second, we have committed to an investigation of musicking beyond the "simple" act of music making, but an investigation encompassing everyone and everything that contributes to the making of a musical event (Small 1988). Third, we all have placed extra attention on the development of communication strategies that reflect the values of each place-based local team (in Bangladesh, China, Haiti, Tanzania, and the Salish Sea), and that are guided by local research partners. We have followed our respective partners' leads on how to connect to viewing audiences in relevant and culturally resonant ways about the environmental or ecological issues that they perceive as being most urgent to address, recalling the feedback cycle of Jean Rouch's (2003) shared anthropology. Fourth, we all hold Freirean pedagogy as a fundamental principle to learning, discerning, and communicating through collective, collaborative processes—not as a coherently defined set of politically neutral acts, but rather more as a locus for action, in some cases perhaps even moving toward revolutionary action (McLaren 2000).

Some of the distinctions between the projects described in this chapter include the following. Pedelty emphasizes bimusicality as a feature of his work in the Salish Sea region. Formally defined as studying music from a tradition other than one's own by learning to perform in the tradition of interest in order to directly experience the technical demands as well as the conceptual and aesthetic aspects of its creation (Hood 1960), bimusicality, in Pedelty's hands, becomes a critical method of performing music with coproducers in the studio and field. It is a practice of learning to hear, interpret, and articulate from an intersubjective perspective. Roy and Dirksen highlight the leadership of activists—both musical and otherwise—who have long been invested in struggles against environmental injustice. For her contributions, Roy joined forces with a cultural collective and musical group that was already active in opposing plans for an energy plant development project that would compromise the biodiversity of the Sundarbans' cherished mangroves. In Dirksen's case, this work is about aligning politically attuned environmental activism with the long tradition of Haiti's *mizik angaje* (politically and socially engaged music; Dirksen 2020) and the practice of applied ethnomusicology, which aims to reach far beyond academia's institutional setting (Dirksen 2012). In contrast, Pang choreographs, composes, and dances with artistic collaborators on her way to staging a visually and sonically arresting interpretive account of a river and its contamination. She draws on her childhood in southwestern China and is influenced by a transnational range of genres, from Sichuan opera to Western classical music of the European concert hall to breaking and other street dance forms that initially rose in the United States.

Because this methodology is premised on direct experience, participant observation, interviewing, experimental filmmaking, and what has effectively turned out to be extended focus group study, it falls firmly within the realm of empirical ecocriticism that this volume is geared toward developing. In these projects, our empirical explorations are mostly qualitative as we explore individual and collective articulations of environmental activism via filmmaking with local colleagues. Through our conversations about and participation in making eco–music videos, we are driven to better understand context, meaning, phenomenology, and critical epistemology. These understandings are acquired experientially and sensorially.

This emphasis on experiential and sensorial learning draws us back to Ofer Gal and Raz Chen-Morris (2010), who remind us that early empiricists such as Locke (1690) foreground the human senses. Later empiricists, however, are highly suspect of our direct, sensory experience. As exploratory technologies developed—telescope, microscope, camera, and so on—most empiricists shift to the view that "accurate scientific observation meant that we are always wrong" (Gal and Chen-Morris, 2010, 121). Although Gal and Chen-Morris may overstate the case, it is helpful to observe how the increasing use of extrasomatic technologies—that is, apparatuses that enhance and extend the human senses—to explore the world took place alongside the development of exogenous techniques to validate knowledge. That dialectical shift in technology and practice led to a radical change in the very definition of empiricism.

While our purpose here is to provide a thick description of our individual experiences with this methodology, rather than reflect on empirical epistemology as a whole, it is worth noting that our film-as-fieldwork methodology integrates both notions of empiricism: it is deeply based on direct sensory experience (active participant observation fieldwork), and it is mediated by extrasomatic technologies that integrate and extend the senses. In fact, our field-based experiences are more heavily augmented by extrasomatic technologies, such as the camera and microphone, as tools for getting at cocreative ways of knowing than is the ethnographic norm. Such technologically enhanced interactions parallel the ways that many people live in an increasingly digitally mediated world. In other words, our methodology is matched to our ontology—perhaps more so than the Malinowskian (1922) methods of participant observation that many of us learned during our graduate training. Cognizant of the complex history of visual anthropology that has long informed ethnomusicology (Baily 1989; Feld 1976; Zemp 1988), we purposefully add camera, microphone, and editing software to the traditional armamentarium of pen and paper, both while in the field and when representing that experience later. Yet the real distinction is that our collaborators are on both sides of the camera and microphone, rather than simply being viewed and recorded through the ethnographer's eye (Grimshaw 2001). In other words, while arguing for a deeply intersubjective and interpretivist field-based methodology, we also claim that the ethical, conscious, creative, and collaborative utilization of media production techniques can be particularly useful when engaging complex environmental challenges at the community level.

Furthermore, in line with ecocritical analysis, with these film projects we have been simultaneously concerned with studying the narrative strategies and techniques being used in the texts (songs and music videos) that we are helping to create in real time. As a team with research partners spread around the globe, we are part of the conversation as decisions about representation get made, and we can interpret and readily discuss related experiences, points of view, attitudes, goals, concerns, and hopes with our colleagues, many of whom have long been involved with environmental activism. In most of the individual case studies, the music videos we have created together are another step forward in our and our collaborators' efforts toward environmental justice. In each case, this has opened up space for cross-cultural partnerships around specific issues of local and global concern.

With the broad range of geographic, cultural, and environmental contexts represented across the research team's individual projects, we have each encountered different needs and concerns and have had to adapt our process-oriented approach accordingly. In figuring out how to proceed in each case, we have devoted considerable time and effort to reflexively examining the conversations and actions behind the scenes, recognizing that these spaces offer possibilities to study the creation of meaning and complex negotiations of social, cultural, and political values. Sometimes behind-the-scenes dialogue has been recorded (through film or audio) to facilitate returning to the conversations for analysis. We therefore use the remainder of this chapter to illustrate our practice via a series of field-based vignettes. U.S.-based team members present their experiences with this research process in turn.

Project Site: The Salish Sea (Reflections from Mark Pedelty)

On Thursday morning, September 20, 2018, a group of us filmed a scene for "LOUD," a music video about noise pollution in the Salish Sea region of western Washington and British Columbia (Pedelty 2018). Karl Demer served as the sound engineer, camera operator, and editor, while I produced and directed the video. Karl and I had arrived in Sooke, British Columbia, the night before, so we were not able to scout the setting during daylight. You can imagine our delight when waking to one of the most idyllic seaside settings a production team could hope for.

After breakfast, Karl and I went to work setting up the day's film shoot. One by one, the talent started showing up: professional singer-songwriter Felicia Harding, body artist Catriona Armour, Victoria-based vocalist Nadine "Let it Shine" Langelo, and Wunderkind singer Hazel MacPherson (Figure 3.1). While Karl set up the audio and video equipment, I dealt with an angry property manager: the property owners had not informed the manager that we had their permission to shoot that day. I ordered and picked up food for the talent and crew, then welcomed and oriented each singer as they arrived. Such are a community-based media producer's typical tasks.

Figure 3.1. Hazel McPherson and Nadine Langelo perform in the music video "LOUD." Image credit: Karl Demer.

Upon arrival, Catriona started the laborious process of applying makeup to each performer. She transformed Felicia into a mermaid-like creature, signifying the space between humans and sea creatures. We had originally decided on a design that would evoke the orca, but that presented unanticipated challenges. So Catriona chose a more abstract set of visual references to the sea, including a kelp-like motif. Felicia's character performed the video's central appeal for quietude in the Salish Sea's increasingly loud and decreasingly biodiverse soundscape.

Despite Catriona's exceptional artistry, my choice to feature body art did not really work. That was mainly my fault as producer and director. However, an international border and university accounting policies played their parts as well, confounding our original plan to gather the binational talent in one place. We originally planned on having a single set, on Orcas Island in Washington State, with all of the characters costumed and painted so that a more singular and coherent vision would extend across the entire video. They were to be liminal characters interacting and performing together in the littoral zone between land and sea. Unfortunately, we were not able to bring the Canadian artists the ten miles from British Columbia to Orcas Island. Having the entire collaborative team in one place, working together, had succeeded in previous Ecosong.net productions, but it was not to be in this case.[4] Visa troubles, a postal strike, an Indigenous drummer's illness, and a university bureaucracy conspired to frustrate our original plans, which made bringing that coherent vision into view impossible. Yet by experiencing these challenges, I learned several things about the area, the soundscape, and topic that might have otherwise remained hidden to me. Although the resulting music video was somewhat disjointed, making "LOUD" was a revelatory field experience.

One of the most rewarding aspects of the "LOUD" project was the way that it revealed the process of community in the making. Communities use ritual to form, organize, communicate, and mobilize. Community filmmaking is therefore a ritual process (Koch 2019). By singing, acting, and filming together, the intentional community formed in Sooke expressed their collective feelings about noise pollution, even if the entire video cast could not convene that day. Undertaken in a creative, ritually charged context, arts-based research can reveal tacit cultural truths (Takach 2016). Two vignettes from the Sooke shoot will be used to illustrate the point.

The first has to do with interspecies communication and, if it is not too much of a stretch, coproduction. As we filmed atop the cliff, Hazel noticed that we were not alone. About fifty feet down the cliff, a couple of Steller sea lions had come to see what we were doing. Seagulls swirled above their heads, hoping for the giant pinnipeds to chum up a salmon meal. After a while the massive mammals went back to hunting, but they returned throughout the day to remind us of what was at stake.

The second vignette illustrates one of the main affordances of coproduction. As mentioned above, various financial and logistical challenges affected the production (Pedelty 2020a). These factors conspired to make it difficult to hire a professional costumer, for example, and my attempts to secure an affordable professional or community volunteer all failed. Moreover, I was reticent to give advice to any of the singers when it came to costuming, especially to the women. For example, I asked Hazel and her mother to make their costuming decisions autonomously.

This led to a simple yet transformative interaction when Hazel called on the collective wisdom of the group to make her final choices. "Would it be OK if I wore this feminist power button?" she asked. Catriona, Nadine, and Felicia answered yes, enthusiastically and in unison. That moment actualized the collective decision-making principles on which the project was predicated. In such moments, the spirit of coproduction are viscerally felt and the intent of arts-based community research realized. An ecofeminist thread was introduced at that point to the project, with Hazel at the lead, which has since been worked more fully into the warp and weft of the project as a whole. That ethos helped propel the production of a follow-up documentary concerning sound pollution in the same region, featuring the environmental organizing efforts of Donna Sandstrom and Sorrel North, women who have challenged the whale-watching industry's sonic harassment of southern resident killer whales. *Sentinels of Silence? Whale Watching, Noise, and the Orca* (Pedelty 2020b) resulted directly from the film-as-fieldwork experience outlined here, and features the voices of community leaders that I met as a result of shooting and distributing "LOUD." Such revelatory experiences have led me to believe that fieldwork based on coproduction can be as useful as traditional forms of participant observation. Film as fieldwork—and field to media—is empirical research in the most literal sense, predicated on direct sensory experience in and within a community.

Project Site: Bangladesh (Reflections from Elja Roy)

"Dwellers of the Forest Arise" is in itself a call for action (Roy 2019). The song presented in the music video, a live performance, describes the mangrove forest of the Sundarbans and how a proposed coal-burning power plant (the Rampal power plant) will harm its flora and fauna. "Dwellers of the Forest Arise" urges listeners to be aware of the situation and to join the movement in saving the forest. The song is composed and performed by Samageet, a Bangladesh-based performing arts group (Figure 3.2). *Sama* means "equal" and *geet* means "songs," so the band's name means "songs of equality."

Sundarbans is an archipelago and the world's largest mangrove forest. It is shared between the Indian state of West Bengal and Bangladesh. One third of the archipelago is in West Bengal, and the rest is in Bangladesh. Growing up in West Bengal, I learned not only about the unique biodiversity of the forest but also how the rising sea level and encroaching oil and coal business threaten its existence. I had been involved with projects concerning the preservation of the forest since 2014, but I went to Bangladesh

Figure 3.2. Samageet performs in "Dwellers of the Forest Arise" music video. Image credit: Ishanu Chakrabarty.

for the first time in summer 2018 as part of the HWW collaboration. Filming "Dwellers of the Forest Arise" was a result of months of high-quality dialogues (Freire [1970] 2000) between me, videographer Ishanu Chakrabarty, and Samageet. I first established contact with the group via Minnesota-based environmental journalist Jeremy Hance. Digital tools such as emails and later social media sites (Facebook and Instagram) allowed me to continue the dialogues and stay connected with my coproducers during the pre- and postproduction stages. Similar to observations made by Linda J. Seligmann and Brian P. Estes (2020), I found that social media and messaging (emails, WhatsApp, and Messenger) created a sense of hyperconnectivity between me and my interlocutors, who soon became my coproducers, in a still-ongoing exchange.

After our initial correspondence in English, I continued the dialogue—as my collaborators recommended—in Bengali, which is my native language and the national language of Bangladesh. My identity as a Bengali played a crucial role in facilitating conversation and collaboration. Although I speak Bengali, I couldn't completely identify as a "native" ethnographer (Ugwu 2017) because I was from a foreign country, India. Thus, my positionality is simultaneously that of a native (because Bengali is my native language) and a nonnative (because I am an Indian) ethnographer. This complicated positioning presented some challenges: where the language instinctively united us, the geopolitics divided us. Although the proposed Rampal power plant is a joint venture between India and Bangladesh, public sentiment in Bangladesh sees most of the anticipated political and economic advantages accruing to India, so most of the protests were against Indian foreign policies. Hence, my role as a Bengali helped me, but at the same time, my Indian identity initially created discomfort, as Samageet's most popular song demands directly in its secondary title, "Go Back, Get Out India." (This same song is also separately titled with a call to action for local residents: "Dwellers of the Forest Arise.") The native ethnographer in me constantly had to defamiliarize myself from what I had taken for granted, and the nonnative one had to embrace the new.

Upon arrival, I met my coproducers before the film shooting day. During these meetings, I discussed the nature of my project and the HWW grant, which was to film the songs they have written and performed to save the forest. With a warm welcome from the group, a cordial sense of coproduction was created. The preproduction discussions included information about the nitty-gritty of logistics and transportation to the shooting location. The meetings also gave me a better sense of what the Bangladeshi Sundarbans movement looks like. Its history, length, and breadth keyed me into the fact that cultural support for the environmental movement was not limited to performing arts, as there were art installations too. It was during these preproduction meetings that I learned about the role of Dhaka University as a forum where all these artists-activists come together to meet, rehearse, and perform, since most of them are current or former students of the university. A clear set of shared goals and expectations were also set. We talked about what the research means to me, how I planned to circulate the video (e.g., via international conferences and festivals), and what they think they will get out of the experience. I consulted with my interlocutors and/or coproducers on "not only in how research is done—including what kinds of data are gathered" (Seligmann and Estes 2020, 183), but also on how the project will be shaped and how it will be presented and/or distributed.

Even though I was the director and producer of the music video, Samageet was free to interpret the aesthetics and execution of the songs. Instead of directing the group's members according to filmic conventions, the videographer and I decided to approach the process the other way around: the performers took the lead. We took an inductive approach and stood at the center of their performance. I used a GoPro to get wide-angle footage, and the videographer used a high-quality digital DSLR camera to focus on individual performers and instruments. It not only helped us to capture the embodied experiences of the performers but also allowed us to be integrated into the performance instead of being mere observers. We moved, crouched, and walked along with the performers, embodying the action as they sang and danced. The idea was to keep the spirit of a street protest alive: we would not stay away from the performers and keep the camera static, but would participate as media makers.

The editing table gave me and the editor, Karl Demer, a unique opportunity to represent a live performance with footage ranging from extreme close-ups to long shots. While Ishanu's camera was particularly helpful in capturing the minute details, individual expressions, and extraordinary moments created during the performance, the GoPro's footage caught the overall energy of the performance with its wide-angle long shots.

"Dwellers of the Forest Arise" is a stand-alone piece with HWW, but it is also a part of a wider agenda, my dissertation, as well as a documentary film that speaks about the forest at large and the emerging musical environmental movement surrounding it, in both India and Bangladesh. The longer documentary film *Musical Mangrove* (Roy 2020) that followed up on the work of the "Dwellers of the Forest Arise" music video involves musical performances by both Bangladeshi and Indian artists (Roy 2021). There are performances by three Indian groups from the forests of Sundarbans (Sundarban Adibashi Loko Sanskriti Sampradaya, Adibashi Tusu Sampradaya, and Bhai Bon Sampradaya) and three urban groups from Bangladesh (Samageet, Bhatial Sohure, and Mabhoi), along with individual performers such as Farhad Syed and Kafil Ahmed.

The project has made me realize the importance of field to media as empirical ecocriticism, which otherwise might be missing from the traditional textual and historical ecocriticism. For example, the music video shoot took place on an open roof in the city of Narayanganj, Bangladesh, which ranks among the top cities for noise pollution. Hence, as expected, the noise from honking and nearby traffic interfered with the audio recording. While I was agonizing over the audio quality, my coproducers from Samageet reminded me that the noise added to the story and symbolized the country's presence in the performance. The final video still has that noise in it; there would have been no other way to capture that distinct Bangladeshi characteristic in the project if the field to media methodology were not in practice. Thus, I would echo Anna Keller, Laura Sommer, Christian A. Klöckner, and Daniel Hanss's (2020) claim that both the researchers and practitioners need to contextualize the environmental arts while exhibiting because this context impacts the audience's emotional response. "Dwellers of the Forest Arise" intends to provide that spatial context to its audiences by embodying and describing Bangladesh's distinct sonic qualities through the performance.

Project Site: China (Reflections from Yan Pang)

My first professional composition lesson was with Professor Mingzhu Song at the Sichuan Conservatory of Music more than a decade ago. My task was to write a song about the elegance of the flowing Funan river (锦江) in Sichuan, China. I contemplated the fast-moving body of water.

"Why is the Funan dark and green?" I asked.

Professor Song raised his eyebrows. "The people of Sichuan do not value this gift. They put their garbage next to it . . . and sometimes into it."

His words stayed with me over the years and inspired me to create socially conscious art. In 2018, I joined the HWW grant collective to make a difference by composing an artwork to bring attention to the pollution of my "Mother River."[5] My contribution to the group echoes the words of Song, a composer who showed me the importance of clean water for my people (Pedelty et al. 2020).

When I returned to China for this HWW project some years after that first lesson with Song, I thought I would be helping to create a straightforward piece about the environment. However, my expectations about the balance between artistry and accessibility shifted after visiting the artists at the Sichuan Conservatory of Music. As part of my field study, I conducted a field trip to the Funan River and met with several contacts in southwestern China who gave me insights into the community in Sichuan province (including in the cities of Chengdu and Neijiang) and their neighbors (including within Guizhou province). The individual who contributed the most to shaping my understanding was Mingzhu Song, who has lived in this community for his whole life. From him, I learned about his Sichuan opera heritage, contextualized within the environment of the community.

While my collaborators were interested in working on this project, I encountered resistance when I asked them to speak for themselves in front of a camera. This resistance was the result of two main challenges grounded in the country's political tensions: regulated speech and interviewees' restricted travel. Because my collaborators knew the project was funded by an American institution, and therefore would be widely distributed and would perhaps have some international influence, they expressed concern about appearing critical of the Chinese government's lack of effort to address environmental issues in China.

While filming individual perspectives carried personal risk to my collaborators, the threat of repercussions also extended to artistic performances. For example, when I was conducting my fieldwork in China, my collaborators and I discussed the possibility of filming Chinese performers on the stage in China. However, my Chinese collaborators alerted us that this type of performance and speech is restricted in China. Local governments try to create an idealized image of the country, so addressing pollution issues and speaking in public about environmental problems are considered controversial. Most of my interviewees were willing to share their perspectives on the reality and impact of chemicals being dumped into the Funan River, but they did not want to be filmed or have their names disclosed. The artists I worked with did not feel comfortable freely expressing their political views because of the possibility of repercussions from the Chinese government.

Conversely, another significant challenge I faced during the field study was collecting video materials. As an American citizen, I was not allowed to record Chinese testimonies in polluted areas. The restrictions and the sensitive political climate around criticizing the government's ineffectual efforts to keep the river clean made it challenging to acquire the necessary legal permissions to film.

To counter these challenges, I used what Mingzhu Song taught me about Sichuan opera musical elements and gestures to create my performance on the stage back in the United States. I aimed to choose the sonic elements and physical gestures that would be understood as the best symbols for conveying the environmentalist message I wanted to share. Although the Chinese participants' voices and bodies were not directly filmed, I pay homage to the people in the polluted areas by referencing their musical traditions and cultural gestures within my performance.

Moreover, while working with my team, I realized that I needed to slow down and stay grounded regarding why I was there with Song to begin with. We wanted to call attention to how the pollution affected people living there, and they wanted their voices (and music) to be heard. Centering the team's motivation—to have local artists more involved—helped all of us continue the process.

Song's work uses musical elements of Sichuan opera to embrace the cultural styles of the province, combined with contemporary compositional values. Experiencing Sichuan opera informs the listener about Chinese art, society, history, and people; it is not just music. Singing, dancing, and playing instruments convey the emotional and dramatic content of this genre. These performative elements correlate to the ways in which a river flows over rapids in the still, deep places. As I listen, the memory of cold water and questions return, push and pull my thoughts, and form the approach to the HWW project. Song's voice echoes like the running water sounds of my youth; both movement and music are necessary to tell the story of environmental awareness. Sichuan opera, created by the Indigenous people of Bashu (巴蜀), the culture that I grew up with, is at risk of dying out. Using this genre to illuminate the situation is my attempt to familiarize other cultures with Sichuan opera while simultaneously working to save the river.

Back in the United States, I worked to interweave the styles of Chinese opera and contemporary music composition to address the chemical pollution of the river as I reflected on the China team's contribution to the HWW project.[6] The goal was to raise the awareness of the local Sichuan population to the condition of the Funan River by combining Chinese, European, and American cultural forms, as well as the disciplines of music, theater, and dance (thus making this an interdisciplinary exploration), with myself as composer; pianist Jason "J-Sun" Noer as choreographer and dancer; and Stefon "Bionik" Taylor as music producer and percussionist (thus resulting in a collaborative process). In "Dancing Upstream: Current Issues of

Environmental Awareness as Performance," I used Song's approach of pure traditional single-line melodies placed in comparatively complicated harmonies and contexts based on pitch collections and figure sequences. This piece embodies China's continual struggle to be a globally connected and industrious nation, which has contributed to conditions and attitudes that have resulted in unacceptable environmental damage.

Delicate like seasonal streams, in the stage performance, J-Sun and I mirror each other's movements with arms and legs, responding to minuscule changes. I remember the strength of the Funan's current at its widest place, and I embody its fierce intensity with a burning glare at whoever would try to control me. Through memory and performance, this project takes an approach that describes important moments of moving bodies informed by lived experience, which impacts music and movement (Pang 2018b, 23). J-Sun developed a strategy to use break(danc)ing techniques to explore and think through water pollution that resulted in a choreographic framework for the artwork. J-Sun's framework follows three steps: "(1) using specific dance techniques to engage with issues like circulations of power, (2) making movement choices which are not part of the accepted canon of techniques to theorize, and (3) documenting the new knowledge produced from the approach in written, spoken, or bodily form" (Mazzola et al. 2020, 171–72).

This process resulted in a performance that blended traditional Chinese music and Western classical-influenced music performed alongside street dance forms (Pang 2018b). The performance consists of playing the piano and dancing on stage with J-Sun (Figure 3.3). A grand piano takes center stage at the beginning of the dance piece; J-Sun sits on a chair while I rest on a bench. An electronic rig surrounds Bionik, who is placed upstage right; he crafts a soundscape of running water and frog croaks. Metalworking sounds, such as the pounding of iron, transform a natural environment into an industrial one (Pedelty et al. 2020).

Along with specific techniques, like the six-step, J-Sun uses a gestural language. Such embodiments include curled fingers sharply pulling into the dancer's body to symbolize purifying the water by taking on the pollutants (Zhao 2013). The choreographic framework pushed us to research how these pollutants, like phosphorus, can be processed to make fertilizer and

Figure 3.3. Yan Pang and J-Sun starting "Dancing Upstream." Photo credit: Bill Cameron.

thus reinvigorate the natural environment (Tweed 2009). Armed with this knowledge, J-Sun adds a pushing-off gesture, a sloughing, to show the distribution of transformed phosphorus to farmlands far away from the river, while I drum on a trash can. Then, I am sitting on the bank of the river again, feeling the sunshine, and listening to the river talk. Now, I am sitting in the spotlight on stage, remembering the memories of my youth. I am the creator of waves washing away the poison while humanity clings to toxic substances. My water is clean and clear, and my music and movement perfectly align to evoke nature. At the end of the piece, I return to the overlapping moments of memory and performance. The sun lowers and the stage lights darken.

All in all, it is important to understand the uniqueness of Sichuan opera as art as well as the environmental threats to the river in Sichuan. I believe the project provides a blueprint for informing other cultures through familiar and unfamiliar art forms by identifying a common concern, ethically using art forms from the affected region, and joining the forces of regional arts with the aesthetics of globally familiar art forms. I suspect that our obstacles in relation to censorship are not rare and that ecocriticism sometimes carries risks. Researchers and art makers must be aware of this possibility and be committed to ensuring that participants are safe from negative repercussions.

Project Site: Haiti (Reflections from Rebecca Dirksen)

The Haiti-based team was the last to complete our music video, titled "M pral plante yon pyebwa" (I will plant a tree), which was ready in rough cut in November 2019 (Vérilus 2020). Featuring popular singer-songwriter BIC (Roosevelt Saillant) and noted journalist/activist Kompè Filo (Anthony Pascal) and filmed by Kendy Vérilus of Verilux Films, the song closely follows its title in encouraging listeners to care for the environment around them and plant trees. Making this music video was not easy. Between spring 2018 and fall 2019, precisely corresponding with the grant period, we faced the extraordinary circumstances of a country effectively shut down for weeks at a time as a result of a still-unfolding movement against corruption in the national government complicated by forceful interventionist international actors. Several periods of Peyi Lòk (country lock) have flared up in the course of protesting (in part) what has become known as the PetroCaribe scandal, in which numerous well-known politicians were implicated in the misdirection of funds from an oil deal with Venezuela that had been earmarked for national development projects (Beauplan et al. 2017). This situation has been confused and compounded by an increase in politically connected gang control of major roads and zones across the country. As a result, at times it has been impossible for anyone—residents and visitors alike—to circulate through certain neighborhoods or to travel between rural and urban spaces.

These conditions had a direct impact on our research project and impeded our process toward timely completion of our music video, as we found ourselves repeatedly devising and then scrapping plans when they proved unfeasible as a result of the political situation. We initially hoped to film at Lakou Souvnans, a historic sacred Vodou yard about ninety miles to the northwest of the capital Port-au-Prince that is a recognized site of national patrimony. Our objective was to highlight the traditional ecological knowledge transmitted through Vodou, the Afro-Haitian sacred worldview and metaphysical practice (Dirksen 2018), and feature well-known musicians and dancers affiliated with Souvnans. When a clear window of opportunity to travel safely through the countryside did not open up and an alternative course that would have taken us to the southern coastal city of Jacmel could not be realized for similar reasons of insecurity, with our project timeline increasingly squeezed, we pivoted to a third plan that would keep us closer to Port-au-Prince and solicited the collaboration of the artist BIC, who is widely celebrated for his politically and socially engaged lyrics (Dirksen 2020). This turned out to be the winning combination at this particular juncture, although we hope to revisit our earlier plans for eco–music videos in the future.

Given this challenging terrain, how did we make this video happen, methodologically speaking? Although frequently diminished or left out of scholarly discussion of methods and procedure, I note the ties that bind: among the few constants through many months of shifting circumstances were journalist Kompè Filo (Anthony Pascale), filmmaker Kendy Vérilus, and me. Over the course of several years, the three of us have held extended conversations about trees and ecological matters more broadly, environmental activism, and the sacred ecology of Vodou—and the best ways to communicate consciousness, meaning, and value to others. In fact, Kompè Filo, a dedicated Vodou practitioner, had for decades before his sudden death in July 2020 led a personal campaign to plant sacred mapou (kapok, or silk cotton) trees across Haiti (Dirksen and Wilcken 2021). Kompè Filo would pay neighborhood children to collect mapou seeds for him, which he cultivated into saplings until they reached several meters in height and were ready to plant into the ground. Then he sought good homes for the trees, with people who would care for them well. We wanted to find a way to complement and honor the spirit of this work through this music video project. Thus, long before the HWW project was ever conceived, we began laying the foundation for efforts like "M pral plante yon pyebwa" to be built organically through our shared interests and concerns. This music video gave us an avenue to channel and coordinate our environmental activism (Dirksen 2022).

That long-term connection and deep commitment are precisely what make such short-term projects possible as more equitable and balanced ventures. Bluntly, ethnography done collaboratively and on the basis of long-term collegial relationships allows for vastly more ethical approaches to research in disciplines wherein the primary methodologies are premised on colonial legacies and imperialist claims to knowledge.[7] Especially when the time frame is short and a product is expected, as scholars, we must continually examine and work to eliminate our involvement in the "hit-and-run" and "extractivist" activities that sustain academia as an institution. Especially for scholars employed by universities, where these legacies are not in the past and remain embedded in ongoing practices and norms, this often means shifting our stance away from our customary roles as star players to facilitators and behind-the-scenes supporters.

Our next steps in completing this music video involved introducing our musical partner BIC to our objectives for the project and our conceptual and theoretical approach. On the basis of these conversations, BIC wrote an inspired song about planting trees as a call for action. He opted for stark, minimalistic lyrics to convey the urgency of this plea. Referencing Kompè Filo's tree-planting activism, at one point BIC issues a word of caution to listeners: "Pou peyi a pa kouche, fòk nou kanpe tankou yon pye mapou" (For the country not to fall, we must stand like a mapou tree). As it is handled in the song and music video, the tree becomes a symbol of strength, dignity, and pride that the Haitian population must hold on to.

In arriving at this mobilizing message, we were well aware of the weight behind this ecoactivist route. The advanced state of deforestation in Haiti is a matter of deep concern. As a team envisioning the possibilities for this music video, we considered the complex layers of context and history: while a punishing lack of access to energy sources has led to overreliance on charcoal among those who simply cannot afford other means to cook their food, deforestation is far more accurately attributed to centuries of land overuse extending back to settler colonial plantations, with their ecologically demanding cash crops of sugar and coffee that required clearing vast numbers of trees. After independence (1804), a rapacious international logging industry flourished from the 1820s to the 1930s, stripping the western third of Hispaniola of most of its remaining old-growth hardwood forests (Bellande 2015). These subsequent but interrelated acts of resource extraction are structurally rooted in a slow violence that perniciously unfolds over time and incrementally increases pressure on those with the least power and resources (Nixon 2011). Yet with the blinding immediacy of the PetroCaribe scandal and other entangled issues that have brought extreme economic strain for most of the population, simultaneously balancing the (often) slower-developing processes of environmental degradation is extremely difficult. The long-term view—for example, of the importance of trees for human health and well-being—is hard to cultivate. With the song "M pral plante yon pyebwa," our goal was to build commitment to planting trees for the benefit of future generations.

The music video was designed to mirror the stark lyrics. It opens with BIC contemplatively gazing out across a range of mostly denuded mountaintops obscured in thick fog. It is a view from Kenscoff, far above the bustling Port-au-Prince metropolis. As the artist mourns the scene, he reflects on other matters of environmental disregard: the piles of trash that persistently collect in waterways, the extraction of the earth's minerals through uncontrolled mining. But this is about moving toward solutions rather than stagnation and further inaction. At the song's refrain (in which BIC vows, "If one day I should cut down a tree, I'll plant three in its place"), a recurring shot places Kompè Filo and BIC back to back as the camera swings around them. Both hold up first one and then three fingers in a gesture of insistence. BIC is next shown raising a pickaxe to dig a hole, into which he places a tree sapling. He is aided by an earnest group of children, who have proudly carried saplings growing from coconut shells to the site before helping place them in the ground (Figure 3.4). These are trees that we, the filmmaking team, cultivated from seeds to saplings, and which have actually been planted. One symbolic element emphasized throughout this video is the representation of multiple generations affected by environmental decline: elder, middle age, youth.

Overall, the experience of collaboratively producing this music video threw into stark relief the inherent tensions between institutional timelines and on-the-ground realities, particularly given excruciatingly difficult circumstances—besides responding in small ways to larger institutional structures with the decolonizing efforts behind this project. As the U.S.-based scholar representing the Haiti team within academic spaces (i.e., to the granting institution, via written publications and conference presentations), I found gracious patience, support, and solidarity from my HWW team members and the HWW consortium leadership, with no pressure to finish up when political realities hindered progress. Moreover, because I am

Figure 3.4. The young residents of Kenskoff and an older neighbor who starred in and helped to plant trees for the music video "M pral plante yon pyebwa." Image credit: Kendy Vérilus.

on the faculty at Indiana University and teach classes as part of my job, I am tied to the university calendar. With the significant delays and obstacles to filming—contrary to our plans, and despite repeated attempts over sixteen months, when I spent long stretches of time in Haiti—I wound up not being able to be present for the actual filming of the video in October 2019. I was hugely disappointed.

In retrospect, however, this led to a radical application of the methodology we have attempted to develop, which was ultimately largely carried to completion by local research partners. Drawing on our previous extended conversations about objectives, motivations, process, and methodology, my Haiti-based colleagues had complete control over the representation of challenging issues facing their country and compatriots. They determined the exact scenes to be filmed depending on the specific circumstances and possibilities they found on the days spent filming. They put together the narrative structure according to the sequences that were filmed. Besides adding English-language subtitles to the video, my role shifted to offering feedback during the editing process. Moreover, and more important, this project for me came to be about taking resources and the process of knowledge-based exploration out of the institutional setting and placing funding and a general orientation toward the project concept squarely into the hands of a community group, thus allowing its members to determine what to do with them, as they/we debated how to address pressing environmental concerns. This goes a long way toward reducing the "ethnographic spectacle" that has inflicted so much of the history of Western scholarly regard, especially as this gaze has intersected with constructions of race and as it has been portrayed through film (Rony 1996).

Concluding Thoughts: Activating Field to Media toward Empirical Justice

Ultimately, we are striving toward a collaborative film-based methodology premised on PAR that leads to an activist ecomusicology (Pedelty 2016). In doing so, we have explored cocreation and coauthorship not only across academic disciplines but also, and more important, beyond the academy with our respective local research teams. We cast this experimental and experiential process as empirical and ethnographic, even as we are committed to extending beyond the bounds of ethnography as usual by drawing our research partners into thinking ethnographically and filmically with us. The camera, symbolic of the entire process, literally helps us focus on defining effective narrative strategies and audiovisual techniques to convey meaning about environmental and ecological matters to local and global audiences.

This methodology is at once empirical and ecocritical. It is empirical in the sense that our process of coming to understand derives from direct experience and deep engagement in sensorial learning. It is ecocritical in the sense that we are collaboratively designing environmental texts to analyze the process of imagining and then making those texts, then reading them after they are made. One advantage that we see in comparison with more typical takes on ecocriticism is that we are intentionally getting tied up in epistemological matters at a much earlier point in the process of knowledge making. That is, we are not studying texts already authored and produced to interpret from a distance the relationships between literature and the environment. Rather, we start by setting the stage—for example, by securing funding and instigating opportunities—for new debates on conversations already underway and then taking a front-row seat as the negotiations over how to make that text unfold with and around us. Once the text is produced, we can go back and reread, searching for new meanings and connections and things we missed along the way, but equipped with our experiences participating in the creation.

Among the most important achievements of this project has been fostering leadership and ownership of the resulting music videos by our collaborators in the field, who are also positioned as researchers, curators, and producers. In encouraging this deeper level of involvement and cocreation, we have aimed to depart from work patterns more typically characteristic within ethnographically oriented disciplines—for example, of simply capturing audiovisual documentation to be analyzed and edited after the period of fieldwork is done. Rather, this is a simultaneous endeavor, a process of discussion, negotiation, and decision-making about how to create a film about something all of us involved see as being important. As a research team, we have actively tried to figure out exactly how everything gets balanced and performed. As such, we might think of this methodology as working toward empirical justice.

Notes

Funding for this project came from the Humanities without Walls consortium program "Humanities in a Changing Climate," a joint initiative of the Mellon Foundation and University of Illinois. The authors thank their respective teams in Bangladesh, China, Haiti, Tanzania, and the U.S./Canadian border region of the Salish Sea.

1. The project specs and completed videos can be viewed online at Ecosong.net (https://www.ecosong.band/#/field-to-media/). As a result of scheduling conflicts, team member Tara Hatfield was unable to join us for the writing of this chapter, but her contributions to the project have been essential and are described in greater detail elsewhere (Pedelty et al. 2020). Her team's music video is "Zaniko Atoko (Bring Me an Axe)" (Kilimanjaro Media and Okoa Mtaa Foundation 2019).

2. In fact, coming to a uniform consensus on what field to media might look like has posed challenges for at least three reasons. First, we five scholars based in the United States, at the time comprising two faculty and three advanced graduate students, are trained in different disciplines, including communications studies, anthropology, ethnomusicology, and composition. Second, the interests, skills, requirements, and motivations of our collaborators in our respective research sites were also highly varied, leading to quite different conversations and activities in the course of completing this project. Third, we encountered complicated bureaucracy in crossing international borders—in at least one case—and rapidly shifting political and economic conditions—in another

case—that necessitated constantly altering plans. Even as we were united in the overarching goal of ecomusical video production, these contrasting circumstances have led to a more loosely defined process than we might have liked.

3. Before beginning our individual field trips, we sought to establish some sense of shared protocol. To this end, in April 2018, the U.S.-based researchers gathered at Atomic K Productions (https://www.atomick.com/) in Minneapolis for a technical skills workshop led by multimedia producer Karl Demer, who served as our HWW collaboration technological partner. Over the course of two days, we received training on the audiovisual equipment and editing tools that we expected to use. While some members of the team had extensive experience with field technologies, others were experimenting with the equipment for the first time. The workshop session thus covered camera settings, lighting, and audio techniques while also addressing the specific technical challenges that we anticipated might arise. This session together as a group gave us a common point of departure from which to launch our respective video productions.

4. Previous productions are available online (https://www.ecosong.band/).

5. In Chinese culture, people consider the rivers near their home to be a mother who raised them. This relationship to the river is powerful because it brings them food, water, and people.

6. The resulting music video is "Dancing Upstream" (Pang 2018a). Dance videographer V. Paul Virtucio, who is highly experienced in capturing movement, filmed the work.

7. Consider the many urgent demands for decolonizing methodology (Smith [1999] 2012), decolonizing ethnomusicology (Rosenberg 2016), and decolonizing listening practices and music studies (Robinson 2020).

References

Baily, John. 1989. "Filmmaking as Musical Ethnography." *World of Music* 31 (3): 3–20.

Beauplan, Evallière, Nenel Cassy, Antonio Cheramy, Richard Lenine Hervé Fourcand, and Onondieu Louis. 2017. Rapport de la Commission Sénatoriale Spéciale d'Enquête du Fonds Petro Caribe couvrant les périodes annuelles allant de Septembre 2008 à Septembre 2016. Sénat de la République, October 2017. https://www.scribd.com/document/364151103/Rapport-Petro-Caribe-Octobre-2017.

Bellande, Alex. 2015. *Haïti déforestée, paysages remodelés.* Montréal: Les Éditions du CIDIHCA.

D'Amico, Leonardo. 2020. *Audiovisual Ethnomusicology: Filming Musical Cultures.* Bern: Peter Lang.

Dirksen, Rebecca. 2012. "Reconsidering Theory and Practice in Ethnomusicology: Applying, Advocating, and Engaging beyond Academia." *Ethnomusicology Review* 17. http://www.ethnomusicologyreview.ucla.edu/journal/volume/17/piece/602.

Dirksen, Rebecca. 2018. "Haiti, Singing for the Land, Sea, and Sky: Cultivating Ecological Metaphysics and Environmental Awareness through Music." *MUSICultures* 45 (1–2): 112–35.

Dirksen, Rebecca. 2020. *After the Dance, the Drums Are Heavy: Carnival, Politics, and Musical Engagement in Haiti.* New York: Oxford University Press.

Dirksen, Rebecca. 2022. "Reinvoking Gran Bwa (Great Forest): Music, Environmental Justice, and a Vodou-Inspired Mission to Plant Trees across Haiti." In *The Routledge Companion to Music and Human Rights,* edited by Julian Fifer, Angela Impey, Peter G. Kirchschlaeger, Manfred Nowak, and George Ulrich, 228–45. London: Routledge.

Dirksen, Rebecca, and Lois Wilcken. 2021. "The Drum and the Seed: A Haitian Odyssey about Environmental Precarity." In *Performing Environmentalisms: Expressive Culture and Ecological Change,* edited by John McDowell, Katey Borland, Rebecca Dirksen, and Sue Tuohy, 132–62. Urbana: University of Illinois Press.

Feld, Steven. 1976. "Ethnomusicology and Visual Communication." *Ethnomusicology* 20 (2): 293–325.

Freire, Paulo. (1970) 2000. *Pedagogy of the Oppressed.* 30th anniversary ed. Translated by Myra Bergman Ramos. New York: Continuum.

Gal, Ofer, and Raz Chen-Morris. 2010. "Empiricism without the Senses: How the Instrument Replaced the Eye." In *The Body as Object and Instrument of Knowledge: Embodied Empiricism in Early Modern Science,* edited by Charles T. Wolfe and Ofer Gal, 121–47. Dordrecht: Springer.

Grimshaw, Anna. 2001. *The Ethnographer's Eye: Ways of Seeing in Anthropology.* Cambridge: Cambridge University Press.

Gubner, Jennie. 2018. "The Music and Memory Project: Understanding Music and Dementia through Applied Ethnomusicology and Experiential Filmmaking." *Yearbook for Traditional Music* 50:15–40.

Harbert, Benjamin J. 2018. *American Music Documentary: Five Case Studies of Ciné-Ethnomusicology.* Middletown, Conn.: Wesleyan University Press.

Hood, Mantle. 1960. "The Challenge of 'Bi-musicality.'" *Ethnomusicology* 4 (2): 55–59.

Johnson, Jeffrey C., Christine Avenarius, and Jack Weatherford. 2006. "The Active Participant-Observer: Applying Social Role Analysis to Participant Observation." *Field Methods* 18 (2): 111–34.

Keller, Anna, Laura Sommer, Christian A. Klöckner, and Daniel Hanss. 2020. "Contextualizing Information Enhances the Experience of Environmental Art." *Psychology of Aesthetics, Creativity, and the Arts* 14 (3): 264–75.

Kilimanjaro Media and Okoa Mtaa Foundation, dir. 2019. "Zaniko Atoko (Bring Me an Axe)." Produced by @Kichakani Mobile Studio and Tara Hatfield. Yaeda Chini, Tanzania: Kilimanjaro Media. Film, 4 minutes. https://youtu.be/MuHeK1h6900.

Koch, Julia. 2019. "Fieldwork as Performance: Being Ethnographic in Film-making." *Anthropology Southern Africa* 42 (2): 161–72.

Locke, John. 1690. *An Essay Concerning Human Understanding.* Volume 1. London: Eliz. Holt, for Thomas Basset. https://www.gutenberg.org/files/10615/10615-h/10615-h.htm.

MacDonald, Michael B. 2020a. "CineMusicking: Ecological Ethnographic Film as Critical Pedagogy." In *Research Handbook on Childhoodnature,* edited by Amy Cutter-Mackenzie-Knowles, Karen Malone, and Elisabeth Barratt Hacking, 1735–52. Cham, Switz.: Springer.

MacDonald, Michael B. 2020b. "Thanks for Being Local: CineMusicking as a Critical Pedagogy of Popular Music." In *The Sage Handbook of Critical Pedagogies,* edited by Shirley R. Steinberg and Barry Down, 1242–54. London: Sage.

Malinowski, Bronisław. 1922. *Argonauts of the Western Pacific: An Account of Native Enterprise and Adventure in the Archipelagoes of Melanesian New Guinea.* London: Routledge and Kegan Paul.

Mazzola, Guerino, Jason Noer, Yan Pang, Shuhui Yao, Jay Afrisando, Christopher Rochester, and William Neace. 2020. *The Future of Music: Towards a Computational Musical Theory of Everything.* Cham, Switzerland: Springer.

McLaren, Peter. 2000. *Che Guevara, Paulo Freire, and the Pedagogy of Revolution.* Lanham, Md.: Rowman & Littlefield.

Ndaliko, Chérie Rivers. 2016. *Necessary Noise: Music, Film, and Charitable Imperialism in the East of Congo.* New York: Oxford University Press.

Nixon, Rob. 2011. *Slow Violence and the Environmentalism of the Poor.* Cambridge, Mass.: Harvard University Press.

Pang, Yan, dir. 2018a. "Dancing Upstream: Current Issues of Environmental Awareness as Performance." Produced by Yan Pang. Minneapolis: Ecosong.net. Film, 15 minutes. https://youtu.be/G9pxe4ckr90.

Pang, Yan. 2018b. "Rural and Urban Explorations of Transnational Performance: A Dialogue between Traditional Southwestern Chinese and Contemporary North American Music and Dance." PhD diss., University of Minnesota.

Pedelty, Mark. 2016. *A Song to Save the Salish Sea: Musical Performance as Environmental Activism.* Bloomington: Indiana University Press.

Pedelty, Mark, dir. 2018. "LOUD." Produced by Mark Pedelty. Minneapolis: Ecosong.net. Film, 10 minutes. https://www.ecosong.band/#/loud-1/.

Pedelty, Mark. 2020a. "Singing across the Sea: The Challenge of Communicating Marine Noise Pollution." In *Water, Rhetoric, and Social Justice: A Critical Confluence,* edited by Casey R. Schmitt, Theresa R. Castor, and Christopher S. Thomas, 293–312. Lanham, Md.: Lexington.

Pedelty, Mark, dir. 2020b. *Sentinels of Silence? Whale Watching, Noise, and the Orca.* Produced by Mark Pedelty. Minneapolis: Ecosong.net. Film, 26 minutes. https://www.ecosong.band/#/sentinelsofsilence/.

Pedelty, Mark, Rebecca Dirksen, Tara Hatfield, Yan Pang, and Elja Roy. 2020. "Field to Media: Applied Ecomusicology in the Anthropocene." *Popular Music* 39 (1): 22–42.

Ranocchiari, Dario, and Eugenio Giorgianni. 2020. "Doing Ethnographically Grounded Music Videos." *Visual Ethnography* 9 (1): 7–18.

Reason, Peter, and Hilary Bradbury. 2007. *The Sage Handbook of Action Research: Participative Inquiry and Practice.* 2nd ed. London: Sage.

Robinson, Dylan. 2020. *Hungry Listening: Resonant Theory for Indigenous Sound Studies.* Minneapolis: University of Minnesota Press.

Rony, Fatimah Tobing. 1996. *The Third Eye: Race, Cinema, and Ethnographic Spectacle.* Durham, N.C.: Duke University Press.

Rosenberg, Davin, ed. 2016. "Decolonizing Ethnomusicology." Special issue, *SEM {StudentNews}* 12 (2): 1–45.

Rouch, Jean. 2003. *Ciné-Ethnography.* Edited and translated by Steven Feld. Minneapolis: University of Minnesota Press.

Roy, Elja, dir. 2019. "A Song for the Sundarbans: Dwellers of the Forest Arise." Minneapolis: Ecosong.net. Produced by Elja Roy. Film, 8 minutes. https://youtu.be/JoKzouroepM.

Roy, Elja, dir. 2020. *Musical Mangrove.* Produced by Elja Roy. Minneapolis: Ecosong.net. Film, 30 minutes. https://youtu.be/JoKzouroepM.

Roy, Elja. 2021. "Art, Activism and Sundarbans: A Case Study of Ecomusical Environmental Movement through Film." PhD diss., University of Minnesota.

Seligmann, Linda J., and Brian P. Estes. 2020. "Innovations in Ethnographic Methods." *American Behavioral Scientist* 64 (2): 176–97.

Small, Christopher. 1988. *Musicking: The Meanings of Performing and Listening.* Middletown, Conn.: Wesleyan University Press.

Smith, Linda Tuhiwai. (1999) 2012. *Decolonizing Methodologies: Research and Indigenous Peoples.* London: Zed Books.

Steinberg, Shirley R., and Barry Down, eds. 2020. *The Sage Handbook of Critical Pedagogies.* London: Sage.

Takach, Geo. 2016. *Scripting the Environment: Oil, Democracy, and the Sands of Time and Space.* Cham, Switzerland: Palgrave Macmillan.

Tweed, Katherine. 2009. "Sewage Industry Fights Phosphorus Pollution." *Scientific American,* November 1, 2017. https://www.scientificamerican.com/article/sewages-cash-crop/.

Ugwu, Chidi. 2017. "The 'Native' as Ethnographer: Doing Social Research in Globalizing Nsukka." *Qualitative Report* 22 (10): 2629–37.

Vérilus, Kendy, dir. 2020. "M pral plante yon pyebwa." Produced by Rebecca Dirksen. Port-au-Prince: Verilux Films. Film, 4 minutes. https://youtu.be/KuKpDKKAw2U.

Zemp, Hugo. 1988. "Filming Music and Looking at Music Films." *Ethnomusicology* 32 (3): 393–427.

Zhao, Rui. 2013. "Solving the Problem of Urban River Pollution: Protect the River from the Headwater and Restore the Ecosystem." *Huamin Research Center China NGO Case Study Series* 3:1–7. https://socialwork.rutgers.edu/sites/default/files/china_ngo_case_study_3.pdf.

PART II

Case Studies

Chapter 4

Does Climate Fiction Work?

An Experimental Test of the Immediate and Delayed Effects of Reading Cli-Fi

MATTHEW SCHNEIDER-MAYERSON, ABEL GUSTAFSON, ANTHONY LEISEROWITZ, MATTHEW H. GOLDBERG, SETH A. ROSENTHAL, AND MATTHEW BALLEW

Increasingly, scholars of environmental communication are joining ecocritics, activists, authors, and other cultural producers in calling for more creative forms of climate communication (Boykoff 2019), including films, television shows, theater, art, and literature. Nearly half of Americans report reading a novel or short story at least once a year (NEA 2015), but environmental literature has received surprisingly little attention in empirical environmental communication research. This research gap is not due to a lack of content for analysis; there are now hundreds of widely read works of so-called climate fiction. Although many of them are written with an activist desire to raise awareness and generate behavioral and political change (Schneider-Mayerson 2017), the effects of reading climate fiction have not yet been studied in controlled experiments.

Given the pervasiveness and potential persuasiveness of climate fiction, together with the growing and increasingly undeniable urgency of climate change, empirical research on this subject is needed. This chapter reports on the results of the first experimental test of the effects of reading climate fiction on readers' beliefs and attitudes about climate change. To measure these effects, we conducted a classic (pre–post–post) experimental study, gathering information about participants before they read a story, immediately after, and then again two weeks later. We used two different types of climate fiction—a speculative dystopian story ("The Tamarisk Hunter" by Paolo Bacigalupi [2008]) and a realist story exploring the psychological dynamics of climate change awareness and denial ("In-Flight Entertainment" by Helen Simpson [2011])—so that we could measure, in a controlled experiment, not only whether climate fiction has an influence on readers but also how consistent these effects are across multiple different stories representing different modes of storytelling.

We review the relevant literature, present the design and results of this experiment, and discuss the implications for future research and practice.

The Significance of Climate Fiction

Climate change fiction, also known as cli-fi, has become one of the most significant trends in English-language literature over the last decade (Schneider-Mayerson 2017). In one sense, climate fiction is not new; authors have been imagining climate futures for almost as long as climate change has been a topic of public concern (Trexler 2015). However, in the 1990s, the number of novels that featured anthropogenic climate change as a central element began to grow. By the late 2000s, as the impacts of climate change became more evident and public concern intensified, dozens of such novels were published in English (Schneider-Mayerson 2017; Trexler 2015). Today, the genre is growing so quickly that any comprehensive list would soon become outdated. These works are no longer relegated to the margins of literature or viewed as mere genre fiction; instead, they represent some of the most well-respected and widely read English-language authors writing today, including Margaret Atwood, T. C. Boyle, Barbara Kingsolver, Amitav Ghosh, N. K. Jemisin, Ian McEwan, Cormac McCarthy, David Mitchell, and Kim Stanley Robinson. Fictional works that focus on climate change are also beginning to win major literary awards. Lauren Groff's collection of short stories, *Florida* (2018), was a finalist for the National Book Award for fiction, Diane Cook's *The New Wilderness* (2020) was short-listed for the Booker Prize, Richard Powers was awarded the Pulitzer Prize for fiction for *The Overstory* (2018), and Jericho Brown was awarded the Pulitzer Prize for poetry for *The Tradition* (2019).

Countless ecocritics have speculated about the potential impact of climate fiction on individual readers (Gaard 2014; Goodbody 2014; Mehnert 2016), and this consideration is often implicit in the growing attention to climate fiction in literary criticism (Johns-Putra 2016; Schneider-Mayerson, Weik von Mossner, and Malecki 2020). Scholars of environmental communication and environmental psychology argue that the arts can play a critical role in influencing beliefs, attitudes, and behaviors related to climate change (Boykoff 2019; Gabrys and Yusoff 2012; Milkoreit 2016). For example, Julia B. Corbett and Brett Clark (2017) assert that the arts allow "the so-called invisibility of climate change to be seen, felt, and imagined in the present and the future" and "encourage critical reflection on existing social structures and cultural and moral norms." Among authors, artists, screenwriters, and filmmakers, this belief in the power and importance of "climate storytelling" is so pervasive that it has led to the emergence of climate media consultants to help creative workers craft more effective climate narratives (Good Energy Stories, n.d.).

The extant empirical scholarship on the influence of environmental art on its audiences has primarily focused on film (Bilandzic and Sukalla 2019; Howell 2011; Leiserowitz 2004) and fine art (Curtis 2009; Sommer and Klöckner 2019). Accordingly, environmental communication researchers have called for more study of the effects of communicating through the environmental arts (Moser 2016), suggesting that stories might engage people in new and persuasive ways.

The Persuasive Power of Narrative

Narrative persuasion involves embedding a persuasive message within a story that has both a plot and identifiable characters (Kreuter et al. 2007). A growing body of empirical research indicates that narrative storytelling is an effective mode of communicating persuasive messages, particularly in the context of science communication (Dahlstrom 2014; Shen, Sheer, and Li 2015). Scholars theorize two central mechanisms by which a narrative format enhances the persuasive effects of a message: identification and transportation (Cohen et al. 2015; Hinyard and Kreuter 2007).

Identification refers to the connection the audience feels to the characters in a narrative. Several studies have operationalized this construct (and analogous constructs) with measures that include elements such as perceived similarity between the reader and character, the reader's understanding of the character, and their shared emotional experience (Busselle and Bilandzic 2009; Cohen 2001; de Graaf et al. 2012; Moyer-Gusé 2008; Tal-Or and Cohen 2010). Because identification with a narrative's characters is motivated in part by perceived similarity, it is likely that the alignment of worldviews, values, or ideology between the audience and the character can affect the level of identification that the audience feels. For narratives presenting a message about risk—as is the case with many stories about environmental issues and climate change in particular—the evidence suggests that identification increases risk perceptions by decreasing perceived social and psychological distance between the audience and the narrative's characters (So and Nabi 2013).

Narrative persuasion is also considered to be effective as a result of increased transportation, which is defined by Melanie C. Green and Timothy C. Brock (2000) as a state in which "all mental systems and capacities become focused on events occurring in the narrative" (701). A meta-analysis of seventy-six studies (Van Laer et al. 2013) found that transportation into narratives has large effects on emotions and moderate effects on beliefs, attitudes, and behavioral intentions (Appel and Richter 2010; Green 2004; Green and Brock 2000; Vaughn et al. 2009). Scholars argue that transportation can enhance persuasion by reducing counterarguing—the conscious consideration of rebuttals to the message—and, as with identification, by decreasing the psychological distance between the audience and the content described in the narrative (Moyer-Gusé 2008; So and Nabi 2013).

While empirical research has found narratives to be persuasive for prosocial, noncontroversial messages such as public health campaigns (Murphy et al. 2013; Niederdeppe, Shapiro, and Porticella 2011), several other experiments have also found narratives to be persuasive for polarizing, contentious topics such as immigration, stem cell research, gay rights, affirmative action, and the Israeli–Palestinian conflict (Boswell, Geddes, and Scholten 2011; Cohen et al. 2015; Mazzocco et al. 2010; Zhang and Min 2013). Many of these studies specifically tested the explanatory mechanisms of transportation and identification. For example, Juan-José Igartua and Isabel Barrios (2012) found that identification mediates (explains) the relationship between narrative appeals and attitude change about one's own religion.

In addition to being effective across myriad topics, narrative persuasion also appears to be effective across multiple types of narratives. Studies indicate that narratives presented as being fictional are no more or less effective than narratives presented as nonfiction (Appel and Malečkar 2012; Strange and Leung 1999; Wheeler, Green, and Brock 1999). Together, the large body of research on narrative persuasion supports the speculations of ecocritics, public intellectuals, activists, and authors that climate fiction could have a significant persuasive effects on readers' beliefs, attitudes, and behaviors.

The Understudied Influence of Climate Fiction

Only a few studies have investigated the effects of literature on environmental topics. For example, studies indicate that reading environmental narratives about animals can lead to a boost in pro-animal attitudes in student samples (Malecki et al. 2018, 2019). Further, Kathryn E. Cooper and Eric C. Nisbet (2016) report that narratives told through news stories, documentaries, and fictional television shows about genetically modified organisms and fracking result in increased risk perceptions and support for environmental policy via increased narrative involvement and negative affect. Similarly, Markus Appel and Martina Mara (2013) find that narratives about sustainable driving behavior increase intentions toward fuel-efficient driving, particularly when the narrative's characters are portrayed (and perceived) as trustworthy. However, another study (Jones 2014) observes that narratives about climate change policy are no more persuasive than a list of facts. W. P. Malecki and colleagues (2021) compare the impact of extinction narratives presented via print, video, and audio. And Anandita Sabherwal and Ganga Shreedhar (2022) examine the impact of a protagonists' motivation for taking pro-environmental actions on readers' support for climate policies and intentions to take individual and collective action.

Despite the growing corpus of climate fiction and the evidence indicating the persuasive power of communicating about the environment through narratives, only two studies on the influence of these literary works of climate change fiction have been published. Using an exploratory qualitative survey, Matthew Schneider-Mayerson (2018) finds that climate fiction readers report that reading climate fiction causes them to perceive climate change as more psychologically proximate, to explore the psychological and social dimensions of climate change, and to appreciate the scale and gravity of climate change. A second survey (Schneider-Mayerson 2020) suggests that a climate fiction novel can make readers more aware of climate injustice and lead them to empathize with climate migrants. However, both studies are surveys that ask readers to describe the ways in which they are (or are not) affected, rather than to directly measure the effects of reading climate fiction in a controlled experiment. In the present study, we address this gap in the literature with a controlled experiment testing the influence of reading climate change fiction on a national sample of fiction readers.

Given the existing evidence of the effectiveness of narratives in persuasion, we expected to find that (H1) participants who read climate fiction stories will report more pro-climate beliefs and attitudes relative to participants in a control group.[1] Following from the empirical and theoretical research on the mechanisms of narrative persuasion, we expected that (H2) the attitudinal and behavioral effects of reading climate fiction would be mediated (explained) by identification and transportation.

Most persuasion research assesses the immediate effects of a treatment message and thus ignores the critical question of the longevity of those effects. Some tests of narrative persuasion have found long-lasting effects on audiences—such as when attitudes are measured a few weeks after the narrative experience (Appel and Mara 2013; Appel and Richter 2007; Jensen et al. 2011)—whereas others have found that effects last for a week but disappear within a month (Malecki et al. 2019). Because this research is sparse and our understanding of the longevity and stability of message effects is still in its infancy, we did not offer hypotheses but instead asked (RQ1): After one month, how do the climate change beliefs and attitudes of participants who read climate fiction stories compare to those who did not?[2]

Methods

Population of interest. The population of interest for this experiment was fiction readers who tend to believe that global warming is happening and anthropogenic, but do not consider the risks to be severe and do not consider the issue to be highly important to them personally. These are characteristics of the "Concerned" and "Cautious" segments of Global Warming's Six Americas (Maibach, Roser-Renouf, and Leiserowitz 2009), an established instrument that segments Americans according to their climate beliefs and attitudes. This focus on the Concerned and Cautious heightens this study's ecological validity. That is, we were not interested in measuring the effect of reading climate fiction on those who tend to doubt or dismiss the reality of anthropogenic climate change (the "Doubtful" or "Dismissive" segments), or on those who do not have an interest in reading fiction, because such individuals are less likely to read climate fiction outside the current experimental context (Schneider-Mayerson 2018). We also did not test the effect of climate fiction on individuals who already have high levels of worry and perceived risk regarding climate change (the "Alarmed" segment of the Six Americas), because they are already highly concerned about the issue.

Design and procedure. To test the effect of reading climate fiction on beliefs and attitudes about climate change, this experiment was constructed with two treatment groups and one control group for between-groups comparisons, using within-subjects measurements that spanned three time points (pre–post–post). First, a time 1 (T1) screener survey identified participants who are fiction readers and belong to the Cautious or Concerned segments. Those who satisfied these criteria were eligible to participate in the time 2 (T2) main study. About two weeks after T1, eligible participants were invited into the T2 portion, which contained a pretest, the treatment stimuli (reading a climate fiction short story), and an immediate posttest. About one month later, participants who completed T2 were invited to participate in a delayed posttest (T3) to assess the longevity of the effects of reading a climate fiction story at T2.

At T2, participants first completed a battery of pretest measures of beliefs and attitudes about climate change. Then participants were randomly assigned to one of three experimental conditions. In two of these conditions, participants were assigned to read one of two climate fiction short stories. In the third condition (the control), participants read a fictional short story that did not touch on climate change or related issues. After reading their story, participants were given the same measures of beliefs and attitudes, some additional measures of their experience with and perceptions of the stories, and several demographic questions. Participation in the delayed T3 involved completing, for a third time, the battery of measures of beliefs and attitudes about global warming.

At each time point, participants provided informed consent after being made aware of the confidentiality of their response data, the protection of their anonymity, and the optional nature of their participation. All participants were compensated immediately after each portion of the study that they completed.[3]

Sample. Participants were recruited from Amazon's Mechanical Turk (MTurk), an online service popular with social scientists. It offers access to a pool of workers who perform small tasks, such as taking surveys or doing data entry, for monetary compensation. Samples recruited from opt-in online panels like MTurk are not nationally representative, but they are more diverse than student samples, and they perform similarly to nationally representative samples on many tasks (Bartneck et al. 2015; Hauser and Schwarz 2015; Kees et al. 2017). MTurk also affords the ability to select participants who gave particular answers on a prior survey and recontact them with an invitation to participate in another survey, which was necessary for the longitudinal design of this study.

When recruiting participants for T1, we used the TurkPrime interface to automatically exclude all MTurk workers who had previously completed any of the research team's prior surveys and experiments about climate change, to exclude all IP addresses flagged by MTurk as suspicious, and to exclude all MTurk workers located outside of the United States. In total, 5,655 American adults entered the T1 survey. Of these, 18 cases were removed for incomplete responses, and 78 cases were removed because of 39 instances of duplicate IP addresses. This left 5,561 valid cases for analysis. Of these, 1,984 (35.7 percent) satisfied the inclusion criteria for T2: they were Concerned or Cautious according to the Six Americas segmentation, they indicated that they read "occasionally," they were at least "moderately" interested in reading fiction, and they were willing to participate in T2.

After about a two-week delay, these 1,984 qualified T1 participants were invited to participate in T2. The reparticipation rate was 84 percent (N_{T2} = 1,671). To reduce the potential for selective participation biases and demand effects, the T2 recruitment posting on MTurk targeted T1 participants without revealing to them that T2 was related to T1. Of the participants who entered T2, a total of 1,507 (90 percent) completed the survey. After data cleaning (see the online supplement for details), we had 1,294 valid T2 cases for analysis, as follows: n = 434 in the control group, n = 469 in the "In-Flight Entertainment" group (referred to as "Flight"), and n = 391 in the "Tamarisk

Hunter" group ("Tamarisk"). This final sample was predominantly white (77 percent) and female (65 percent), and skewed younger and more liberal than the American population (full sample demographics are provided in the online supplement). The skew of the sample toward liberal and female individuals reflects the population of interest, as Americans in the Concerned and Cautious segments tend to be more liberal than the national population, and Americans who read literature at least once per year are more likely to be female (62 percent) (NEA 2015).

Approximately one month later, the participants who completed T2 were invited to participate in T3. As with the T2 recruitment, the T3 recruitment posting on MTurk did not mention that the task was connected to the prior T2 or T1 tasks. Of the 1,284 valid T2 cases, 744 (57 percent) completed T3 after the one-month delay.[4] This attrition did not substantially alter the demographic or ideological sample characteristics.[5]

Stimulus materials. Participants in the Flight treatment condition read Helen Simpson's "In-Flight Entertainment" (2011), and participants in Tamarisk treatment condition read Paolo Bacigalupi's "The Tamarisk Hunter" (2008). We chose to test two very different short stories (instead of just one) primarily to enable greater generalizability of observed effects. Additionally, these short stories fit into different categories of Matthew Schneider-Mayerson's (2017) typology of American climate fiction, which allowed us to observe whether different kinds of fiction narratives might have significantly different persuasive effects.

"In-Flight Entertainment" is a realist narrative that explores the psychological and moral dimensions of the "denial, avoidance, and acceptance" of the gravity of climate change in the present (Schneider-Mayerson 2017, 312). The story does so by staging a lengthy conversation about the anthropogenic nature, reality, and future impacts of climate change between a skeptical, self-centered, wealthy businessman and a retired climate scientist in the first class section of an intercontinental airplane, while they complain about the inconvenience of an elderly passenger having a heart attack nearby. After he dies, the plane is forced to make an emergency stop as the former scientist describes, in surprisingly calm and resigned terms, the inevitability of a socioecological collapse that will unfold because "nothing [is] going to stop people flying" (Simpson 2011, 15) and acting selfishly. The businessman is the story's focalizer, with a third-person narrator offering an inside perspective on his thoughts, dreams, and emotions. While the story can be described as realist, set in a world that readers are likely to see as a familiar (or at least recognizable) present-day setting, the tone is caustically satirical, describing the pettiness of the businessman's concerns (the luxurious food, the in-flight movie) compared to the encroaching reality of climate change and the death of a fellow passenger. By literally and metaphorically dramatizing the lack of appropriate concern and urgency about and response to climate change, "In-Flight Entertainment" indirectly implicates the reader.

In contrast, "The Tamarisk Hunter" is a prototypical "cautionary fable of the Anthropocene" (Schneider-Mayerson 2017, 313) that places the reader in a dystopic, climate-changed future. The speculative fiction narrative is set decades in the future in an American Southwest characterized by climate apartheid, in which cities are drying up and disappearing and individual states, fighting (literally) for scarce water, have closed their borders to migrants from other states. California has laid claim to the Southwest's water, using forced labor to construct a concrete waterway that carries water west and aggressively protecting it with military force. The third-person focalizer of "The Tamarisk Hunter" is a stubborn, clever, impoverished holdout who subsists on a homestead in desiccated Arizona long after most have emigrated. For removing tamarisks, which suck up an extraordinary amount of water, he receives a small compensation from California officials. Through the consciousness of his protagonist, Bacigalupi describes the onset of the permanent drought—and, it is suggested, the dissolution of the United States. In the story's conclusion, the narrator is notified that the "water bounty" (124) program is being discontinued, leaving the protagonist without any means of survival. "The Tamarisk Hunter" describes the future as a wasteland. The struggle to mitigate climate change has been lost, and with it the possibility of a unified nation and, for people in many places, a decent life.[6]

These stories were selected for their engaging storytelling, their equivalent length,[7] and their ecological validity; Bacigalupi and Simpson are two widely read and respected authors of environmentally engaged fiction. Both stories were lightly edited to enhance clarity, and "In-Flight Entertainment" was edited to make its protagonists American because perceived similarity has been shown to affect character identification (Brown 2015).

Participants in the control condition read the short story "Good People" by David Foster Wallace (2007). It portrays a young couple in a tense scene of introspection and conversation as they ponder their futures and the available courses of action in light of their recent pregnancy and their plan to have an abortion. "Good People" was selected as a control because it is similar in length to the treatments[8] and was also written by an established author of short stories, but the story does not touch on environmental or climate issues. The full texts of all three stories as presented to the participants are provided in the online supplement.

Measures

Global warming beliefs and attitudes. At the beginning of the T2 pretest, participants were first shown a definition of global warming (see the online supplement for the full text).[9] Then we administered self-report measures of beliefs and attitudes about climate change, which were adapted from those used in the biannual Climate Change in the American Mind survey (Leiserowitz et al. 2019). The full text of each question and all response options is presented in the online supplement. We measured belief in the existence of global warming using a seven-point scale with anchors of "I strongly believe global warming is *not* happening" (1), "I am unsure whether or not global warming is happening" (4), and "I strongly believe global warming *is* happening" (7). Next, we measured belief in the anthropogenic nature of global warming using a seven-point scale with anchors of "I believe global warming is caused entirely by natural changes in the environment" (1), "I believe global warming is caused equally by natural changes and human activities" (4), and "I believe global warming is caused entirely by human activities" (7).

Next, we measured worry about climate change, two types of risk perceptions, and personal importance. These four items produce the Six Americas segmentation (Chryst et al. 2018). Participants indicated their worry about global warming on a four-point scale from "Not at all worried" (1) to "Very worried" (4), and indicated how much they think global warming will harm themselves and future generations, respectively, on separate four-point scales ranging from "Not at all" (1) to "A great deal" (4), with an additional "Don't know" option. We measured personal importance of global warming by asking "How important is the issue of global warming to you personally?" with responses given on a five-point scale ranging from "Not at all important" (1) to "Extremely important" (5).

Because both of the climate fiction stimuli described or portrayed the future effects of climate change, we also measured the perceived severity of these future effects by asking how global warming would affect the frequency of "Droughts and water shortages," "People living in poverty," "Refugees," and "Floods" over the next forty years. Responses to these items were given on five-point scales with response options ranging from "Many less" (1) to "Many more" (5).

Empirical research shows that persuasive messages about threats such as climate change should not just focus on the threat but should also promote self-efficacy, the perception that one is capable of performing the recommended solution, and response efficacy, the perception that the recommended action will be effective (Nabi, Gustafson, and Jensen 2018; Witte and Allen 2000). Importantly, perceptions of efficacy are predictors of support for climate change solutions (Roser-Renouf et al. 2014). Therefore, we measured perceptions of efficacy by asking, "In your opinion, is it possible to reduce global warming enough to prevent catastrophic future harm to [People in the United States] [The stability of earth's climate] [Future generations of people]?" Responses to each item were provided on a five-point scale ranging from "No, definitely not" (1) to "Yes, definitely" (5), with an additional option of "N/A because global warming will not cause catastrophic harm."

A by-product of beliefs and risk perceptions about climate change is the overall perception of its priority for political decision makers. To assess this, participants were asked, "Do you think the following should be low, medium, high, or very high priority for the president and Congress?" with two separate items of "Global warming" and "Developing sources of clean energy." Responses were recorded on four-point scales corresponding to the question stem (low, medium, high, very high).

All of the above measures were administered both before and after participants read their assigned story. The following section describes measures that were only administered after the stimuli because they refer to participants' experience while reading their specific story.

Narrative transportation and identification. The literature review indicates that the effects of narrative persuasion depend on experiences of transportation and identification. To measure transportation, we consolidated the core dimensions and items of several prior lengthy transportation scales (Busselle and Bilandzic 2009; de Graaf et al. 2012; Green and Brock 2000; Kim and Biocca 1997; Tal-Or and Cohen 2010) into a short scale of three items. These three items asked "To what degree did you [Have a vivid image of the events in the story] [Become fully absorbed in the story] [Feel as if you were present in the world that the story created]?" Responses were given on a four-point scale of "Not at all" (1), "Only a little" (2), "Moderately" (3), and "A lot" (4). An initial pilot test on an independent sample ($N = 803$) found that this three-item measure was internally consistent, and the main study corroborated these findings. A description of the methods of the pilot test and exploratory factor analysis is presented in the online supplement.

Similarly, to measure identification, we synthesized the primary dimensions and items from several prior measures of character identification (Busselle and Bilandzic 2009; Cohen 2001; de Graaf et al. 2012; Green and Brock 2000; Tal-Or and Cohen 2010) into a short scale of two items. These two items asked, "To what degree did you [Feel the emotions the characters were feeling] [Imagine what it would be like to be in the position of the characters]?" Responses were given on a four-point scale of "Not at all" (1), "Only a little" (2), "Moderately" (3), and "A lot" (4). This two-item measure demonstrated adequate internal consistency in the pilot test sample and in the main study sample.[10]

Analysis

For empirical ecocritics, social scientists, and readers who are interested in conducting their own empirical research, we provide a brief description of our analysis.

The Six Americas segmentation is calculated from the measures of worry, personal risk, risk to future generations, and personal importance. A publicly available tool allows users to upload a data set of responses to these four items.[11] The tool then automatically calculates the Six Americas segment for each respondent and appends them to the data set as a new variable. The methodological specifications of the segmentation analysis are described by Chryst and colleagues (2018).

H1 predicted that reading climate fiction stories would positively affect global warming beliefs and attitudes of Concerned and Cautious fiction readers. To test this hypothesis, we used an analysis of covariance (ANCOVA) for each outcome variable to compare T2 posttest means across experimental groups while controlling for the corresponding T2 pretest variable. This mixed design increases statistical power relative to between-groups comparisons of posttest means that ignore pretest values (Charness, Gneezy, and Kuhn 2012; Goldberg et al. 2019). In instances where the omnibus test indicated significant differences across conditions, least-square difference post hoc comparisons were used to compare each treatment group to the control group.

H2 predicted that the effects of the treatments will be mediated (explained) by participants' reported transportation into the story and identification with the characters. To test this hypothesis, we entered the transportation index and the identification index as mediators in separate mediation models (Figure 4.1) using the PROCESS macro in SPSS software. Transportation and identification were tested in separate models because they are highly correlated (r = .69). We also used separate models for the two treatment conditions (Flight and Tamarisk), such that each model was coded as X = condition (control = 0 and treatment = 1), M = mediator (transportation or

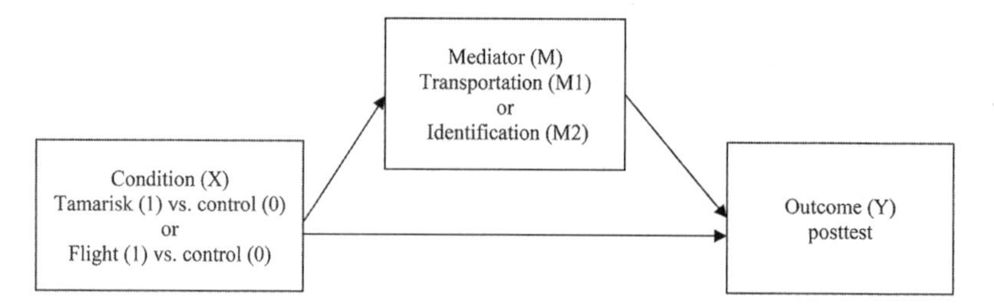

Figure 4.1. Parallel mediation with transportation (M1) and identification (M2) as mediators of the effect of X (treatment/control condition pairs) on Yi (each dependent variable). The corresponding pretest variable is included as a covariate.

identification), and Yi = a T2 posttest outcome variable. Each model also included Yi's corresponding T2 pretest variable as a covariate, as in the ANCOVAs described above.

RQ1 asked whether the total (main) effects hypothesized in H1 would be observed after a one-month delay. To inform this question, the same analyses corresponding to H1 were performed using the delayed posttest (T3) values of each outcome variable instead of the immediate posttest (T2) values. Like the tests of H1, the tests of RQ1 controlled for the T2 pretest values of the corresponding outcome variable.

Results

Immediate effects: Tests of H1. As Table 4.1 in the appendix shows, our results partially supported H1. "In-Flight Entertainment" had significant positive effects on multiple dependent variables: the belief that climate change is anthropogenic, risk perceptions (personal risk and risk to future generations), climate change issue priority, and beliefs that climate change will cause more droughts, more poverty, more refugees, and more floods. "The Tamarisk Hunter" also had significant positive effects on multiple dependent variables: risk perceptions (personal risk and risk to future generations), climate change issue priority, and beliefs that climate change will cause more droughts, more poverty, and more refugees.

However, neither story significantly affected clean energy issue priority, or perceived efficacy to reduce climate change enough to prevent harm to people in the United States and to the stability of the earth's climate.

There were two unexpected negative results. "The Tamarisk Hunter" had a significant negative effect on perceptions of future flooding, and "In-Flight Entertainment" had a significant negative effect on efficacy regarding preventing harm to future generations of people.

Overall, the effects of the two climate fiction stories were highly similar. There were significant differences between participants in the Flight and Tamarisk conditions on only three outcome variables. The first was belief that climate change is anthropogenic, which was significantly higher in the Flight condition than the Tamarisk condition.[12] The second was in the belief that climate change will increase floods, which was lower in the Tamarisk condition than the Flight condition.[13] The third significant difference was the perception of efficacy regarding preventing harm to future generations, which was lower for the "In-Flight Entertainment" readers than readers of "The Tamarisk Hunter."[14]

As an exploratory analysis, we investigated whether gender, age, political ideology, and Six Americas segment might moderate (influence) any of the treatment effects. The tests of these interactions revealed that the treatment effects did not differ significantly across levels of any of these potential moderators. That is, there was no significant difference in the effect of the story on readers based on gender, age, political ideology, or climate concern (Concerned versus Cautious). These results are presented in the online supplement.

Immediate effects: Tests of H2. The immediate posttest also supported H2. As Table 4.2 in the appendix shows, for both the Tamarisk and the Flight treatments, participants' level of transportation and identification (respectively) mediated several of the significant X–Y relationships. This is to say that there was a strong, statistically significant relationship between transportation into the narrative and identification with the characters and the dependent variables that were tested, such as the belief that climate change will harm future generations and the personal importance of climate change. However, most of these significant indirect effects involved partial mediation, such that the residual direct effect of X on Y was still significant when accounting for the mediator. The only instances of full mediation of a significant main effect were, respectively, transportation and identification's mediation of the effect of "The Tamarisk Hunter" on perceived personal harm.

Transportation seemed to consistently be a stronger mediator of the effects of "In-Flight Entertainment" than identification (Table 4.2).[15] This means that transportation into the narrative seemed to explain the effects of the story more than identification with the characters. Regarding "The Tamarisk Hunter," the indirect effects of transportation and identification were similar across most outcome variables. This is partly because "The Tamarisk Hunter" (relative to the control) did not have a significantly different effect on transportation compared to identification.[16]

Delayed effects: Tests of RQ1. Our analyses revealed that after a one-month delay, there were only two significant effects of reading a climate fiction story, neither of which was positive (see Table 4.3 in the appendix). At T3, participants who had read "The Tamarisk Hunter" reported a significantly lower belief that climate change is happening and expected significantly fewer floods due to climate change compared to those in the control condition. All of the significant positive treatment effects that were observed in the immediate posttest had diminished to statistical nonsignificance. However, this is not conclusive evidence that these effects disappeared entirely, because the difference between the significant effects and nonsignificant effects was not significant (Gelman and Stern 2006).

Discussion

This study was the first randomized controlled experiment on the persuasive effects of reading a climate fiction story, and one of very few randomized controlled experiments on the effects of literature on readers' environmental attitudes and beliefs. The results indicate significant though small effects on a set of beliefs and attitudes that in most cases were partially mediated by an experience of narrative transportation and in some cases were partially mediated by identification with the story's characters. This evidence is consistent with a common interpretation that transportation and identification are mechanisms by which narrative persuasion takes place (Cohen et al. 2015; Hinyard and Kreuter 2007), though the correlational nature of mediation analyses does not enable causal claims.

The effects of the two climate fiction stories on beliefs and attitudes were observed immediately after the treatments, but they decreased to nonsignificance after a one-month delay. While the two stories were quite different in content and style, their respective effects were surprisingly similar in size and transience, which may suggest some tentative implications about the generalizability of these results to climate fiction short stories at large. These findings build on a large body of literature indicating the persuasive power of storytelling in environmental issues, and they add important detail about the longevity of the effects that occur from a single exposure to a message.

"The Tamarisk Hunter" likely had a significant negative effect on perceptions of future flooding for an obvious reason: the story is set in a parched dystopian landscape where water is a scarce commodity, making future floods seem improbable. "In-Flight Entertainment" likely had a significant negative effect on efficacy regarding preventing harm to future generations because the two major characters in the story agree that protesting climate change is "a waste of time" (Simpson 2011, 9) given its advanced stage. One might hope that the story's satirical tone would lead to a desire for action on the part of readers—that it would disturb readers to witness two people casually giving up on the future, just as it (ironically) disturbs the wealthy businessman to hear the climate scientist's fatalism (Garrard 2013), and thereby challenge readers to take action. But on the perceived efficacy of preventing future harm, "In-Flight Entertainment" seems to have had a more straightforward effect. Readers experienced a conversation in which two people agree that it's too late to do anything meaningful about climate change, which conveys a demobilizing message. Especially given the lack of conversations about climate change in some places (Maibach et al. 2016)—which means that for some readers, this is a rare opportunity to observe or vicariously participate in such a conversation—it might not be surprising that this narrative experience had a negative impact on readers. Moreover, the story portrayed a climate scientist discounting the efficacy and value of working to mitigate climate change. Climate scientists possess a great deal of authority, but most readers will never meet one personally, so this might be readers' only experience witnessing a climate scientist's private opinions about the issue, even if it is fictional.

Some ecocritics have advocated for more irony and irreverence in climate communication (Seymour 2018), but this result demonstrates that such an approach can also have a counterproductive effect on audiences. While environmental social scientists have identified perceptions of efficacy as a critical attribute of environmental communication, it has been more common for ecocritics to focus on the awareness of negative environmental outcomes. But awareness only goes so far, especially if action is perceived as futile. As noted above, empirical research finds that persuasive messages about environmental threats should not just focus on the threat but should also promote self-efficacy, the perception that one is capable of performing the recommended action, and response efficacy, the perception that the recommended action will be effective (Nabi, Gustafson, and Jensen 2018; Witte and Allen 2000). Especially because climate change is increasingly being publicly acknowledged in most countries as a grave and urgent threat that requires immediate action, ecocritics ought to evaluate works not only for their ability to generate awareness but also for their ability to highlight and even expand the reader's perception of self-efficacy and response efficacy.

While some ecocritics, activists, and authors might be disappointed to learn that the effects of these stories diminished to nonsignificance after two weeks, the details and the context of this experiment should be taken into account. The immediate treatment effects observed in our study were small to medium in size at the immediate posttest measurement and were nonsignificant after one month. This is not surprising to scholars of environmental communication because prior research has shown that immediate changes in response to a single stimulus usually do not result in (empirically measurable) permanent changes. However, this should not be interpreted as evidence that these messages failed, or that climate fiction, and environmental literature in general, can only have small and transient effects on readers' beliefs and attitudes. Research indicates that repeated exposure and reinforcement from other media and social influences can lead to compounding effects and to behavioral change (Abelson 1985; Funder and Ozer 2019). For example, Dustin Carnahan, Daniel E. Bergan, and Sangwon Lee (2021) have demonstrated that the effects of a single message tend to dissipate, but repeated messages can cause the effects to stick. One of the limitations of this study—as with most experiments on media effects—is its single-exposure message design, but finding significant immediate effects from a single exposure in an artificial setting is still a valuable insight. This is partially because the effects of a single exposure in an artificial setting may represent a lower bound of the real-world effects. Given the dominance of the novel in many countries, reading climate fiction in the real world generally involves significantly longer narratives. It also frequently involves multiple exposures because readers who enjoy reading climate fiction will experience many works of cli-fi. Both longer narratives and multiple exposures might be expected to result in larger and longer-lasting impacts. Most empirical research on the influence of literary narratives has used single short stories or excerpts, but a novel might be expected to have larger and longer-lasting effects because its greater length and immersive detail could cause greater transportation and identification and also provide more information about environmental issues.

Lessons, Limitations, and Practical Implications

This study offers experimental evidence about a question that has been at the center of the widespread academic and popular interest in climate fiction, and it opens up multiple directions for future research. For example, further research is needed to test other important questions such as how different genres, plot elements, and styles of climate narratives might have different effects. In addition, readers' literary preferences, prior worldviews, demographics, and psychological traits might affect how they process and respond to literary works with environmental themes. While the present study focuses on a specific and practically important subset of the population (American fiction readers in the Concerned or Cautious segments of Global Warming's Six Americas), future research could investigate the effects of reading climate fiction among other groups.

One limitation of this study is that many readers of climate fiction might belong to the Alarmed segment of Global Warming's Six Americas, not the Concerned or Cautious segments. We excluded the Alarmed from the present study because they already score very high on the outcome variables, indicating that they do not need to be persuaded that climate change is a significant threat. Still, future research should investigate the effects of climate fiction among the Alarmed. Qualitative empirical ecocritical research suggests that climate fiction can lead already concerned readers to take action (Schneider-Mayerson 2018), and related research has demonstrated that narratives can strengthen the attitude–behavior relationship (Rhodes, Toole, and Arpan 2016). Reading climate fiction may help activate pro-environmental behavior in individuals who already have strong pro-environmental attitudes.

Another important consideration when interpreting these results is that MTurk samples are not representative of the broader population. We recruited from this pool of participants because it enabled us to study a highly specific subpopulation, to cover a wide geographic range within the United States, to recontact participants after a long interval, and to recruit participants into T2 and T3 without revealing that it was associated with a prior survey they had completed (thus reducing the risk of selective attrition). Despite these advantages, the nonrepresentative nature of this sample is a barrier to generalizability.

The findings of this study have widespread practical implications. They highlight the effects that a growing and important category of literature is likely having on the beliefs and attitudes of Concerned and Cautious readers, who are likely to represent a large portion of the readers of climate fiction. This means that activists seeking persuasive climate change communication content should note the growing body of research indicating the effectiveness of narratives and storytelling, and of climate fiction specifically. It also means that authors, editors, and publishers who want to contribute to the ongoing efforts to mitigate and adapt to climate change might consider writing and publishing climate fiction. This study also found that the effects of these stories are often explained (at least in part) by transportation into the story and sometimes by identification with the characters. Thus, a practical recommendation for strategic communicators, authors, and cultural producers is to seek to maximize these two experiences when making decisions about format, style, and content.

Closer to home for many readers of this volume, it means that college instructors and secondary school teachers might do their part by assigning (more) climate fiction. While this research focused on short stories, there are many existing works of climate fiction and nonfiction storytelling that could be used in diverse contexts.

Online Supplement

The online supplement is available at https://figshare.com/articles/journal_contribution/Environmental_Literature_as_Persuasion_An_Experimental_Test_of_the_Effects_of_Reading_Climate_Fiction/12957574

Funding

This research was supported by a Yale-NUS College internal grant (IG17-LR104).

Table 4.1. Immediate Main Effects of Climate Fiction Stories

Outcome	*Control*		*Flight*		*Tamarisk*		*Omnibus Test*			*FvsC*	*TvsC*
	EMM	SE	EMM	SE	EMM	SE	F	*p*	η^2	*d*	*d*
Happening	6.01	.024	6.01	.023	6.06	.025	1.56	.211	.002	0.01	0.05
Human caused	5.20	.024	**5.40**	.023	5.25	.026	19.46	<.001	.029	**0.20*** **	0.05
Worried	3.02	.019	3.05	.019	**3.08**	.020	2.27	.104	.004	0.06	**0.10***
Harm personally	2.47	.020	**2.60**	.020	**2.54**	.022	9.45	<.001	.015	**0.18*** **	**0.10***
Harm future generations	3.43	.019	**3.55**	.018	**3.53**	.022	12.43	<.001	.019	**0.19*** **	**0.15*** **
Personal importance	3.36	.023	**3.47**	.022	**3.45**	.024	6.47	.002	.010	**0.13*** **	**0.11** **
More droughts	4.40	.021	**4.51**	.020	**4.51**	.022	8.44	<.001	.013	**0.14*** **	**0.17*** **
More poverty	3.96	.025	**4.18**	.024	**4.20**	.026	30.13	<.001	.045	**0.28*** **	**0.31*** **
More refugees	3.99	.023	**4.13**	.022	**4.13**	.024	13.21	<.001	.020	**0.17*** **	**0.17*** **
More floods	4.32	.026	**4.50**	.025	**4.14**	.027	46.21	<.001	.067	**0.24*** **	**0.22*** **
GW priority	2.83	.021	**2.99**	.020	**2.95**	.022	16.98	<.001	.026	**0.20*** **	**0.15*** **
CE priority	3.34	.018	3.39	.018	3.36	.019	1.84	.159	.003	0.07	0.02
Prevent harm to people in United States	3.78	.031	3.70	.030	3.80	.033	2.78	.063	.004	0.08	0.15
Climate stability	3.70	.032	3.63	.031	3.67	.034	1.16	.313	.002	0.07	0.04
Future generations	3.85	.032	**3.73**	.031	3.87	.034	5.83	.003	.009	**0.12** **	0.03

Note. Measurements taken in immediate posttest at T2. Results are from ANCOVA controlling for T2 pretest. Omnibus indicates effect of experimental condition in overall ANCOVA; FvsC, pairwise comparison of standardized marginal means between Flight and control conditions; TvsC, pairwise comparison of standardized marginal means between Tamarisk and control conditions; EMM, estimated marginal means; SE, standard error; η^2, partial eta square; and *d*, Cohen *d* standardized effect. Bold text indicates instances of the treatment's being significantly different than the control: $^{***}p < .001$; $^{**}p < .01$; $^{*}p < .05$.

Table 4.2. Indirect Effects of Transportation and Identification

Outcome (Y)	*X = Flight (1), Control (0)*				*X = Tamarisk (1), Control (0)*			
	M = Transportation		*M = Identification*		*M = Transportation*		*M = Identification*	
	β	*LLCI, ULCI*	*β*	*LLCI, ULCI*	*β*	*LLCI, ULCI*	*β*	*LLCI, ULCI*
Happening	**.034**	.014, .056	.006	−.001, .015	.015	−.001, .033	**.022**	.002, .045
Human caused	**.030**	.014, .048	**.011**	.002, .022	.010	−.001, .023	.011	−.002, .025
Worried	**.065**	.037, .099	**.014**	.003, .031	**.044**	.021, .070	**.047**	.022, .076
Harm personally	**.035**	.013, .058	**.010**	.001, .023	**.031**	.012, .053	**.043**	.021, .066
Harm future generations	**.029**	.010, .051	.006	−.001, .017	**.023**	.007, .043	.020	−.001, .043
Personal importance	**.040**	.019, .064	**.013**	.003, .027	**.028**	.009, .051	**.048**	.026, .075
More droughts	.038	.018, .061	.006	−.001, .015	**.034**	.015, .056	**.032**	.012, .055
More poverty	**.021**	.001, .042	.000	−.008, .008	**.039**	.020, .061	**.031**	.011, .052
More refugees	**.025**	.007, .047	.005	−.001, .013	**.033**	.016, .053	**.030**	.012, .052
More floods	**.037**	.015, .062	**.008**	.001, .021	.010	−.009, .030	.009	−.011, .031
GW priority	**.028**	.012, .046	**.008**	.001, .017	**.034**	.016, .054	**.049**	.029, .075
CE priority	**.040**	.020, .064	**.012**	.002, .024	**.029**	.012, .050	**.046**	.026, .068
Prevent harm to people in the United States	.007	−.016, .029	.007	−.000, .018	.006	−.016, .030	.002	−.024, .030
Climate stability	−.004	−.028, .020	.005	−.002, .016	.002	−.019, .025	−.001	−.026, .024
Future generations	−.001	−.022, .019	.006	−.001, .016	−.003	−.024, .019	−.007	−.032, .016

Note. β indicates standardized indirect effects; LLCI, bootstrapped lower limit 95% confidence interval (CI); and ULCI, bootstrapped upper limit 95% CI. Variables entered as X = condition (Flight/Tamarisk = 1, control = 0), M = mediator (transportation/identification), and Y = outcome variable. The corresponding pretest variable was included as a covariate. Significant indirect effects are indicated by bold and by confidence intervals that do not cross zero.

Table 4.3. Delayed Main Effects of Climate Fiction Stories

Outcome	*Control*		*Flight*		*Tamarisk*		*Omnibus Test*			*FvsC*	*TvsC*
	EMM	SE	EMM	SE	EMM	SE	*F*	*p*	η^2	*d*	*d*
Happening	6.05	.042	5.99	.042	**5.88**	.044	4.27	.014	.010	0.06	**0.17****[**]
Human caused	5.14	.045	5.18	.045	5.07	.047	1.65	.192	.004	0.04	0.07
Worried	2.98	.026	3.03	.026	3.00	.027	0.81	.444	.002	0.08	0.04
Harm personally	2.48	.031	2.49	.031	2.43	.033	0.96	.384	.002	0.01	0.08
Harm future generations	3.50	.029	3.51	.029	3.48	.031	0.19	.823	.002	0.02	0.03
Personal importance	3.34	.035	3.39	.035	3.31	.037	1.21	.300	.003	0.06	0.04
More droughts	4.44	.035	4.50	.035	4.43	.036	1.25	.286	.003	0.10	0.01
More poverty	4.00	.041	4.09	.041	4.07	.043	1.58	.206	.004	0.12	0.09
More refugees	4.05	.041	4.11	.041	4.14	.044	1.08	.339	.003	0.08	0.10
More floods	4.40	.036	4.45	.036	**4.27**	.038	5.93	.003	.014	0.08	**0.17**[*]
GW priority	2.81	.036	2.85	.036	2.82	.038	0.32	.725	.001	0.05	0.01
CE priority	3.32	.034	3.33	.034	3.31	.036	0.09	.915	.000	0.02	0.01
Prevent harm to people in the United States	3.73	.049	3.84	.049	3.78	.052	1.29	.277	.003	0.12	0.06
Climate stability	3.61	.051	3.71	.051	3.61	.054	1.33	.265	.003	0.11	0.00
Future generations	3.82	.049	3.81	.049	3.81	.052	0.04	.996	0.00	0.00	0.00

Note. Measurements taken in the delayed posttest at T3. Results are from ANCOVA controlling for T2 pretest. Omnibus indicates the effect of experimental condition in the overall ANCOVA; FvsC, pairwise comparison of standardized marginal means between the Flight condition and the control condition; TvsC, pairwise comparison of standardized marginal means between the Tamarisk condition and the control condition; EMM, estimated marginal means; SE, standard error; η^2, partial eta squared; *d*, Cohen *d* standardized effect. Bold text indicates instances of the treatment's being significantly different than the control: $^{***}p < .001$; $^{**}p < .01$; **p is an invitation to this new area of research. It* < .05.

Notes

This chapter is derived from Matthew Schneider-Mayerson, Abel Gustafson, Anthony Leiserowitz, Matthew H. Goldberg, Seth A. Rosenthal, and Matthew Ballew, "Environmental Literature as Persuasion: An Experimental Test of the Effects of Reading Climate Fiction," *Environmental Communication* 2020, copyright Taylor and Francis, https://doi.org/10.1080/17524032.2020.1814377.

1. H1 stands for hypothesis 1, emphasizing the importance of using quantitative methods to test specific hypotheses.

2. RQ stands for research question and is more appropriate when there is uncertainty about a specific issue.

3. The compensation was as follows: T1 = \$0.20; T2 = \$2.50; T3 = \$0.50.

4. Sample sizes are as follows: 300 in control, 299 in Flight, and 269 in Tamarisk.

5. See the supplemental materials for a detailed comparison.

6. "The Tamarisk Hunter" takes place in the same fictional world as Bacigalupi's novel *The Water Knife* (2015), which was the subject of a separate empirical study by one of the authors (Schneider-Mayerson 2020). However, the two works of fiction are quite different. For example, "The Tamarisk Hunter" does not contain the cynical noir tone that characterizes *The Water Knife,* and its protagonists are not climate migrants who turn to violence. As such, while the results of these two studies might appear contradictory, they are in fact complementary.

7. For the three stimuli, we attempted to select texts that would have similar length, and measured the reading time of our participants. For "In-Flight Entertainment," word count was 4,418, with reading time of mean 18:07 minutes and median 17:27 minutes. For "The Tamarisk Hunter," word count was 5,095, with reading time of mean 18:10 minutes and median 17:05 minutes.

8. For "Good People," word count was 3,211, with reading time of mean 13:53 minutes and median 12:46 minutes.

9. For this experiment, we used the Yale Program on Climate Change Communication's previously validated questions, which consistently use the term "global warming" instead of "climate change." In describing our results in this chapter, we use the two phrases interchangeably.

10. The Spearman-Brown coefficients were .75 and .70, respectively.

11. It can be found at https://climatecommunication.yale.edu/visualizations-data/sassy/.

12. $M_{diff} = 0.15$; $SE_{diff} = .035$; $p < .001$.

13. $M_{diff} = 0.36$; $SE_{diff} = .036$; $p < .001$.

14. $M_{diff} = 0.15$; $SE_{diff} = .046$; $p < .001$.

15. This may be explained by the fact that the "In-Flight Entertainment" (relative to the control) had a larger effect on transportation ($\beta = .54$, SE = .07, 95% confidence interval [CI] 0.41, 0.67) than on identification ($\beta = .17$, SE = .07, 95% CI 0.04, 0.31),

$Z = 3.89$, $p < .001$. For the formula used to compare regression coefficients, see Paternoster et al. (1998).

16. The transportation effect was as follows: $\beta = .48$, SE = .07, 95% CI 0.34, 0.62; identification effect, $\beta = .51$, SE = .07, 95% CI 0.37, 0.65, $Z = 0.34$, $p = .733$. In theory, it is possible to evaluate the unique indirect effects of each mediator while accounting for the other (e.g., via parallel mediation). However, as mentioned above, transportation and identification were not included as mediators in the same models as a result of high intercorrelation.

References

Abelson, Robert P. 1985. "A Variance Explanation Paradox: When a Little Is a Lot." *Psychological Bulletin* 97 (1): 129.

Appel, Markus, and Barbara Malečkar. 2012. "The Influence of Paratext on Narrative Persuasion: Fact, Fiction, or Fake?" *Human Communication Research* 38 (4): 459–84.

Appel, Markus, and Martina Mara. 2013. "The Persuasive Influence of a Fictional Character's Trustworthiness." *Journal of Communication* 63 (5): 912–32.

Appel, Markus, and Tobias Richter. 2007. "Persuasive Effects of Fictional Narratives Increase over Time." *Media Psychology* 10 (1): 113–34.

Appel, Markus, and Tobias Richter. 2010. "Transportation and Need for Affect in Narrative Persuasion: A Mediated Moderation Model." *Media Psychology* 13 (2): 101–35.

Bacigalupi, Paolo. 2008. "The Tamarisk Hunter." In *Pump Six and Other Stories*. San Francisco: Night Shade Books.

Bacigalupi, Paolo. 2015. *The Water Knife*. New York: Penguin Random House.

Bartneck, Christoph, Andreas Duenser, Elena Moltchanova, and Karolina Zawieska. 2015. "Comparing the Similarity of Responses Received from Studies in Amazon's Mechanical Turk to Studies Conducted Online and with Direct Recruitment." *PLoS One* 10 (4): e0121595.

Bilandzic, Helena, and Freya Sukalla. 2019. "The Role of Fictional Film Exposure and Narrative Engagement for Personal Norms, Guilt and Intentions to Protect the Climate." *Environmental Communication* 13 (8): 1069–86.

Boswell, Christina, Andrew Geddes, and Peter Scholten. 2011. "The Role of Narratives in Migration Policy-making: A Research Framework." *British Journal of Politics and International Relations* 13 (1): 1–11.

Boykoff, Max. 2019. *Creative (Climate) Communications: Productive Pathways for Science, Policy and Society*. Cambridge: Cambridge University Press.

Brown, William J. 2015. "Examining Four Processes of Audience Involvement with Media Personae: Transportation, Parasocial Interaction, Identification, and Worship." *Communication Theory* 25 (3): 259–83.

Busselle, Rick, and Helena Bilandzic. 2009. "Measuring Narrative Engagement." *Media Psychology* 12 (4): 321–47.

Carnahan, Dustin, Daniel E. Bergan, and Sangwon Lee. 2021. "Do Corrective Effects Last? Results from a Longitudinal Experiment on Beliefs toward Immigration in the U.S." *Political Behavior* 43 (3): 1227–46.

Charness, Gary, Uri Gneezy, and Michael A. Kuhn. 2012. "Experimental Methods: Between-Subject and Within-Subject Design." *Journal of Economic Behavior and Organization* 81 (1): 1–8.

Chryst, Breanne, Jennifer Marlon, Sander van der Linden, Anthony Leiserowitz, Edward Maibach, and Connie Roser-Renouf. 2018. "Global Warming's 'Six Americas Short Survey': Audience Segmentation of Climate Change Views Using a Four Question Instrument." *Environmental Communication* 12 (8): 1109–22.

Cohen, Jonathan. 2001. "Defining Identification: A Theoretical Look at the Identification of Audiences with Media Characters." *Mass Communication and Society* 4 (3): 245–64.

Cohen, Jonathan, Nurit Tal-Or, and Maya Mazor-Tregerman. 2015. "The Tempering Effect of Transportation: Exploring the Effects of Transportation and Identification during Exposure to Controversial Two-sided Narratives." *Journal of Communication* 65 (2): 237–58.

Cooper, Kathryn E., and Erik C. Nisbet. 2016. "Green Narratives: How Affective Responses to Media Messages Influence Risk Perceptions and Policy Preferences about Environmental Hazards." *Science Communication* 38 (5): 626–54.

Corbett, Julia B., and Brett Clark. 2017. "The Arts and Humanities in Climate Change Engagement." In Oxford Research Encyclopedia of Climate Science. https://oxfordre.com/climatescience/?url=%2Fclimatescience%2Fview%2F10.1093%2Facrefore%2F9780190228620.001.0001%2Facrefore-9780190228620-e-780.

Curtis, David J. 2009. "Creating Inspiration: The Role of the Arts in Creating Empathy for Ecological Restoration." *Ecological Management and Restoration* 10 (3): 174–84.

Dahlstrom, Michael F. 2014. "Using Narratives and Storytelling to Communicate Science with Nonexpert Audiences." Proceedings of the National Academy of Sciences of the United States of America 111 (suppl. 4): 13614–20.

De Graaf, Anneke, Hans Hoeken, José Sanders, and Johannes W. J. Beentjes. 2012. "Identification as a Mechanism of Narrative Persuasion." *Communication Research* 39 (6): 802–23.

Funder, David C., and Daniel J. Ozer. 2019. "Evaluating Effect Size in Psychological Research: Sense and Nonsense." *Advances in Methods and Practices in Psychological Science* 2 (2): 156–68.

Gaard, Greta. 2014. "What's the Story? Competing Narratives of Climate Change and Climate Justice." *Forum for World Literature Studies* 6 (2): 272–88.

Gabrys, Jennifer, and Kathryn Yusoff. 2012. "Arts, Sciences, and Climate Change: Practices and Politics at the Threshold." *Science as Culture* 21 (1): 1–24.

Garrard, Greg. 2013. "The Unbearable Lightness of Green: Air Travel, Climate Change, and Literature." *Green Letters* 17 (2): 175–88.

Gelman, Andrew, and Hal Stern. 2006. "The Difference between 'Significant' and 'Not Significant' Is Not Itself Statistically Significant." *American Statistician* 60 (4): 328–31.

Goldberg, Matthew H., Sander van der Linden, Matthew T. Ballew, Seth A. Rosenthal, and Anthony Leiserowitz. 2019. "The Role of Anchoring in Judgments about Expert Consensus." *Journal of Applied Social Psychology* 49 (3): 192–200.

Good Energy Stories. n.d. https://www.goodenergystories.com/. Accessed April 20, 2022.

Goodbody, Axel. 2014. "Risk, Denial, and Narrative Form in Climate Change Fiction: Barbara Kingsolver's 'Flight Behaviour' and Ilija Trojanow's 'Melting Ice.'" In *The Anticipation of Catastrophe: Environmental Risk in North American Literature and Culture*, edited by Sylvia Mayer and Alexa Weik von Mossner, 39–58. Heidelberg: Universitatsverlag Winter.

Green, Melanie C. 2004. "Transportation into Narrative Worlds: The Role of Prior Knowledge and Perceived Realism." *Discourse Processes* 38 (2): 247–66.

Green, Melanie C., and Timothy C. Brock. 2000. "The Role of Transportation in the Persuasiveness of Public Narratives." *Journal of Personality and Social Psychology* 79 (5): 701.

Hauser, David J., and Norbert Schwarz. 2016. "Attentive Turkers: MTurk Participants Perform Better on Online Attention Checks than Do Subject Pool Participants." *Behavior Research Methods* 48 (1): 400–7.

Hinyard, Leslie J., and Matthew W. Kreuter. 2007. "Using Narrative Communication as a Tool for Health Behavior Change: A Conceptual, Theoretical, and Empirical Overview." *Health Education and Behavior* 34 (5): 777–92.

Howell, Rachel A. 2011. "Lights, Camera . . . Action? Altered Attitudes and Behaviour in Response to the Climate Change Film *The Age of Stupid*." *Global Environmental Change* 21 (1): 177–87.

Igartua, Juan-José, and Isabel Barrios. 2012. "Changing Real-World Beliefs with Controversial Movies: Processes and Mechanisms of Narrative Persuasion." *Journal of Communication* 62 (3): 514–31.

Jensen, Jakob D., Jennifer K. Bernat, Kari M. Wilson, and Julie Goonewardene. 2011. "The Delay Hypothesis: The Manifestation of Media Effects over Time." *Human Communication Research* 37 (4): 509–28.

Johns-Putra, Adeline. 2016. "Climate Change in Literature and Literary Studies: From Cli-Fi, Climate Change Theater, and Ecopoetry to Ecocriticism and Climate Change Criticism." *Wiley Interdisciplinary Reviews: Climate Change* 7 (2): 266–82.

Jones, Michael D. 2014. "Communicating Climate Change: Are Stories Better than 'Just the Facts'?" *Policy Studies Journal* 42 (4): 644–73.

Kees, Jeremy, Christopher Berry, Scot Burton, and Kim Sheehan. 2017. "An Analysis of Data Quality: Professional Panels, Student Subject Pools, and Amazon's Mechanical Turk." *Journal of Advertising* 46 (1): 141–55.

Kim, Tacyong, and Frank Biocca. 1997. "Telepresence via Television: Two Dimensions of Telepresence May Have Different Connections to Memory and Persuasion." *Journal of Computer-Mediated Communication* 3 (2): JCMC325.

Kreuter, Matthew W., Melanie C. Green, Joseph N. Cappella, et al. 2007. "Narrative Communication in Cancer Prevention and Control: A Framework to Guide Research and Application." *Annals of Behavioral Medicine* 33 (3): 221–35.

Leiserowitz, Anthony A. 2004. "Day after Tomorrow: Study of Climate Change Risk Perception." *Environment: Science and Policy for Sustainable Development* 46 (9): 22–39.

Leiserowitz, Anthony, Edward W. Maibach, Seth Rosenthal, et al. 2019. "Climate Change in the American Mind: April 2019." Yale Program on Climate Change Communication and the George Mason University Center for Climate Change Communication. New Haven, Conn.: Yale Program on Climate Change Communication.

Maibach, Edward, Anthony Leiserowitz, Seth Rosenthal, Connie Roser-Renouf, and Matthew Cutler. 2016. "Is There a Climate 'Spiral of Silence' in America?" Yale Program on Climate Change Communication and the George Mason University Center for Climate Change Communication. New Haven, Conn.: Yale Program on Climate Change Communication. http://climatecommunication.yale.edu/publications/climate-spiral-silence-america/.

Maibach, Edward, Connie Roser-Renouf, and Anthony Leiserowitz. 2009. "Global Warming's Six Americas, 2009: An Audience Segmentation Analysis." Yale Program on Climate Change Communication and the George Mason University Center for Climate Change Communication. New Haven, Conn.: Yale Program on Climate Change Communication.

Malecki, Wojciech, Bogusław Pawłowski, Marcin Cieński, and Piotr Sorokowski. 2018. "Can Fiction Make Us Kinder to Other Species? The Impact of Fiction on Pro-animal Attitudes and Behavior." *Poetics* 66: 54–63.

Malecki, Wojciech, Piotr Sorokowski, Bogusław Pawłowski, and Marcin Cieński. 2019. *Human Minds and Animal Stories: How Narratives Make Us Care About Other Species*. London: Routledge.

Malecki, W. P., Alexa Weik von Mossner, Piotr Sorokowski, and Tomasz Frackowiak. 2021. "Extinction Stories Matter: The Impact of Narrative Representations of Endangered Species across Media." ISLE: Interdisciplinary Studies in Literature and Environment. https://doi.org/10.1093/isle/isab094.

Mazzocco, Philip J., Melanie C. Green, Jo A. Sasota, and Norman W. Jones. 2010. "This Story Is Not for Everyone: Transportability and Narrative Persuasion." *Social Psychological and Personality Science* 1 (4): 361–68.

Mehnert, Antonia. 2016. *Climate Change Fictions: Representations of Global Warming in American Literature*. London: Palgrave Macmillan.

Milkoreit, Manjana. 2016. "The Promise of Climate Fiction: Imagination, Storytelling, and the Politics of the Future." In *Reimagining Climate Change*, edited by Paul Wapner and Hilal Elver, 171–91. London: Routledge.

Moser, Susanne C. 2016. "Reflections on Climate Change Communication Research and Practice in the Second Decade of the 21st Century: What More Is There to Say?" *Wiley Interdisciplinary Reviews: Climate Change* 7 (3): 345–69.

Moyer-Gusé, Emily. 2008. "Toward a Theory of Entertainment Persuasion: Explaining the Persuasive Effects of Entertainment-Education Messages." *Communication Theory* 18 (3): 407–25.

Murphy, Sheila T., Lauren B. Frank, Joyee S. Chatterjee, and Lourdes Baezconde-Garbanati. 2013. "Narrative versus Nonnarrative: The Role of Identification, Transportation, and Emotion in Reducing Health Disparities." *Journal of Communication* 63 (1): 116–37.

Nabi, Robin L., Abel Gustafson, and Risa Jensen. 2018. "Framing Climate Change: Exploring the Role of Emotion in Generating Advocacy Behavior." *Science Communication* 40 (4): 442–68.

NEA (National Endowment for the Arts). 2015. "A Decade of Arts Engagement: Findings from the Survey of Public Participation in the Arts, 2002–2012." NEA Research Report 58. https://www.arts.gov/impact/research/publications/decade-arts-engagement-findings-survey-public-participation-arts-2002-2012.

Niederdeppe, Jeff, Michael A. Shapiro, and Norman Porticella. 2011. "Attributions of Responsibility for Obesity: Narrative Communication Reduces Reactive Counterarguing among Liberals." *Human Communication Research* 37 (3): 295–323.

Paternoster, Raymond, Robert Brame, Paul Mazerolle, and Alex Piquero. 1998. "Using the Correct Statistical Test for the Equality of Regression Coefficients." *Criminology* 36 (4): 859–66.

Rhodes, Nancy, Jennifer Toole, and Laura M. Arpan. 2016. "Persuasion as Reinforcement: Strengthening the Pro-environmental Attitude–Behavior Relationship through Ecotainment Programming." *Media Psychology* 19 (3): 455–78.

Roser-Renouf, Connie, Edward W. Maibach, Anthony Leiserowitz, and Xiaoquan Zhao. 2014. "The Genesis of Climate Change Activism: From Key Beliefs to Political Action." *Climatic Change* 125 (2): 163–78.

Sabherwal, Anandita, and Ganga Shreedhar. 2022. "Stories of Intentional Action Mobilise Climate Policy Support and Action Intentions." *Scientific Reports* 12 (1): 1–8.

Schneider-Mayerson, Matthew. 2017. "Climate Change Fiction." In *American Literature in Transition*, 2000–2010, edited by Rachel Greenwald Smith, 309–21. Cambridge: Cambridge University Press.

Schneider-Mayerson, Matthew. 2018. "The Influence of Climate Fiction: An Empirical Survey of Readers." *Environmental Humanities* 10 (2): 473–500.

Schneider-Mayerson, Matthew. 2020. "'Just as in the Book'? The Influence of Literature on Readers' Awareness of Climate Injustice and Perception of Climate Migrants." *ISLE: Interdisciplinary Studies in Literature and Environment* 27 (2): 337–64.

Schneider-Mayerson, Matthew, Alexa Weik von Mossner, and W. P. Malecki. 2020. "Empirical Ecocriticism: Environmental Texts and Empirical Methods." *ISLE: Interdisciplinary Studies in Literature and Environment* 27 (2): 327–36.

Seymour, Nicole. 2018. *Bad Environmentalism: Irony and Irreverence in the Ecological Age*. Minneapolis: University of Minnesota Press.

Shen, Fuyuan, Vivian C. Sheer, and Ruobing Li. 2015. "Impact of Narratives on Persuasion in Health Communication: A Meta-analysis." *Journal of Advertising* 44 (2): 105–13.

Simpson, Helen. 2011. "In-Flight Entertainment." In *In-Flight Entertainment*, 6–21. London: Vintage.

So, Jiyeon, and Robin Nabi. 2013. "Reduction of Perceived Social Distance as an Explanation for Media's Influence on Personal Risk Perceptions: A Test of the Risk Convergence Model." *Human Communication Research* 39 (3): 317–38.

Sommer, Laura Kim, and Christian Andreas Klöckner. 2021. "Does Activist Art Have the Capacity to Raise Awareness in Audiences? A Study on Climate Change Art at the ArtCOP21 Event in Paris." *Psychology of Aesthetics, Creativity, and the Arts* 15 (1): 60–75.

Strange, Jeffrey J., and Cynthia C. Leung. 1999. "How Anecdotal Accounts in News and in Fiction Can Influence Judgments of a Social Problem's Urgency, Causes, and Cures." *Personality and Social Psychology Bulletin* 25 (4): 436–49.

Tal-Or, Nurit, and Jonathan Cohen. 2010. "Understanding Audience Involvement: Conceptualizing and Manipulating Identification and Transportation." *Poetics* 38 (4): 402–18.

Trexler, Adam. 2015. *Anthropocene Fictions: The Novel in a Time of Climate Change*. Charlottesville: University of Virginia Press.

Van Laer, Tom, Ko De Ruyter, Luca M. Visconti, and Martin Wetzels. 2013. "The Extended Transportation-Imagery Model: A Meta-analysis of the Antecedents and Consequences of Consumers' Narrative Transportation." *Journal of Consumer Research* 40 (5): 797–817.

Vaughn, Leigh Ann, Sarah J. Hesse, Zhivka Petkova, and Lindsay Trudeau. 2009. "'This Story Is Right On': The Impact of Regulatory Fit on Narrative Engagement and Persuasion." *European Journal of Social Psychology* 39 (3): 447–56.

Wallace, David Foster. 2007. "Good People." *New Yorker*, January 28, 2007. https://www.newyorker.com/magazine/2007/02/05/good-people.

Wheeler, Christian, Melanie C. Green, and Timothy C. Brock. 1999. "Fictional Narratives Change Beliefs: Replications of Prentice, Gerrig, and Bailis (1997) with Mixed Corroboration." *Psychonomic Bulletin and Review* 6 (1): 136–41.

Witte, Kim, and Mike Allen. 2000. "A Meta-analysis of Fear Appeals: Implications for Effective Public Health Campaigns." *Health Education and Behavior* 27 (5): 591–615.

Zhang, Lihong, and Young Min. 2013. "Effects of Entertainment Media Framing on Support for Gay Rights in China: Mechanisms of Attribution and Value Framing." *Asian Journal of Communication* 23 (3): 248–67.

Chapter 5

The Roles of Exemplar Voice, Compassion, and Pity in Shaping Audience Responses to Environmental News Narratives

JESSICA GALL MYRICK AND MARY BETH OLIVER

Not everyone has an equal chance of being negatively affected by environmental problems, including climate change, deforestation, and health issues due to environmental toxins. Individuals who have experienced racial and ethnic discrimination or oppression as well as those living in low-income areas are far more likely to encounter health problems due to environmental hazards (Morello-Frosch et al. 2011; Muller, Sampson, and Winter 2018). This inequity in risk, as well as the inequality in participation in the political processes that can help mitigate environmental hazards, are serious problems for those concerned about environmental justice, including frontline communities, activists, and policy makers (Schlosberg 2004). Media and communication scholars and ecocritics alike are concerned with the ways narratives about environmental harms may either perpetuate or help mitigate these negative outcomes that disproportionately affect some social groups more than others.

In this chapter, we examine the role of narratives in affecting individuals' awareness of and concern about environmental injustice. Narratives are not just novels or films. Indeed, many news stories about environmental problems are narratives, telling stories about these environmental dilemmas to readers and viewers using the typical narrative tools of plot and character development. How audiences respond to environmental news narratives may affect subsequent perceptions, intent to learn more, and willingness to get involved in the struggle against environmental injustice that adversely and disproportionately affects racial and ethnic minorities in the United States and beyond. As such, the effects of these environmental news narratives on public consciousness are more than objects of scholarly and theoretical intrigue; they are crucial to understand and consider in efforts to shape a more just, egalitarian society that grapples with the consequences of poor decisions that affect our environment.

In considering how environmental news narratives might be shaping public opinion and public willingness to think more deeply about environmental injustice, it is helpful to critically examine how journalists are constructing these stories and how they are presenting the victims of environmental injustices to their audiences. There are many aspects of how the ways that information is presented that ecocritics and environmental communication scholars might investigate. In the present moment, during which we are increasingly focused on hearing the voices of previously marginalized and oppressed people, one interesting and important issue is how narrative news-reporting techniques portray victims of environmental injustice—and if different portrayals of these individuals result in different emotional, cognitive, and behavioral responses in audiences. It is possible that some narrative news-reporting techniques evoke compassion for victims of environmental injustices, whereas others evoke pity, a different emotion that has been described as an ambivalent feeling that can foster negative stereotypes of victims (Fiske et al. 2002). The stereotype content model suggests that when we make downward social comparisons with other groups of people whom we like but view as incompetent, we feel pity (Cuddy, Fiske, and Glick 2007). Unlike the power of compassion to connect dissimilar people, pity may encourage self-interested emotion regulation, like avoiding unpleasant feelings, as well as defensive responses. That is, compassion can bring us closer to others, but pity may perpetuate a barrier.

One narrative technique that might be related to compassion versus pity is the use of more or less exemplar voice—direct quotes from victims of environmental injustice. By exemplar voice, we mean the amount of the message in which the people directly affected are quoted in telling their story in their own words, as opposed to being paraphrased by a reporter. Although most news reports include quotations or testimony from affected individuals, the amount of content directly from those individuals varies greatly. In considering a news report as a narrative, we can think of this as the amount of character dialogue versus narrator exposition that differs from story to story. The empirical question at hand is whether the amount of dialogue, or exemplar voice, results in different patterns of emotional, cognitive, and behavioral responses from audiences. That is, hearing more from individuals directly affected may help change public perceptions of them from individuals to be pitied to individuals to be assisted in their quest for equity and health justice. If hearing more of the stories about environmental injustice directly from the individuals most affected by them can shift perceptions, then reporters and advocates ought to consider changing how they tell these stories.

To empirically assess the roles of exemplar voice in eliciting compassion versus pity after reading an environmental news narrative, we conducted two experiments. Before exploring the methods and results of those experiments, we first outline existing literature on emotions, news reporting, and narratives in the context of a particular environmental injustice: the Flint, Michigan, water crisis. We then detail the results of our empirical studies and conclude by offering a discussion of the insights these data provide for future work and the integration of empirical and critical perspectives in thinking about how news narratives shape public understanding of the reality of environmental injustice and willingness to support and join the struggle for environmental justice.

News Narratives and Effects on Audiences

Empirical ecocriticism aims to empirically assess how environmental texts may shape audiences' environmental thoughts, feelings, beliefs, and behaviors (Schneider-Mayerson, Weik von Mossner, and Malecki 2020). Whereas most of the work in empirical ecocriticism has examined fictional environmental texts (Schneider-Mayerson et al. 2020), a prominent source of environment-related texts consumed by the public is news media. Many environmental news stories are narratives, telling stories about environmental problems through the lenses of those directly affected. In particular, issues of environmental justice offer a context where a narrative storytelling approach may be crucial for fostering a deeper public understanding of the ways in which systemic injustices lead to environmental and related health hazards for minority groups (Bullard 1999). Environmental justice can be defined as seeking equity in three realms: the distribution of environmental risk, the recognition of diverse participants and experiences, and participation in the political processes related to environmental policy (Schlosberg 2004).

Because structural causes and systemic ills such as racism, sexism, and classism typically underlie these inequities in environmental and related health outcomes, it is important for the public to understand that affected individuals are not to blame for their circumstances and that policy changes are needed to implement necessary changes (Bullard 1999). Exactly how audiences respond to environmental narratives may affect audience perceptions and willingness to seek more information or actively involve themselves in the pursuit of environmental justice.

Why, exactly, might narratives be crucial to fostering deeper public understanding of environmental justice issues? Why might they motivate audiences to take action to ameliorate these injustices? Research on narratives and the ways in which audiences develop connections with mediated others offers us important insights. First, social science research demonstrates that people tend to be insensitive to group or mass suffering, whereas they respond with compassion when presented with an individual under duress (Cameron and Payne 2011; Slovic 2010). As such, narratives may encourage audiences of environmental news to experience feelings such as compassion, thereby increasing their intentions to assist victims of environmental problems. For instance, Scott R. Maier, Paul Slovic, and Marcus Mayorga (2016) report that the narrative technique of personification—that is, putting a human face on an issue—evoked stronger emotional responses than nonpersonified news stories about violence in Africa among respondents in their Mechanical Turk sample. Research has also found that news narratives about the environment can increase empathy for individuals negatively affected by drilling for natural gas, which in turn decreases audience support for drilling (Shen, Ahern, and Baker 2014).

Making contact with others through media. There is a substantial body of empirical evidence that media provide a psychological meeting place whereby individual audience members can get to know people unlike them with whom they may not have regular physical contact. That is, people can come into parasocial contact with individuals in their out-groups via media exposure, and by consuming media featuring positive portrayals of individuals from an out-group, they are more likely to develop positive attitudes toward the out-group as a whole (Schiappa, Gregg, and Hewes 2005). Furthermore, these relationships that audiences form with individuals featured in the media have been shown to motivate them to take action to help others and to act in prosocial ways (Myrick 2017a, 2017b; Myrick and Willoughby 2019). In the case of environmental injustice, news stories about individuals dealing with contaminated water might enable audiences who are not directly affected and/or who are members of different racial or socioeconomic groups to better connect with the affected individuals and overcome negative stereotypes that they might associate with the affected groups.

This research does not suggest that media exposure is a panacea for bringing disparate groups together and motivating audiences to reconsider their thoughts and behaviors regarding environmental justice issues. Importantly, if the mediated portrayal of individuals from racial minorities affected by environmental injustices is not positive in nature, then white audiences are less likely to develop more positive attitudes from mediated exposure to out-group members (Ramasubramanian 2013). Even if the mediated experience is aimed at evoking empathy or a sense of understanding of the barriers facing individuals, these portrayals can potentially backfire and cause audiences to distance themselves from dissimilar others. For example, in their study of the effects of virtual reality representations of mental illness, Sriram Kalyanaraman and colleagues (2010) find that after individuals without schizophrenia wore a virtual headset so they could experience the world as if they had the condition, participants actually desired greater social distance from individuals with schizophrenia. This study provides a cautionary tale to environmental justice storytellers. Some audience members who see the reality of those who are worse off than them will recoil at the portrayal of suffering instead of diving into the topic and seeking ways to redress the injustice.

Exemplification and exemplar voice. In order to better understand the features of environmental news narratives that could foster rather than inhibit deeper engagement with environmental injustice, we now turn to the role of exemplars, or characters, in these reports. Most literary critics, narrative scholars, storytellers, and even science communication experts would agree that characters and protagonists are an essential facet of any narrative (Hassabis et al. 2014; Jamieson 2018), including environmental narratives (Caracciolo 2019; James 2015; Weik von Mossner 2019). In the context of news stories, the characters are actual flesh-and-blood people featured by journalists. In research on news media effects on news audiences, these interviewees are typically referred to as exemplars. Theorizing on exemplars centers around exemplification theory, which argues that when individuals are featured in news reports, audiences are more strongly influenced than when they see base-rate statistics about the same phenomena (Zillmann 2006; Zillmann and Brosius 2000). One reason exemplification effects occur is that audiences can make better emotional connections with exemplars, who are actual people, than with statistics. For example, Maria E. Grabe and colleagues (2017) note that personalizing news accounts of policy issues by including emotional testimony from individuals directly affected results in audiences reporting greater feelings of empathy toward and identification with those people affected by social issues, which increased their perceptions that the social issues discussed in the news stories were important. Given the vital role of emotional responses to news stories in shaping subsequent audience reactions and behaviors, the role of exemplars in determining how audiences will respond to environmental news narratives is an important consideration for this area of research.

Although numerous studies have documented the existence of exemplification effects in a variety of news contexts (Gibson and Zillmann 1994, 2000; Zillmann 2006), this work typically tests the effects of the *number* of exemplars used in a story and not on the *amount* of news content voiced directly by the exemplars. To gain additional insight into the nuances of how exemplification may shape audience responses to environmental news narratives, we offer an investigation of the role of exemplar voice in this arena. We define exemplar voice as the amount of the news narrative in which the people directly affected are quoted and are featured telling their story in their own words, as opposed to their experiences being paraphrased by a reporter. If the number of exemplars leads to stronger affective reactions in news audiences, we posit that the amount of direct voice given to exemplars will shift the nature of the affective responses to news narratives.

Emotional Responses to Suffering

Exemplification theory and research points out the ability of exemplars to amplify the emotional responses that audiences experience after consuming news stories. However, research and theory regarding the psychology of emotions underscores the idea that the specific type of emotion experienced by audiences is a crucial consideration in determining what and how they will think and behave (Nabi 2010). Below we discuss the literature surrounding two relevant emotions for the context of narratives about environmental injustices: compassion and pity.

Compassion. Compassion is defined as a state of being moved by someone else's suffering and wanting to help (Lazarus 1991). It is a social emotion, meaning it is a feeling that involves thinking about others and the connections we share. Martha Nussbaum (1996) argues that compassion is a "central bridge between the individual and the community" (28). The experience of compassion has been tied to having an altruistic concern for another's suffering (Blum 1980). This emotion is often categorized as a prototypical moral emotion (Haidt 2003), meaning it is tied to notions of right and wrong. As such, experiencing compassion can motivate individuals to perform altruistic, prosocial behaviors in hopes of alleviating the suffering of others (Stellar et al. 2017). In a media context, feelings of compassion after viewing media related to cancer victims have been linked to the reduction of stigma perceptions and an increase in prosocial behavioral intentions (Myrick 2017b). The emotion has also been shown to motivate audiences to support climate change mitigation policies (Lu and Schuldt 2016).

Pity. Pity can be thought of as another type of empathic feeling (Bartsch et al. 2018; Haidt 2003). It has been described as an ambivalent emotion that can promote negative stereotypes (Fiske et al. 2002). Pity has also been the focus of much literature and philosophy concerning the role of emotions in our social lives (Nussbaum 1996). In particular, Aristotle's work refers to pity, alongside fear, as a typical emotional response to tragedy (Nussbaum 1996; Punter 2014). In 350 BCE, Aristotle (2015) writes in *Rhetoric* that "what we fear for ourselves excites our pity when it happens to others" (88), suggesting that while fear is close and personal, pity is felt when something bad occurs in a less personally relevant context.

Pity and compassion are similar in that both arise from viewing the suffering of others, but the two are not identical affective states. During the Victorian era, philosophers began differentiating these emotions (Nussbaum 1996). Feeling pity implies a power differential such that the individual worthy of pity is of lower status than the individual experiencing the emotion (Punter 2014). According to Nietzsche (1968), "to show pity is felt as a sign of contempt because one has clearly ceased to be an object of fear as soon as one is pitied" (119). Pity has also been associated in many works of literature with a feeling of finality; that is, pity is experienced when hope is absent and there is nothing one can do to ameliorate the suffering of another (Punter 2014).

These philosophical and literary viewpoints on pity align with the limited existing empirical work on pity as a response to media. For example, in a study of audience responses to viewing media about Paralympic athletes, Anne Bartsch and colleagues (2018) find that when these videos elicited feelings of pity in able-bodied audiences, pity was negatively related to prosocial attitudes. That is, feeling pity after watching differently abled athletes was associated with a decrease in the likelihood of taking action to help others. This suggests that people who feel pity may feel as if there is nothing they can or need to do to help the recipients of their pity.

Although this study of audience responses to Paralympic sports coverage aligns with theoretical and philosophical perspectives on pity, there is little empirical research on pity as an emotional response to media. Moreover, research has yet to empirically compare compassion and pity as different emotional responses to the suffering of victims of an environmental injustice, leaving a gap in the existing literature on compassion and pity. Below we discuss how the ways in which an environmental news narrative is reported may evoke more or less compassion versus pity. We then offer the results of two studies that aimed to test the proposed differences in audience responses to different types of environmental news narratives.

Water Narratives and Audience Responses

Given the theoretical differences in behavior associated with feelings of compassion versus pity, we argue that environmental narrative news coverage that evokes more or less of these emotions will have differential effects on news audiences. In particular, we predict that when the victims are given more voice in a news narrative about an environmental injustice, readers will experience more compassion and less pity for those directly affected. Then, these emotional responses, as well as other cognitive responses to the news story, will shape policy support and information-seeking behavior, with compassion more likely to motivate increases in these outcomes. To assess the role of exemplar voice in eliciting compassion versus pity after reading a news story about an environmental injustice, we empirically test how the amount of exemplar voice—in this case, direct quotes from victims of poor water quality—affects audiences in two experiments. These experiments use narrative content based on news coverage of the Flint, Michigan, drinking water crises.

First, a quick summary of the Flint crisis. In 2011, the U.S. state of Michigan took over Flint's finances after an audit revealed the city was $25 million in debt (CNN 2021). The governor of Michigan at the time, Rick Snyder, appointed a series of financial managers to take control of Flint. In 2014, the emergency manager of Flint, appointed by Snyder, decided to switch the city's water supply from the Detroit water system to the Flint River. In violation of federal law, water drawn from the Flint River was not being treated with an anticorrosive agent, which resulted in lead from aging service lines leaching into the public water supply.

Almost immediately after the city's switch to drinking Flint River water, Flint residents reported serious problems to government officials, including discolored, sewage-smelling water, rashes, and unexplained illnesses (CNN 2021). The city of Flint issued multiple boil water advisories during 2014 but told residents afterward that the water was safe to drink. In January 2015, the Detroit water and sewage department offered to reconnect Flint with water from Lake Huron and to waive the reconnection fee, but the appointed Flint officials declined the offer, citing the expense of switching back. A week later, Flint residents brought jugs of discolored water to a community forum, and news outlets across the country reported on the rashes and illnesses experienced by Flint residents. Importantly, the story of the Flint water crisis is one of environmental injustice. The majority of the residents of Flint are not white, and racial biases that have become structurally entrenched in state politics played a significant role in the policy decisions that resulted in lead leaching into the residents' water (Pulido 2016).

During the course of 2015, researchers from the federal Environmental Protection Agency and Virginia Tech University found that Flint repeatedly violated water quality standards regulating bacterial contamination, disinfectant by-products, and lead (CNN 2021). As reported by CNN, a Virginia Tech study released in September 2015 found that 40 percent of homes in Flint had elevated lead levels. News reports detailed how additional research spearheaded by a local pediatrician revealed that after the change in the water source, the proportion doubled of children citywide who had high levels of lead in their blood.

Lead poisoning will have lasting consequences for the people of Flint. According to the U.S. Centers for Disease Control and Prevention (CDC), even low levels of lead in the blood can affect children in the midst of brain development, and lead exposure can affect nearly every system in the body (CDC, n.d.). Chronic exposure to low amounts of lead in children has been linked with decreased intelligence, impaired neurobehavioral development, decreased hearing acuity, and decreased ability to maintain a steady posture or growth.

A qualitative textual analysis of news coverage of the Flint water crisis in the *New York Times* and the *Wall Street Journal* found that the descriptions reporters used about the people of Flint typically focused exclusively on their problems and largely ignored the agency of these residents, who were actively organizing and petitioning government officials (Carey and Lichtenwalter 2019). Interviews with Flint residents and journalists revealed that many residents were frustrated with the ways media outlets would shorten their explanations of what was happening, and many journalists reporting on the story had little familiarity with Flint or its people (Rutt and Bluwstein 2017).

By portraying Flint residents as upset but unable to do anything about their plight, and by minimizing the space given to residents' own account of events, news narratives about the Flint water crisis may have caused audiences to look down on the victims, thereby creating a moral distance between themselves and the larger structural issues contributing to Flint's problems (Carey and Lichtenwalter 2019; Eller 2008). The present series of studies aims to assess the possibility that giving the people directly affected by environmental injustices greater voice in news accounts may shift readers' moral calculus so that victims receive more compassion, an emotion associated with greater agency than that of pity. Below, we detail the methods used in our two empirical studies.

Study 1

METHOD

Design and procedures. A two-level between-subjects online experiment was used in this study.[1] All stories featured expository information about the nature of the Flint water crisis as well as direct quotes from those experiencing the crisis (exemplar voice). However, one experimental news story contained exemplar voice only at the beginning of the story, whereas the other news story featured exemplar voice throughout the story.

Participants began the questionnaire by answering several demographic questions. To disguise the purpose of the study, participants were then asked to view and rate their perceptions of the layout of a news website from the Associated Press. They were then presented with a news story about unsafe drinking water. After reading the story, participants rated their responses to the story, reported their attitudes about public policy regarding clean drinking water, and indicated their likelihood of seeking greater information about water quality. Finally, at the end of the questionnaire, participants were given the opportunity to browse a website about clean drinking water from the CDC.

Participants. Participants were recruited from Amazon's Mechanical Turk (N = 59). Participants ranged in age from twenty-three to sixty-six years (mean [M] = 40.53, standard deviation [SD] = 11.24), with the majority being female (61.0 percent) and white (88.1 percent). The largest percentage of the sample identified as Democrats (40.7 percent), and the majority (54.2 percent) reported having earned a four-year college degree or higher.

Stimulus materials. The news stories about the Flint water crisis were formatted from actual materials from news outlets including CNN and the *New York Times.* The real stories from CNN provided factual background information and timelines of events, and the content adapted from the *New York Times* featured an individual who was directly affected by the Flint water crisis (described below). By featuring such an individual, these news reports likely help audiences to see Flint through his eyes, gain a better sense of the situation in Flint, and evoke emotional responses to the individuals' plights (Weik von Mossner 2017).

We edited the stimuli used for the controlled experiment to make them appear as authentic web-based stories published by the Associated Press. This outlet was chosen because it is a relatively neutral yet common news outlet that would be unlikely to evoke partisan or ideological backlash from American audiences. The structure of the news story stimuli was typical of any news report in that it provided background context (i.e., information about the water source switch in Flint, the involvement of different government officials and actors, and the specific health harms to Flint residents) as well as some quotations. As such, the low-voice condition represented rather typical reporting on the Flint case, which was frustrating to Flint residents because of its minimal use of direct testimony from those with first-hand experience of the consequences of government actions and inaction (Rutt and Bluwstein 2017). Conversely, the high-voice version of the stimuli placed a Flint resident as the primary storyteller while containing the same factual content of the narrative as the low-voice version.

Both versions of the stimuli featured the same headline ("Flint Still Dealing with Fallout from Drinking Water Crisis") and the same lede (introductory) paragraph. Further, both began and ended with quotes from a resident of Flint, Sid Booker, who had numerous ailments from drinking contaminated water. The first quote from Booker stated, "I'll use tap water for washing dishes, for mopping floors, for really, really quick showers. But, I only use bottled water for my cooking, drinking and washing my face or hands. The tap water is still disgusting to me. I won't even let the hand towel in my bathroom absorb a single drop of it. Sometimes when I shower I still get chalky white spots on my skin." The ending remark, referencing Booker's dog, was, "I don't even give it to Sparky here. I don't care how many filters they give us or how new the pipes are. I don't care what they say. How can I trust them again?"

In the low-voice condition, these two quotes were the only quotes in the story, with the rest of the information concerning exposition and policy information. In the high-voice condition, Booker's exemplar testimony was featured throughout the narrative alongside expository information. For example, an additional quote from Booker in the high-voice condition states, "'Then, after New Year's 2016, the governor announced we actually did have an emergency on our hands, I marked that day down right here,' Mr. Booker notes as he points to his calendar." That same information was conveyed in the low-voice condition with a sentence from the reporter reading, "On January 5, 2016, Governor Snyder declared a state of emergency in the county surrounding Flint." The low-voice article included 1,279 words and the high-voice article included 1,537 words. In the low-voice article, Booker was quoted three times using 182 words; in the high-voice article, he was quoted nine times using 527 words. Both articles included all of the same factual information about the Flint water crisis and only differed in how much of the information appeared to come from the reporter's paraphrasing of events versus Booker's direct testimony.

Measures. Table 5.1 provides correlations and descriptive information for all of the measured variables in both study 1 and study 2.

Two measures were used as a means of assessing perceptions of expository and exemplar voice in the news story. Perceived focus on the characters included two items: "The news story emphasized a character's experiences" and "The news story emphasized a character's action." Perceived focus on policy also included two items: "The news story emphasized policy information" and "The news story emphasized statistical information."

Measures of compassion and pity were modeled after the measures used by Anne Bartsch and colleagues (2018) in their research concerning perceptions of an advertisement for the Paralympic games. The measure of compassion included the items "compassionate," "caring," and "supportive." The measure of pity included the items "pity," "sympathetic," and "sorry for."

Political support was assessed using two items: "We should immediately increase government regulation on the quality of drinking water" and "We should immediately increase taxes on industries and businesses that contribute to poor drinking water quality." Information seeking was assessed using five items (e.g., "I plan to seek information about drinking water quality in the near future"). Finally, time spent looking at the CDC's website for clean water was measured using Qualtrics' built-in timer (CDC, n.d.).[2]

RESULTS

Before conducting a path analysis, we conducted a preliminary analysis of the effects of story condition on all of the variables examined. A multivariate

Table 5.1. Correlations and Descriptive Statistics for Measured Variables

	1	2	3	4	5	6	7
1. Policy focused	—	**.26***	**.18**	**.18**	**-.15**	**.06**	**-.08**
2. Character focused	.34***	—	**.49*****	**.52*****	**.28***	**.37****	**.15**
3. Pity	.40***	.22**	—	**.78*****	**.35****	**.17**	**.27***
4. Compassion	.31***	.27***	.72***	—	**.41****	**.36****	**.26**
5. Political support	.30***	.12	.26***	.25**	—	**.43*****	**.26***
6. Information seeking	.37***	.23**	.24**	.29***	.21**	—	**.15**
7. Time on website[a]	.11	.04	.22**	.14	.17*	.10	—
Means	**4.57**	**5.66**	**5.89**	**5.77**	**5.32**	**4.68**	**1.25**
	4.51	4.71	5.53	5.42	4.78	3.90	1.17
SD	**1.16**	**0.98**	**1.25**	**1.35**	**1.62**	**1.44**	**0.46**
	1.21	1.46	1.12	1.24	1.35	1.52	0.34
Cronbach's alpha[b]	**0.44**	**0.32**	**0.8**	**0.9**	**0.64**	**0.96**	
	0.48	0.66	0.73	0.85	0.45	0.97	

Note. Correlations for study 1 (n = 59) are above the diagonal and on the first line of descriptive and reliability information in bold. Correlations for study 2 (n = 181) are below the diagonal and on the second line of descriptive and reliability information in nonbold. All variables except for Time on Website were measured on scales ranging from 1 to 7.

[a]All correlations and descriptive information for Time on Website are for $\log_{10}$-transformed variables.

[b]Because measures of Policy Focused and Character Focused consisted of two items each, correlations are reported rather than Cronbach's alpha. No reliabilities are reported for Time on Website because this was a single-item measure.

*$p < .05$; **$p < .01$; ***$p < .001$.

analysis of variance (MANCOVA) revealed a multivariate effect of story condition, with Wilks $\lambda = .72$, $F(7, 51) = 2.74$, $p < .05$, and $\eta^2 = .27$. Table 5.2 reports the means, standard errors, and univariate results, showing that the story featuring the character voice throughout was associated with higher scores on character focus, compassion, pity, and information-seeking intent.

A path analysis was conducted to examine outcomes associated with reading a news story featuring different amounts of exemplar voice. We chose to include perceived character focus and policy focus as immediate outcomes, as subsequent downstream perceptions were thought to depend on participants' perceptions of these story features (O'Keefe 2003). These perceptions were

Table 5.2. Multivariate and Univariate Results for Study 1

	Character Voice at Beginning	*Character Voice Throughout*	*Univariate* F	η^2
Character focus	5.24 (.16)	6.12 (.17)	14.95***	.21
Policy focus	4.61 (.21)	4.52 (.22)	0.10	.00
Compassion	5.33 (.23)	6.26 (.24)	7.74**	.12
Pity	5.57 (.22)	6.24 (.23)	4.49*	.07
Political support	5.11 (.29)	5.55 (.31)	1.10	.02
Information-seeking intent	4.34 (.25)	5.07 (.26)	4.07*	.07
Time on website	1.19 (.08)	1.33 (.09)	1.39	.02

Note. This analysis was associated with a multivariate main effect of Wilks $\lambda = .72$, $F(7, 51) = 2.74$, $p < .05$, $\eta^2 = .27$. Numbers in parentheses are standard errors.
$^{*}p < .05$; $^{**}p < .01$; $^{***}p < .001$.

then used to predict feelings of compassion and pity, with these two affective states then used to predict support for public policy and information-seeking intent. The final dependent variable in the model was time spent on the clean water website. Figure 5.1 shows the statistics associated with this model that evidenced adequate model fit: χ^2 $(df = 13) = 15.12$, $p = .30$; CFI = .98; RMSEA = .05; 95 percent confidence interval .00, .15; PCLOSE = .43; SRMSR = .08.

This analysis found that the story including exemplar voice throughout was significantly associated with greater perceptions of character focus, but not policy focus. In turn, perceived character focus was associated with heightened feelings of both compassion and pity, whereas perceived policy

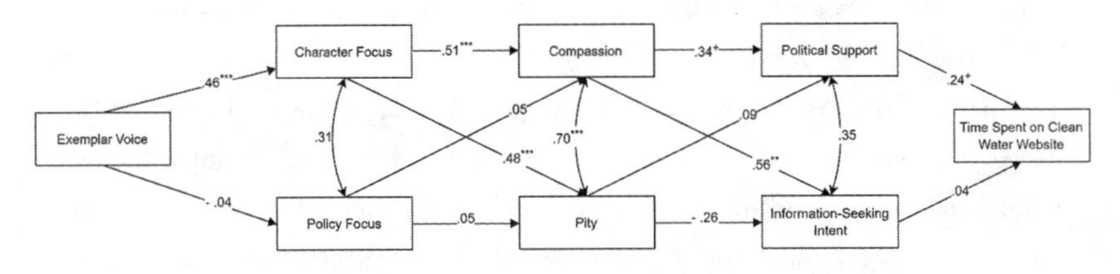

Figure 5.1. Study 1 path analysis.

focus was unrelated to either affective state. Compassion was positively associated with both greater political support and intent to seek more information about safe drinking water. In contrast, pity was associated with lower levels of information seeking intent. Finally, political support (but not information seeking) was associated with greater time spent on a clean water website.

Bootstrapping using 2,000 bootstrap samples and bias-corrected confidence intervals was used to examine the indirect effects of the story condition on all of the mediators and the final, endogenous variable. This analysis showed total indirect of effects of story condition on compassion ($\beta = .23$, $p < .01$), pity, ($\beta = .22$, $p < .01$), political support ($\beta = .10$, $p < .01$), and time on website ($\beta = .03$, $p < .05$). These results indicate that greater exemplar voice did increase audience responses of compassion, pity, political support, and time spent on the website.

Study 2

Although study 1 generally demonstrated the importance of exemplar voice on affective responses and of compassion (versus pity) on indicators of support for the importance of clean water, the study was underpowered—that is, the sample size was small, which makes it difficult to assess how strong or weak the identified effects were. Furthermore, both news stories included exemplar voice, albeit to differing levels. Consequently, a replication and extension of this study was conducted to address these limitations.

METHOD

Participants in this second study were undergraduate students ($N = 181$) at a large public university in the Mid-Atlantic region of the United States ranging in age from 18 to 25 years ($M = 20.34$, $SD = 1.01$). The majority of the sample was female (74.6 percent), white (88.1 percent), and reported identifying as a Democrat (45.9 percent).

Measures and online procedures were the same as in study 1. In this study, however, an additional experimental condition was included by introducing a news story about Flint that featured only expository information and no exemplar voice. Hence, this study was a between-subjects three-way experimental design.[3]

RESULTS

Before performing a path analysis, we conducted a preliminary analysis of the effects of story condition on all of the variables examined. This MANCOVA revealed a multivariate effect of story condition: Wilks $\lambda = .80$, $F(14, 344) = 2.85$, $p < .001$, $\eta^2 = .10$. Table 5.3 reports the means, standard errors, and univariate results, showing that the stories featuring the character voice throughout or at the beginning and end were associated with higher scores on character focus compared to the story featuring no character voice.

Table 5.3. Multivariate and Univariate Results for Study 2

Characteristic	*No Voice*	*Character Voice at Beginning and End*	*Character Voice Throughout*	*Univariate* F	η^2
Character focus	3.97_A (.18)	4.85_B (.18)	5.26_B (.17)	14.12***	.14
Policy focus	4.59 (.16)	4.42 (.16)	4.52 (.15)	0.29	.00
Compassion	5.31 (.16)	5.40 (.16)	5.54 (.16)	0.52	.01
Pity	5.53 (.15)	5.46 (.15)	5.60 (.14)	0.23	.00
Political support	4.66 (.18)	4.63 (.18)	5.03 (.17)	1.71	.02
Information-seeking intent	3.81 (.20)	3.75 (.20)	4.11 (.19)	0.98	.01
Time on website	1.21 (.04)	1.12 (.04)	1.19 (.04)	1.02	.01

Note. This analysis was associated with a multivariate main effect of Wilks $\lambda = .80$, $F(14, 344) = 2.85$, $p < .001$, $\eta^2 = .10$. Numbers in parentheses are standard errors. Within rows, means with no letters in common differ at $p < .05$ using Bonferroni post hoc comparison.
***$p < .001$.

Figure 5.2 provides the output for the path analysis that was conducted, with the resultant model showing adequate levels of model fit: χ^2 ($df = 18$) = 37.13, $p = .005$; CFI = .94; RMSEA = .08; 95 percent confidence interval .04, .11; PCLOSE = .10; SRMSR = .07. The experimental conditions were dummy coded, with the news story with no exemplar voice coded as the omitted category.

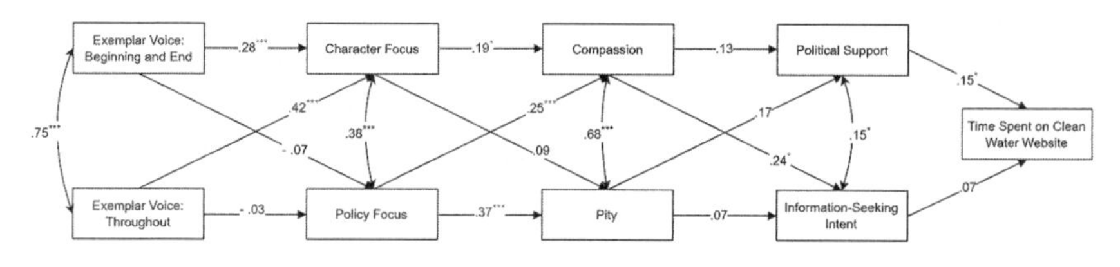

Figure 5.2. Study 2 path analysis.

Similar to study 1, this analysis showed that the presence of exemplar voice was significantly associated with greater perceptions of character focus, although this relationship was stronger for the news story featuring exemplar voice throughout compared to exemplar voice only at the beginning and end of the story. As in study 1, the perception of character focus was positively associated with compassion. However, unlike in study 1, perceived character focus was unrelated to pity. Furthermore, policy focus was associated with both heightened compassion and pity. Compassion, in turn, was associated with greater intent to engage in information seeking but was unrelated to political support. Finally, as in study 1, greater political support was associated with greater time spent on the clean water website.

Bootstrapping using 2,000 bootstrap samples and bias-corrected confidence intervals was used to examine the indirect effects of story conditions on all of the mediators and the final, endogenous variable. This analysis showed only a marginally significant indirect effect of the story with narration throughout on compassion ($\beta = .07, p < .10$). That is, this indirect effect was not significant at the $p < .05$ level, and therefore we cannot draw a strong conclusion that the story condition affected compassion level.

DISCUSSION

This chapter combines approaches from environmental communication and ecocriticism in a number of ways. First, it focuses on the effects of news reporting, a common medium examined in empirical environmental communication–focused studies. Environmental journalism can be considered the first draft of history and a starting point for inspiring more critical, advocacy, and/or activist writing. Yet it has rarely been examined by ecocritics, who have tended to focus instead on fictional texts and, more recently, other forms of media, like film and television. This is a missed opportunity because news is a critical way that environmental issues are communicated to the public. Moreover, individual stories often form the basis for longer book projects written by journalists, which can capture the public's attention in different ways, and perhaps reach different audiences, than their shorter, more episodic narrative news counterparts.

This chapter examines the effects of a fairly nuanced aspect of that form of storytelling—exemplar voice—from the perspective of empirical ecocriticism, which is more concerned with the formal and aesthetic features of narratives than environmental communication. Another way that this case study exemplifies a combination of empirical and critical approaches is through our use of social scientific theory and our analysis of actual texts—in this case, news articles. That is, we developed our hypotheses from deductive theories regarding the psychology of emotions and from existing critical examinations of the shortcomings of news coverage of the situation in Flint (Carey and Lichtenwalter 2019; Rutt and Bluwstein 2017). As such, while this case study leans toward the empirical social science tradition in which we are trained and typically practice, it aims to provide critical audiences with multiple avenues for application.

To summarize the results across the experiments, both studies found that including more exemplar voice in an environmental news narrative about contaminated drinking water has emotional, cognitive, attitudinal, and behavioral implications. The more audiences hear directly from individuals affected by an environmental injustice, the more they focus on the voices of these individuals. This greater character focus in turn generates higher levels of compassion. Theoretically, generating compassion was seen as important; it was predicted to be the emotional response that makes victims of environmental injustice seem more human and more agentic, whereas the generation of pity could cause downward social comparison and do less to inspire audiences to support policy changes or to look for more information about environmental justice issues. The results largely supported this prediction, with compassion being a strong and positive predictor of political support and information-seeking intent, and with political support then predicting actual time spent on a website that provides more information about water contamination.

The narrative factors that facilitate these different emotions also deserve additional attention. Notably, in both studies, perceiving the story as character focused was an important precursor for eliciting compassion. In addition to increasing the amount of exemplar voice in environmental justice narratives, other features may likewise increase perceptions among audiences that a story is character focused. For instance, adding more images or visuals of affected individuals in text-based reporting could also potentially increase perceived character focus. By giving individual, concrete faces to the people affected by environmental injustices, narratives may help counteract our well-documented tendency to avoid empathizing with depictions of mass suffering (Cameron and Payne 2011; Slovic 2010).

The Flint water crisis was hardly an isolated or anomalous event, but is representative of most environmental injustice in that people of different backgrounds were not equally affected. Flint is a majority-Black city, and additional research is needed to understand the interplay of race and exemplar voice. In particular, it would be helpful to have additional research to test how the racial identity of the victims versus the audience affects audience emotional responses and behaviors. In our studies, the news stories were text based and the quotations were edited to mask the racial identities of exemplars. This was done to help isolate the effects of emotions on subsequent political support and behavior, as is the strength of a controlled experiment. However, it does not represent the reality of most environmental justice narratives, in which images and other text-based cues may more explicitly alert audiences that the affected individuals identify with either the same or a different race than themselves.

Another consideration is that the actual reporting of the Flint water crisis often avoided any explicit discussion of the civil rights issues involved (Carey and Lichtenwalter 2019). By not directly addressing the racial and structural inequalities at the heart of environmental justice issues, news coverage may not evoke as much compassion in audiences as it would be minimizing one of the main barriers Flint residents faced when pursuing justice. One of the reasons that narratives have been found to be more persuasive than nonnarrative messages is that they can reduce counterarguing in audiences (Green and Brock 2000). It could be that news narratives with higher levels of exemplar voice likewise prevent white audiences from feeling quite as defensive about the role of white privilege and structural forces shaped by decades of decisions by white individuals in harming the health and well-being of other groups in society.

It is also noteworthy that our focus on audience emotions generated by environmental justice reporting techniques leaves a number of alternative cognitive explanations for our results that future research could explore. First, it is possible that consuming environmental news narratives featuring higher levels of exemplar voice makes people feel more efficacious in their intentions to help the specific people mentioned in the story, as opposed to helping large numbers of people. Therefore, the audience's perceived potential contribution to others may be more salient when they think about helping individuals who spoke in the news story (Bartels and Burnett 2011). Second, by speaking in their own voices instead of being paraphrased, the greater exemplar voice may have fostered the audience's perception that the victims were active and responsible members of the community, and previous research has found that individuals are more willing to help active victims (Skitka 1999).

As with any social scientific endeavor, the present research comes with limitations. The sample sizes in both studies were relatively small, and the results are therefore underpowered. Future work should include larger and more diverse samples that could detect more nuanced effects in different types of audience members. Additionally, these studies only examined one specific context (the Flint, Michigan, water crisis) in one specific country (the United States), and additional work is needed to see how audiences respond to the amount of exemplar voice used to tell the stories of other environmental justice topics in other contexts and other countries. Furthermore, we only examined the amount of exemplar voice; future research should also investigate the type and tone of exemplar voice.

Additionally, there was a strong correlation between our two emotions of interest, compassion and pity. This calls into question the distinctiveness of the two that is argued for by theorists. Although our prediction was supported in that compassion exhibited more downstream effects on political attitudes and information-seeking behavior, these results are presented with caution, as feelings of compassion and pity were highly correlated with each other. Although these two emotions are conceptually different, in a self-report questionnaire, it may be difficult for most audiences to distinguish between the two. Additional work is needed to gain insights into the nuanced differences in these feelings and the best ways to measure them as audience responses. Nonetheless, the stronger relationships between compassion and attitudes and information seeking than between pity and these same outcomes in both of our studies suggests that the emotions deserve further investigation. Many more questions remain around the effects of types and the nature of exemplar voices.

Despite these limitations, the present work offers a strong starting point for gaining a deeper understanding of the ways in which reporting practices may influence how environmental news narratives affect audiences. Across two controlled experiments with different samples, we found that higher levels of exemplar voice contributed to stronger perceptions that environmental news narratives were character focused, which facilitated stronger feelings of compassion. Feeling compassionate, as compared to feeling pity toward those affected negatively by the Flint water crisis, was then associated with stronger political support for policies that would improve drinking water quality and with more time spent on a clean water website. These results may motivate scholars and storytellers who are interested in helping audiences better understand issues of environmental justice and how they affect real people across the globe to pay further attention to compassion—and who exactly gets to tell the stories of environmental injustices.

Notes

1. A two-level between-subjects experiment refers to a procedure in which individuals are randomly assigned to one of two conditions in which some key variables are manipulated (e.g., exemplar voice) but other variables remain constant.

2. Because this time measure was positively skewed (skewness = 2.88), $\log_{10}$ transformations were used in both this study and in study 2 (Tabachnick and Fidell 2001).

3. A three-level between-subjects experiment refers to a procedure in which individuals are randomly assigned to one of three conditions in which some key variables are manipulated (e.g., exemplar voice) but other aspects of the message remain constant.

References

Aristotle. 2015. *Rhetoric.* Translated by W. Rhys Roberts. CreateSpace Indepedent Publishing.

Bartels, Daniel M., and Russell C. Burnett. 2011. "A Group Construal Account of Drop-in-the-Bucket Thinking in Policy Preference and Moral Judgment." *Journal of Experimental Social Psychology* 47 (1): 50–57.

Bartsch, Anne, Mary Beth Oliver, Cordula Nitsch, and Sebastian Scherr. 2018. "Inspired by the Paralympics: Effects of Empathy on Audience Interest in Para-sports and on the Destigmatization of Persons with Disabilities." *Communication Research* 45 (4): 525–53.

Blum, Laurence A. 1980. "Compassion." In *Explaining Emotions,* edited by A. O. Rorty, 507–17. Chicago: University of Chicago Press.

Bullard, Robert D. 1999. "Dismantling Environmental Racism in the USA." *Local Environment* 4 (1): 5–19.

Cameron, C. Daryl, and B. Keith Payne. 2011. "Escaping Affect: How Motivated Emotion Regulation Creates Insensitivity to Mass Suffering." *Journal of Personality and Social Psychology* 100 (1): 1–15.

Caracciolo, Marco. 2019. "Form, Science, and Narrative in the Anthropocene." *Narrative* 27 (3): 270–89.

Carey, Michael Clay, and Jim Lichtenwalter. 2019. "'Flint Can't Get in the Hearing': The Language of Urban Pathology in Coverage of an American Public Health Crisis." *Journal of Communication Inquiry* 44 (1): 26–47.

CDC (U.S. Centers for Disease Control and Prevention). n.d. "Lead FAQs." Accessed January 21, 2019. https://www.cdc.gov/nceh/lead/faqs/lead-faqs.htm.

CNN. 2021. "Flint Water Crisis Fast Facts." January 14, 2021. https://www.cnn.com/2016/03/04/us/flint-water-crisis-fast-facts/index.html.

Cuddy, Amy J. C., Susan T. Fiske, and Peter Glick. 2007. "The BIAS Map: Behaviors from Intergroup Affect and Stereotypes." *Journal of Personality and Social Psychology* 92: 631–48.

Eller, Ronald D. 2008. *Uneven Ground: Appalachia since 1945.* Lexington: University Press of Kentucky.

Fiske, Susan T., Amy J. C. Cuddy, Peter Glick, and Jun Xu. 2002. "A Model of (Often Mixed) Stereotype Content: Competence and Warmth Respectively Follow from Perceived Status and Competition." *Journal of Personality and Social Psychology* 82: 878–902.

Gibson, Rhonda, and Dolf Zillmann. 1994. "Exaggerated versus Representative Exemplification in News Reports Perception of Issues and Personal Consequences." *Communication Research* 21 (5): 603–24.

Gibson, Rhonda, and Dolf Zillmann. 2000. "Reading between the Photographs: The Influence of Incidental Pictorial Information on Issue Perception." *Journalism and Mass Communication Quarterly* 77 (2): 355–66.

Grabe, Maria E., Mariska Kleemans, Ozen Bas, Jessica Gall Myrick, and Minchul Kim. 2017. "Putting a Human Face on Cold, Hard Facts: Effects of Personalizing Social Issues on Perceptions of Issue Importance." *International Journal of Communication* 11: 907–29.

Green, Melanie C., and Timothy C. Brock. 2000. "The Role of Transportation in the Persuasiveness of Public Narratives." *Journal of Personality and Social Psychology* 79 (5): 701–21.

Haidt, Jonathan. 2003. "The Moral Emotions." In *Handbook of Affective Sciences,* edited by Richard J. Davidson, Klaus R. Scherer, and H. Hill Goldsmith, 852–70. Oxford: Oxford University Press.

Hassabis, Demis, R. Nathan Spreng, Andrei A. Rusu, Clifford A. Robbins, Raymond A. Mar, and Daniel L. Schacter. 2014. "Imagine All the People: How the Brain Creates and Uses Personality Models to Predict Behavior." *Cerebral Cortex* 24 (8): 1979–87.

James, Erin, 2015. *The Storyworld Accord: Econarratology and Postcolonial Narratives.* Lincoln: University of Nebraska Press.

Jamieson, Kathleen Hall. 2018. "Crisis or Self-correction: Rethinking Media Narratives about the Well-being of Science." *Proceedings of the National Academy of Sciences of the United States of America* 115 (11): 2620–27.

Kalyanaraman, Sriram, David L. Penn, James D. Ivory, and Abigail Judge. 2010. "The Virtual Doppelganger: Effects of a Virtual Reality Simulator on Perceptions of Schizophrenia." *Journal of Nervous and Mental Disease* 198 (6): 437–43.

Lazarus, Richard S. 1991. *Emotion and Adaptation.* Oxford: Oxford University Press.

Lu, Hang, and Jonathon P. Schuldt. 2016. "Compassion for Climate Change Victims and Support for Mitigation Policy." *Journal of Environmental Psychology* 45: 192–200.

Maier, Scott R., Paul Slovic, and Marcus Mayorga. 2016. "Reader Reaction to News of Mass Suffering: Assessing the Influence of Story Form and Emotional Response." *Journalism* 18 (8): 1011–29.

Morello-Frosch, Rachel, Miriam Zuk, Michael Jerrett, Bhavna Shamasunder, and Amy D. Kyle. 2011. "Understanding the Cumulative Impacts of Inequalities in Environmental Health: Implications for Policy." *Health Affairs* 30 (5): 879–87.

Muller, Christopher, Robert J. Sampson, and Alix S. Winter. 2018. "Environmental Inequality: The Social Causes and Consequences of Lead Exposure." *Annual Review of Sociology* 44 (1): 263–82.

Myrick, Jessica Gall. 2017a. "Identification and Emotions Experienced after a Celebrity Cancer Death Shape Information Sharing and Prosocial Behavior." *Journal of Health Communication:* 1–8.

Myrick, Jessica Gall. 2017b. "Public Perceptions of Celebrity Cancer Deaths: How Identification and Emotions Shape Cancer Stigma and Behavioral Intentions." *Health Communication* 32 (11): 1385–95.

Myrick, Jessica Gall, and Jessica Fitts Willoughby. 2019. "The Role of Media-Induced Nostalgia after a Celebrity Death in Shaping Audiences' Social Sharing and Prosocial Behavior." *Journal of Health Communication* 24 (5): 461–68.

Nabi, Robin L. 2010. "The Case for Emphasizing Discrete Emotions in Communication Research." *Communication Monographs* 77 (2): 153–59.

Nietzsche, Friedrich. 1968. "The Anti-Christ." In *"Twilight of the Idols" and "The Anti-Christ."* Harmondsworth, U.K.: Penguin.

Nussbaum, Martha C. 1996. "Compassion: The Basic Social Emotion." *Social Philosophy and Policy* 13 (1): 27–58.

O'Keefe, Daniel J. 2003. "Message Properties, Mediating States, and Manipulation Checks: Claims, Evidence, and Data Analysis in Experimental Persuasive Message Effects Research." *Communication Theory* 13 (3): 251–74.

Pulido, Laura. 2016. "Flint, Environmental Racism, and Racial Capitalism." *Capitalism Nature Socialism* 27 (3): 1–16.

Punter, David. 2014. *The Literature of Pity.* Edinburgh: Edinburgh University Press.

Ramasubramanian, Srividya. 2013. "Intergroup Contact, Media Exposure, and Racial Attitudes." *Journal of Intercultural Communication Research* 42 (1): 54–72.

Rutt, Rebecca L., and Jevgeniy Bluwstein. 2017. "Quests for Justice and Mechanisms of Suppression in Flint, Michigan." *Environmental Justice* 10 (2): 27–35.

Schiappa, Edward, Peter B. Gregg, and Dean E. Hewes. 2005. "The Parasocial Contact Hypothesis." *Communication Monographs* 72 (1): 92–115.

Schlosberg, David. 2004. "Reconceiving Environmental Justice: Global Movements and Political Theories." *Environmental Politics* 13 (3): 517–40.

Schneider-Mayerson, Matthew, Alexa Weik von Mossner, and W. P. Malecki. 2020. "Empirical Ecocriticism: Environmental Texts and Empirical Methods." *ISLE: Interdisciplinary Studies in Literature and Environment* 27 (2): 327–36.

Schneider-Mayerson, Mathew, Abel Gustafson, Anthony Leiserowitz, Matthew H. Goldberg, Seth A. Rosenthal, and Mathew Ballew. 2020. "Environmental Literature as Persuasion: An Experimental Test of the Effects of Reading Climate Fiction." *Environmental Communication,* 1–16.

Shen, Fuyuan, Lee Ahern, and Michelle Baker. 2014. "Stories That Count: Influence of News Narratives on Issue Attitudes." *Journalism and Mass Communication Quarterly* 91 (1): 98–117.

Skitka, Linda J. 1999. "Ideological and Attributional Boundaries on Public Compassion: Reactions to Individuals and Communities Affected by a Natural Disaster." *Personality and Social Psychology Bulletin* 25 (7): 793–808.

Slovic, Paul. 2010. "If I Look at the Mass I Will Never Act: Psychic Numbing and Genocide." In *Emotions and Risky Technologies,* edited by Sabine Roeser, 37–59. Dordrecht: Springer Netherlands.

Stellar, Jennifer E., Amie M. Gordon, Paul K. Piff, et al. 2017. "Self-Transcendent Emotions and Their Social Functions: Compassion, Gratitude, and Awe Bind Us to Others through Prosociality." *Emotion Review* 9 (3): 200–207.

Tabachnick, Barbara G., and Linda S. Fidell. 2001. *Computer-Assisted Research Design and Analysis.* Needham Heights, Mass.: Allyn & Bacon.

Weik von Mossner, Alexa. 2017. *Affective Ecologies: Empathy, Emotion, and Environmental Narratives.* Columbus: Ohio State University Press.

Weik von Mossner, Alexa. 2019. "Why We Care About (Non) Fictional Places: Empathy, Character, and Narrative Environment." *Poetics Today* 40 (3): 559–77.

Zillmann, Dolf. 2006. "Exemplification Effects in the Promotion of Safety and Health." *Journal of Communication* 56: S221–37.

Zillmann, Dolf, and Hans-Bernd Brosius. 2000. *Exemplification in Communication: The Influence of Case Reports on the Perception of Issues.* Mahwah, N.J.: Lawrence Erlbaum.

Chapter 6

The Reception of Radical Texts

The Complicated Case of Alice Walker's "Am I Blue?"

ALEXA WEIK VON MOSSNER, W. P. MALECKI, MATTHEW SCHNEIDER-MAYERSON, MARCUS MAYORGA, AND PAUL SLOVIC

One of the driving ideas behind empirical ecocriticism is that the combination of humanistic and social science methods will result in a better understanding of the impact of environmental narratives on their audiences. Thorough ecocritical analysis of a given text can yield insights into its narrative structure and main themes, leading to the formulation of hypotheses regarding its potential impact, while social science methods can provide reliable data that will either confirm or reject those hypotheses—so far, so deceptively simple. However, those who engage in empirical research know that the results often generate as many questions as they answer. This is particularly true when the stimulus is as complex as a literary text. The very richness and ambiguity that are cherished by readers and scholars of literature can lead to unexpected and confounding results, and reception is rarely as straightforward as concepts such as the ideal reader (Kress 1985) suggest. Ika Willis (2017) reminds us that "a vast number of interdependent factors are involved in acts of reception, [. . .] from the structure of human brains and the personal histories of individual readers to the material, economic and social structures which influence the production and distribution of texts" (7). In its attempt to grapple with the complexities of both the investigated literary text and its reception, this chapter demonstrates that doing empirical ecocriticism is often not as simple as using social scientific methods to confirm a hypothesis; sometimes empirical audience research uncovers something unexpected that generates further theorizing.

The study we discuss is the latest in a series of attempts to tackle a conundrum posed by Alice Walker's controversial story "Am I Blue?" ([1988] 1996). At its heart, "Am I Blue?" is an autobiographical piece of creative nonfiction that narrates Walker's developing relationship with a horse called Blue. As a Black woman, Walker's narrator is struck when she realizes that not only is Blue heartbroken when he is separated from his mate, but that the reasons for his suffering are analogous to those that inflicted emotional pain on her own ancestors during American slavery, thereby drawing what Marjory Spiegel (1996) calls "the dreaded comparison" between animal husbandry and human enslavement. Moreover, Walker potently reminds readers of the sentience and suffering of other farm animals such as chicken and cattle, who are bred and kept by humans for their eggs, milk, and meat. In 1994, the California State Board of Education banned "Am I Blue?" from a statewide test for tenth graders because it was considered "anti–meat eating" (Holt 1996, 5). Only after a massive public outcry against censoring a story penned by a Pulitzer Prize–winning Black author was the ban retracted. Since then, it has been considered a powerful example of a literary text that successfully raises concern about other species (Hooker 2005) and has frequently been assigned in high school and college courses.

This remarkable reception history made the story an ideal choice for a group of researchers at the University of Wroclaw in Poland (including one of the authors) who conducted a series of experiments to establish whether a literary text makes readers more concerned about animal welfare. But unlike other texts used in their studies, "Am I Blue?" failed to have a positive impact on the participants' attitudes toward animal welfare in general, although it did change readers' attitudes toward horses specifically (Malecki et al. 2019). Additional statistical analysis of the data revealed that the text also had a positive impact on readers' attitudes toward cultural minorities (Malecki, Weik von Mossner, and Dobrowolska 2020).[1] Given the cultural embeddedness of reception (Hall [1973] 1980; Livingstone 1998, 2015), we hypothesized that American readers might respond differently to "Am I Blue?" than readers in Poland. Not only did the participants in the Polish studies read the text in translation, which might affect its impact, but also it seemed likely that the "dreaded comparison" made by Walker's African American narrator might affect readers in culturally specific ways.

The next logical step was to conduct a large survey (N = 800) in the United States; that survey is the focus of this chapter. In addition to testing the potential influence of cultural factors on reception by replicating the first Polish study with American readers, we wanted to find out whether a text manipulation that left out the controversial comparison to human enslavement would lead to a different result in terms of the story's impact on attitudes toward animal welfare.[2] Because the horse in Walker's story does not endure physical violence—whereas the depicted animals in comparable narratives tested by the Polish team did—we added such physical violence in a second text manipulation. Our hypothesis was that the capacity of animals to suffer emotional pain, even though well documented by cognitive ethologists (Bekoff 2008), primatologists (De Waal 2019), and neuroscientists (Panksepp 2004), is still not a universally accepted fact, and that its literary depiction might therefore not have the same impact as a portrayal of physical suffering.

The results of our study, which are as nuanced and complicated as "Am I Blue?" itself, are an excellent example of how empirical research on environmental texts can improve—and at times complicate—our understanding of their reception.

The Narrative Strategies of "Am I Blue?"

The ambiguities within and around "Am I Blue?" begin with the very nature of the text. It has been called an essay (Lioi 2008; Malecki, Weik von Mossner, and Dobrowolska 2020; Ruffin 2008), a story (Benzel 1996), and a short story (Holt 1996), making its position on the fiction–nonfiction spectrum ambiguous.[3] Most commentators assume, however, that it is autobiographical. Walker's first-person narrator starts out by remembering how she once rented "a small house in the country" with her unnamed companion. On the large meadow across from the house, she meets Blue, "a large white horse, cropping grass, flipping its mane, and ambling about" ([1988] 1996, 32). It is an idyllic and beautiful sight, but its pastoral "illusion of peace and harmony" (Marx 1964, 25) hides not only the human labor and machinery necessary to create and maintain it but also the suffering that often results from animal husbandry. Readers soon find out that Blue is neglected and lonely. He is a gregarious animal kept solitary by an owner who lives far away and shows little interest in the well-being of his possession. Only the narrator regularly interacts with Blue, feeding him apples from a tree that grows in her yard, and only the narrator understands how unhappy he is in this condition.

The narrative arc of the story stretches from the narrator's first encounter with the horse to what she perceives as a growing friendship between her and Blue to a moment of happiness for the stallion when he receives a companion, a brown mare that "amble[s] and gallop[s]" with him across the meadow ([1988] 1996, 39). It ends with the sudden separation of the two horses after they have mated, which has a profound effect on Blue, who is once again alone. At first "crazed" (40), he turns apathetic, no longer showing any interest even in his beloved apples. The narrator reads disappointment, pain, disgust, and hatred in the horse's eyes. "What that meant," she explains to the reader, "was that he had put up a barrier within to protect himself from future violence; all the apples in the world wouldn't change the fact" (42). The violence Blue is subjected to is not physical abuse, but he is shown to be in emotional pain.

The narrator conveys that as a Black woman, she can relate to Blue's suffering. "If I had been born into slavery," she writes, "and my partner had been sold or killed, my eyes would have looked like that" ([1988] 1996, 40). From there, she extrapolates to the treatment of enslaved persons in the United States more generally: "The children next door explained that Blue's partner had been 'put up with him' (the same expression that old people used, I noticed, when speaking of an ancestor during slavery who had been impregnated by her owner) so they could mate and she conceive" (40). The comparison implies that Blue has been enslaved by humans and that his feelings are comparable to what a human would feel in such a position. As Walker (1996b) acknowledges in her foreword to Spiegel's *Dreaded Comparison,* this "is a comparison that, even for those of us who recognize its validity, is a difficult one to face. Especially so if we are the descendants of slaves. Or of slave owners. Or both" (13).[4] Being a descendent of enslaved persons, Walker suggests in "Am I Blue?," allows for a deeper understanding of Blue's pain because she can empathize with his situation. It is a case of "situational empathy" as it has been defined by narratologist Patrick Colm Hogan (2003), in this case extending across species.[5]

Importantly, readers never have direct access to Blue's consciousness. Even though the story is centrally concerned with the horse's emotional suffering, it never offers an "insider perspective" on animal subjectivity (Weik von Mossner 2017, 107), like we find, for example, in Anna Sewell's influential novel *Black Beauty* ([1877] 2012). In "Am I Blue?," it is the human narrator who attributes a range of thoughts and feelings to the horse. She does this by looking at Blue's face and body, then using what psychologists call theory of mind (ToM) to draw inferences about his mental states. For example, the narrator observes that "sometimes he would stand very still just by the apple tree, and when one of us came out he would whinny, snort loudly, or stamp the ground," deducing that "this meant, of course: I want an apple" (33). In addition to such cognitive processes of mindreading, it is suggested that there are instances of interspecies emotional contagion—situations in which the narrator involuntarily catches the emotions of the horse and feels along with him. After Blue lost his mate, the narrator recalls, "he looked at me. It was a look so piercing, so full of grief, a look so human" (41). Although there is neuroscientific evidence for such processes of transspecies empathy between humans and animals (Franklin et al. 2013), they run the risk of "empathetic inaccuracy" (Keen 2010, 81) in that the deep empathetic sadness the narrator feels in the story may not really or fully correspond to the actual feelings of the horse. However, this doesn't change the fact that the narrator's feelings are genuine and that the words on the page cue readers' empathy for a suffering animal.[6]

It is thus not only the narrator who engages in transspecies empathy. Readers are also invited to feel along. Walker offers them two mental routes to empathize with Blue's plight. On the one hand, they are invited to share the emotions of the human narrator as she experiences feelings of sympathy, compassion, and pity *for* the horse. On the other hand, the text also invites readers to feel along *with* the horse. They learn that Blue "whinnied until he couldn't. He tore the ground with his hooves. He butted himself against his single shade tree. He looked always and always toward the road down which his partner had gone" (Walker [1988] 1996, 41). This information is given by a potentially unreliable first-person narrator, who herself attributes consciousness to the horse. However, because her narration is in "showing" mode, readers can visualize those scenes and thus engage in their own theory of mind to understand that these are physical expressions of pain and despair. They are encouraged to understand, on an emotional level, that nonhuman animals are sentient beings, and that humans therefore have ethical responsibilities and moral obligations toward them (Bekoff 2010; Harper 2010; Singer 2009; Sunstein and Nussbaum 2005).

That the story advocates this insight becomes most obvious on its final page, where it makes the explicit imaginative leap from horses to other animals. After a visiting friend comments that the white horse in the meadow is "the very image of freedom" (Walker [1988] 1996, 43), Walker's narrator thinks about the difference between misleading animal imagery and the real state of affairs: "We are used to drinking milk from containers showing 'contented' cows, whose real lives we want to hear nothing about, eating eggs and drumsticks from 'happy' hens, and munching hamburgers advertised by bulls of integrity who seem to command their fate" (43). Not only does the narrator make the mental leap from the suffering of an individual horse to the exploitation of other species, but she also shows readers how she personally was affected by it when she later "sat down to steaks. I am eating misery, I thought, as I took the first bite. And spit it out" (43). Those are the final words of Walker's story, presenting readers with the narrator's disgusted response to a piece of meat. It is a surprising and even "shocking turn" (Lioi 2008, 20) in a narrative that is rich in unexpected twists and metaphorical undercurrents. As Anthony Lioi (2008) puts it, Walker's story "is an exercise in holding very different subject-positions at the same time" because the narrator self-identifies "as, literally, an American black woman whose family history is a product of slavery, and, metaphorically, as a white slave-holder, an enslaved horse, and the horse's Abolitionist advocate" (22).

These narrative complexities undoubtedly circumscribe readers' engagement with the text, and it is important to trace and analyze them from an ecocritical perspective. But one of the core lessons of empirical ecocriticism is that the assumptions we make about reader engagement based on textual analysis may not always match up with the responses of actual readers. Since the early days of semiotics (Barthes 1977; Eco 1979) and reception aesthetics (Iser 1976; Jauss 1982), it has become commonplace to assert that readers cocreate literary texts, and that such cocreation will be to some degree subjective. With American reader-response theory (Fish 1980) and British cultural studies (Hall [1973] 1980) came a much stronger focus on the agency of real readers, along with a recognition of the inevitable cultural and social embeddedness of such individual acts of reception. With reference to Hall's ([1973] 1980) influential encoding/decoding model of communication, Willis (2017) reminds us that "real readers are able to take up a range of different positions with respect to implied readers: subordinate, resistant, negotiated, oppositional and so on. For this reason, we cannot know how real readers or audiences respond to the protocols and positions set out in a text simply by looking at the text in isolation" (84).[7]

If we are interested in the reception of environmentally engaged texts, we are thus left with at least two options. First, we can examine "historical evidence left behind by real readers of the past" (Willis 2017, 84) where such evidence is available. Second, we can conduct original empirical research on real readers of the present.[8] In our quest to better understand the narrative impact of "Am I Blue?," we did both.

Historical Evidence and Previous Studies

It was Walker's turn to the topic of industrial meat production at the end of "Am I Blue?," along with the narrator's disgusted response to meat, that led the California State Board of Education to ban the story in 1994. That may seem like a remarkable declaration, given that other than those last few lines, the story is centrally focused on a horse and thus on an animal that isn't typically associated with meat production in the contemporary United States. However, at the time of the board's decision, the production and sale of horsemeat for human consumption was not yet illegal in California, and perhaps that played into its declaration that "Am I Blue?" was inappropriate reading material—because "it seemed to violate rural children's family occupations" (Holt 1996, 5).[9] Even though, as Lioi (2008) notes, Walker makes no attempt "to construct a logical argument against factory farming or slaughterhouses" and "Blue is not in danger of being eaten" (21), the story relates the false perception of "freedom" when looking at the lonely, captive horse to the related false perception of the happiness of animals that are bred and raised for their eggs, milks, and meat.

Walker's other comparison in the story—between the suffering of the horse and that of enslaved Black people, Native Americans, and immigrants from Japan, Korea, and the Philippines—was not mentioned in the board's official statement on its decision. However, the transcript of the relevant board meeting reveals that one member, Diane Lucas, claimed, in defense of Walker's story, to have taught it to sixty tenth and eleventh graders "who thought, with no coaching from [her], that the story was about people . . . mention[ing] ethnic diversity and that the story condemned racism" (qtd. in Walker 1996a 82). In her remarks, Lucas did not even mention the suffering horse. Instead, she focused entirely on what she considered the story's metaphorical dimension: its commentary on the historical suffering of enslaved Black people whose emotional needs, like Blue's, were disregarded and denied by owners who had limitless power over them.

The same is true for several of the letters to the editor of the *San Francisco Chronicle.* After learning of the banning of Walker's story, the *Chronicle* republished it and asked for readers' opinions regarding the board's decision. More than 600 letters were sent in response, "running approximately nine to one 'in support of the story, and against the Board of Education's removal of it'" (Walker 1996a, 67). Sixteen of these letters are reprinted in *Alice Walker Banned,* edited by Patricia Holdt, and it is reported that they are "representative of the range of opinions expressed, not the ratio" (Walker 1996a, 67). They vary from calling the story a "pedestrian, uncreative remembrance" written by a "PC empress" (Sherrell [1994] 1996, 68) to the astute observation that "'Am I Blue?' is a very skillfully drawn emotional poke at many subjects" including "slavery, racism, women's rights, general intensitivity and, of course, meat eating" (Grosjean [1994] 1996, 68). They also include a letter from a reader who "gave up eating meat in 1987" because she "no longer felt comfortable eating [her] friends" (Benzel [1994] 1996, 69). The reader goes on to speculate, "If this story makes people uncomfortable, it's only because it provokes one to think about issues we may have become desensitized to" (69). Overall, the available historical evidence suggests that in 1994, readers in California read the story in a variety of ways, perceiving different aspects of it as salient.

More recently, "Am I Blue?" was included as one of several texts in the aforementioned series of experiments in Poland that tested whether literature can impact attitudes toward general animal welfare. The researchers initially hypothesized that the failure of "Am I Blue?" to have such an impact might have been because the experiment was conducted on uniformly white and Polish citizens, and that the African American perspective presented by the story might have been alien or alienating to them (Malecki et al. 2019). However, this hypothesis was eventually dropped because a later experiment, also conducted on uniformly white Polish citizens, found that the story *did* improve attitudes toward horses. Something else must have been responsible for the fact that "Am I Blue?" does not seem to improve attitudes toward animals in general, whereas other works do.

Together with the historical evidence and our narratological analysis, these results led to the formulation of a new hypothesis: that the story's metaphorical link between the fate of the horse and the fate of enslaved persons might somehow block the metonymic associative link from the individual horse to all animals and thus explain why the story does not improve attitudes toward animals in general (Malecki, Weik von Mossner, and Dobrowolska 2020).[10] Given that Walker relies on a comparison between the oppression of humans and animals, and given that for most people the suffering of humans is more important than the suffering of animals, the associative chain leading from a single animal to the species level and then to animals in general might get redirected. The chain might lead instead to human groups, the result of the pull of most readers' default anthropocentric attitudes. This hypothesis has potentially large implications. If drawing connections between human and nonhuman suffering compromises stories' ability to affect attitudes toward general animal welfare, that would be important information not only for ecocritics but also for animal rights organizations such as People for the Ethical Treatment of Animals (PETA) that employ such comparisons in their outreach (Deckha 2008).

This is especially true when human–animal comparisons make direct reference to the history of enslavement and related racist practices of animalizing African Americans. Because of those practices, Black people sometimes perceive PETA's campaigns as humiliating or exploitative (Rodrigues 2020). As Lindgren Johnson (2018) notes, such a perception, and related protests, "rehearses, in the present, a struggle going back to slavery to articulate African American humanity in the face of racist pressures that sought to reduce blackness to a form of 'animality,' mere carnality, a brute body" (1). That is why the "insistence on contrast rather than comparison in regard to black and animal suffering" is often seen as an attempt "to secure black humanity from the ravages of dehumanization" (1). In a society that continues to be marked by racial discrimination, marginalization, and violence, such insistence is understandable.[11] However, we should not forget that the comparison is based on prior disregard for the lives of nonhuman animals. "That the human/animal opposition makes the abjection of *human* others possible," writes Christopher Peterson (2012), "means that insisting on their humanity as a mode of resistance can only reinscribe the speciesist logic that initiates their exclusion" (2). Walker (1996b) seems to share that perspective, noting, "The animals of the world exist for their own reason. They were not made for humans any more than black people were made for white women and men" (14). But what if this comparison is so offensive to some readers that it becomes ineffective?

Aside from the problems related to the dreaded comparison, there are at least two other plausible reasons why, in the study described above, "Am I Blue?" might have failed to have an impact on attitudes toward animals in general, in contrast to other animal narratives tested in similar experiments. All those other narratives displayed physical suffering of an animal protagonist, whereas the suffering of the horse in "Am I Blue?" is primarily psychological. Perhaps psychological suffering was not enough to affect participants' attitudes toward general animal welfare, especially because many people believe that only humans are capable of higher emotions. We also wondered whether the specific context of U.S. culture on which the parallel drawn in the story relies so heavily might play a significant role in its reception. We therefore conducted an experiment with American participants.

The U.S. Study: Method, Materials, and Design

Eight hundred participants were recruited from Prolific for what was described as an online survey on "Personality and Stories."[12] They were compensated $4.75 for their participation. Their mean age was 34.4 years, and 51.8 percent of them identified as female. A total of 68.3 percent of them identified as white, 7.9 percent as Latino/Hispanic, 6.9 percent as East Asian, 5.9 percent as African American, 1.3 percent as South Asian, 0.8 percent as Native American or Alaska Native, 0.6 percent as Middle Eastern, 0.5 percent as Caribbean, 6.8 percent as mixed, and 0.6 percent as other.[13] Overall, the sample was highly educated, with 81.8 percent possessing some college education or higher. Reading frequency was normally distributed, with most participants indicating that they read fiction and nonfiction "occasionally (5 to 6 books or articles a year)."

The design of our study was supposed to help us answer whether the lack of impact of "Am I Blue?" on attitudes toward general animal welfare observed in the Polish studies was due to (1) divergence between the cultural specificity of the text (about American realities and by an American author) and the cultural specificity of the participants, all of whom were Polish; (2) the parallel Walker draws between human and animal oppression; or (3) the story's focus on the horse's emotional, rather than physical, suffering. In order to be able to answer questions (2) and (3), we needed to compare the impact of the Original version of Walker's story on our participants with the impact exerted by two manipulated versions. The Animal version was modified to remove all story elements that referenced human enslavement; the Physical Abuse version included additional text elements that showed Blue being severely whipped and beaten by two handlers as punishment for his "crazed" behavior after his mate was taken away.[14] In order to establish whether the three versions of the text had a statistically significant impact on attitudes toward general animal welfare, we also needed to have a control group. Following the methodology of the Polish study, we decided that the participants in the control group would read a text that would be of similar length to the Walker story and on a subject entirely unrelated to animal welfare—in this case, a journalistic narrative on the discovery of the Higgs boson particle. We also followed the methodology of the Polish study in not revealing the exact purpose of our experiment to the participants, informing them instead that it concerned "Personality and Stories." This strategy was used to minimize social desirability bias.[15]

After being informed about the ostensible purpose of the study, the participants first read the stimulus, depending on their random assignment to one of our four conditions: (1) the Original version of "Am I Blue?," (2) the Animal version, (3) the Physical Abuse version, or (4) the control text. Once they were done reading, they answered questionnaire items measuring their attitudes toward the welfare of animals in both general and more concrete terms. These included the seven items comprising the Attitudes toward Animal Welfare (ATAW) scale used in the Polish study, as well as two policy-related questions. One concerned support for a hypothetical "meat tax" to ensure humane treatment of animals on factory farms, and the other concerned support for real federal legislation—passed by the U.S. House of Representatives and the Senate at the time when we conducted our study and now signed into law—that makes animal cruelty a federal crime. Because the main reason for the California State Board of Education's banning of the text was a suspected impact on readers' attitudes toward meat consumption, we also included an item measuring participants' consumption of animal products. To camouflage the actual purpose of the study, we buried these questions among dozens of items consistent with its ostensible purpose ("Personality and Stories"). These camouflage items included a Ten Item Personality Questionnaire (Gosling, Rentfrow, and Swan 2003) as well as other questions ostensibly probing the participants' personalities and worldviews and the item on attitudes toward minorities that was used in the Polish study.

Because narrative impact has been shown to depend on how engaged one is in reading a given story (Johnson 2012; Mazzocco et al. 2010), we also included a shortened version of the Transportation Scale (Green and Brock 2000), designed to measure how much readers feel transported into the story world.[16] In order to ensure that they were actually reading the text, we included a simple three-item quiz on content, which varied depending on the story condition. Participants that did not answer any of these three correctly were excluded from analysis. As is standard in experiments on literary reception, we controlled for age, gender, ethnicity, and education. Given the topic of Walker's story, we also controlled for whether the participants kept pets because having a pet might have an influence on how one responds to items concerning animal welfare or reacts to a story concerning that subject.

Approximately two weeks after the release of the first survey, we invited the 800 participants to take part in a second survey, which included the ATAW scale, the meat consumption frequency question, the policy-related questions, and several items testing the recollection of the story from two weeks before. We decided to conduct this second survey for two reasons. First, we wanted to test whether the impact of "Am I Blue?," if any, would last. Our main focus in this study was whether the text has a meaningful influence on attitudes, and in order to know that, it is not enough to test its impact immediately after reading, especially because the limited available data on the long-term influence of stories (Malecki et al. 2018; Schneider-Mayerson et al. 2020; Vezzali, Stathi, and Giovanni 2012) do not permit sound generalizations to be made on this subject. Studying the longitudinal effects of "Am I Blue?" was also important to us because we suspected that the text might be subject to the so-called sleeper effect (Appel and Richter 2007). According to the sleeper effect hypothesis, some aspects of a text may act as a discounting cue that mitigates the persuasive influence of the text's content. Because the reader's memory of the discounting cue may fade faster than the memory of the text's content, this could lead to an increased influence of the text over time (Appel and Richter 2007). Fictionality has been shown to act as a discounting cue, with an increased influence of a text occurring after two weeks (Appel and Richter 2007). While its position on the fiction–nonfiction spectrum is ambiguous, we hypothesized that the literariness of "Am I Blue?" might act in similar ways as a discounting cue and that the sleeper effect might manifest itself after a similar period of time.[17] Comparing the results obtained in both surveys allowed us to test that hypothesis. Overall, 729 participants completed both surveys.

Results

Our statistical analyses revealed that Walker's story (all variants combined) did not have a significantly different effect than the control text on the participants' attitudes toward general animal welfare, as measured by the ATAW.[18] This result is in line with the result of the Polish study, suggesting that the lack of impact reported in that study was not due to the Polish participants' unfamiliarity with the cultural context of the story, which should have been perfectly familiar to the American participants in the present study. However, there were significant effects of the two text manipulations on three individual items in the ATAW scale.

For the Animal condition, there was more support than in the control condition for the statement "The low costs of food production do not justify maintaining animals under poor conditions" (Figure 6.1).[19] In accordance with our dreaded comparison hypothesis, participants who read the

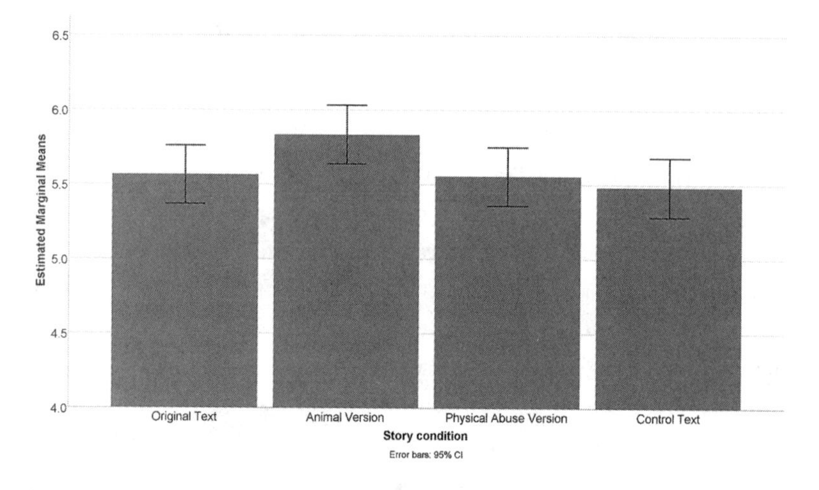

Figure 6.1. Agreement for "The low costs of food production do not justify maintaining animals under poor conditions."

manipulation of the story that excluded all parallels to human slavery thus were more likely to prioritize animal welfare over other concerns, such as food prices. No such effect was observed in the case of people in the other two experimental conditions.

For the Physical Abuse condition, there was significantly less support for the following two statements: "Human needs should always come before the needs of animals"[20] and "Basically, humans have the right to use animals as we see fit"[21] (Figure 6.2). Participants who had read the manipulation of the story that added physical abuse to the emotional suffering of the horse were thus less likely than people in the control group to support statements implying the absolute priority of human needs over animal needs or the rightlessness of animals. No such effect was observed in the two other conditions.

We also received interesting results for our two policy-related items. Regarding the question about supporting a hypothetical small tax to ensure humane treatment of farmed animals, our analyses revealed a significant difference: respondents who were exposed to the Physical Abuse text had significantly more support for the meat tax than the three other stories (Figure 6.3).[22] The results for our other policy question—asking participants whether they

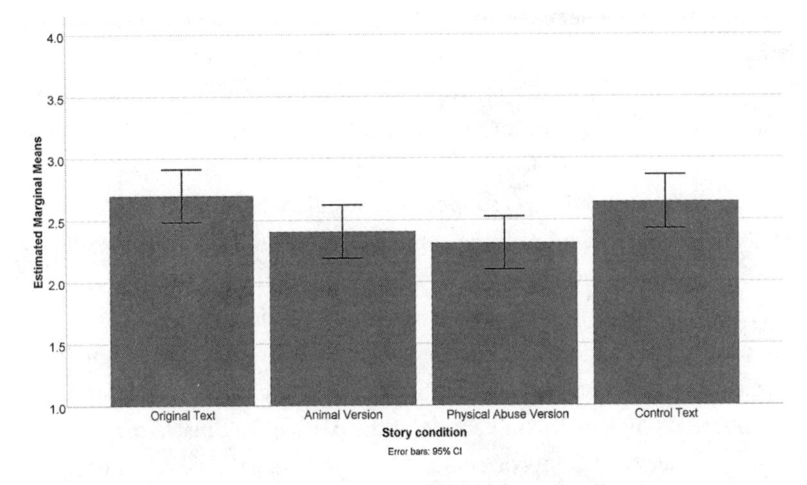

Figure 6.2. Agreement for "Basically, humans have the right to use animals as we see fit."

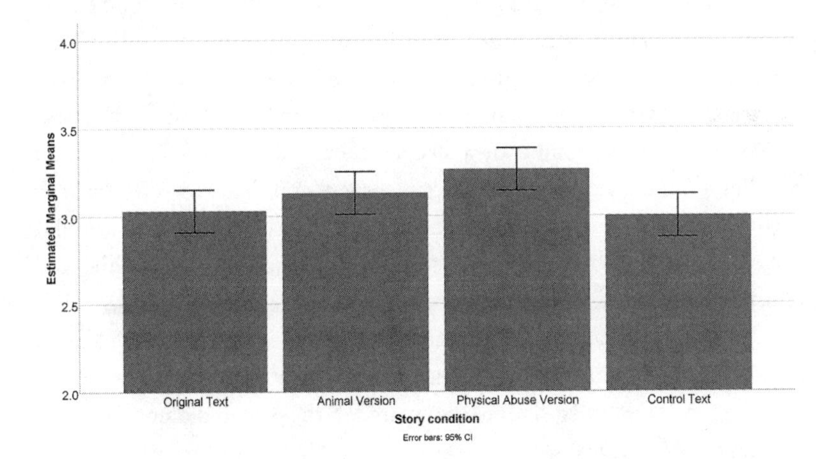

Figure 6.3. Support for tax on meat products to ensure humane treatment of animals.

would support a federal law under congressional consideration that would make animal abuse a federal crime—were counterintuitive.[23] There was a marginally significant and relatively small effect of the story condition (combined "Am I Blue?" conditions versus the control text)[24] on support for the policy, but in the opposite direction as predicted: the three versions of the Walker story garnered lower support for the bill than the control text.[25] That is, the participants who read the control text were more willing to support the bill than those who read any variant of "Am I Blue?"

These comparisons were made while controlling for age, gender (binary), education, pet keeping, reading frequency (both fiction and nonfiction), narrative transportation, and attention. In all the groups, the attention to the story was high, with 81.1 percent of the participants answering all three attention-check questions correctly.[26] These percentages differed little among the conditions. In addition, 92.04 percent self-reported that they read the Original version of "Am I Blue?" "closely, to attend to the detailed information presented in the story," while only 7.96 percent reported reading the story "quickly, to get the general idea." Participants in the control condition indicated slightly faster reading, with 20.8 percent reporting a quick read. We also found that the Original version of "Am I Blue?" had significantly higher means on the transportation scale than the control text, which suggests that it was more engaging.[27] Rates of previous exposure to Walker's story was low, with only four participants reporting familiarity.

No evidence was found for a sleeper effect. Participants who recalled the themes of the story[28] in the second survey (conducted two weeks after the first) did not show increased ratings in the ATAW[29] over those who did not recall the themes of the text.[30] We also found no evidence of a delayed effect of the story in the second survey regarding support for the meat tax[31] or support for legislation making animal abuse a federal crime.[32] People who recalled the themes of the story in the second survey did not show increased support for a tax[33] or legislation[34] compared to those who did not recall the themes of the text.[35]

Given the concerns of the California State Board of Education about the story being anti–meat eating, we also checked whether reading the Original version of "Am I Blue?" had any effect on people's self-reported meat consumption, but we found no effect. The ratings between the two surveys were thus consistent,[36] suggesting that reading Walker's story had no effect on participants' desire to eat meat.

Additionally, we compared the results regarding participants' support for minorities, as measured by a direct question, to the results obtained in the Polish study.[37] Controlling for age, gender, ethnicity, education, and transportation into the story, the story condition did not have a significant effect on support for ethnic minorities in the second survey, thereby failing to replicate the results of the Polish study.[38] Another counterintuitive result was revealed by the analysis of this item in the second survey. A marginally significant difference was observed between story conditions, but not in the way we would have expected. Readers of the Animal version of the text expressed significantly *more* support for ethnic minorities than readers of the other texts, controlling for the wave 1 responses (Figure 6.2).[39] Recall of the text did affect attitudes toward ethnic minorities; people who recalled the themes of the story in the second survey showed more support for ethnic minorities[40] over those who did not recall the themes of the text.[41] However, this general difference was mainly due to higher ratings of support in those who recalled themes in the Animal version.[42]

Finally, we took advantage of the fact that we controlled for a variety of demographic factors in the study and investigated whether any of those affected the dependent variables our study evaluated. Some of them did. We found that participants' attitudes toward animals, as measured by the ATAW, were significantly predicted by gender, with women holding more favorable attitudes toward protecting animals than men.[43] Pet owners held more favorable attitudes toward protecting animals in general.[44] There also was a small but significant relationship between reading fiction and views toward animals, with people who read more fiction holding more favorable views toward animals.[45] Women were significantly more likely to support the tax policy[46] and the anti–animal abuse bill than men.[47] Older adults were also more likely to support the policy than younger adults,[48] and pet owners supported it more than those without pets.[49] Age, gender, and education were also significant covariates (independent variables that can influence the statistical outcome without being of direct interest to the study) for support for ethnic minorities. Participants who were older,[50] female,[51] and more educated indicated greater support for ethnic minorities.[52]

Discussion

One of our main aims in conducting this study was to see whether the lack of impact of "Am I Blue?" on attitudes toward general animal welfare (as measured by the ATAW scale) observed in the two previous studies conducted in Poland might not have been due to a divergence between the cultural specificity of the text and the background and knowledge of the participants, all of whom were Polish; and, more generally, to see whether that lack of impact was culturally specific. Our results suggest that the story's lack of impact on attitudes toward general animal welfare is not specific to the Polish context, and that it cannot be explained by a cultural disconnect between the Polish participants in the earlier study and the text. The American participants in the present study remained equally unmoved in their attitudes toward general animal welfare as measured by the ATAW scale. It would appear, then, that the lack of attitudinal impact is due to the text itself.

Our second hypothesis was that the lack of impact of "Am I Blue?" on attitudes toward animals in general might be due to the parallel Walker draws between human and animal oppression. Here too our results—as measured by the ATAW scale—were negative in that they did not show a significant difference between the Original version of Walker's story and the Animal version. However, things get more complicated when we look at one of the items on the scale in isolation. Given that the Animal version did have a positive impact on readers' agreement with the statement "The low costs of food production do not justify maintaining animals under poor conditions"—and given that this effect was not observed in readers who read the Original version of the text—there is some support for our hypothesis that the human–animal comparison somehow blocks the metonymic chain that leads from the individual horse Blue to animals in general.[53]

Counterintuitively, the Animal version of the story also improved attitudes toward minorities in the second survey. Although this version does not mention the history and experiences of minorities at all, it resulted in better attitudes toward minorities than did exposure to either the control text or both other versions of "Am I Blue?" (the Original and the Physical Abuse version) that did include the human–animal oppression parallel. This is so unexpected that at first glance, one might treat it simply as a result of chance or error. However, rather than rejecting an unexpected result out of hand, we can look for possible explanations for its occurrence. As noted earlier, this analogy has been observed to be controversial, even offensive, to some. It should also be noted that the Original version of Walker's story, which includes that comparison, did have a positive impact on attitudes toward minorities immediately after reading in the Polish study, but did not have any immediate impact in the U.S. study in the first wave of our survey. Therefore, our questions should be rephrased as follows: why did the human–animal oppression parallel have an impact on attitudes toward minorities in the Polish context immediately after reading, whereas it did not have such an impact in the United States? And why did a text manipulation that omits the parallel have no impact on American readers immediately after reading, but did have an impact after two weeks?

One possible explanation for this counterintuitive result is that participants in the U.S. study were more likely to respond negatively to a comparison between animal husbandry and American slavery for the culturally specific reasons outlined above. Beyond simply comparing animal suffering to human suffering, Walker makes concrete references to American history and racial politics—references that are likely to be more salient to, and perhaps controversial for, an American reader than for a Polish reader, who looks at this history as a cultural outsider. About 30 percent of the participants in the U.S. study identified as Black, Indigenous, or people of color, and they would seem particularly likely to object to this comparison. Some of the participants who identified as white may have felt implicated in the history of racial inequality and discrimination in different ways, and they are also more likely than the participants in the Polish study to personally know people belonging to these minority groups. It follows, then, that the dreaded comparison would be more problematic for the participants in the American study. This might have been especially true in 2019 (when the experiment was conducted), a time of growing awareness of racial discrimination among white Americans. This might have been a reason behind the observed difference in the influence of the text on attitudes toward minorities.[54]

Why would the Animal version have an impact in the U.S. context, but only after two weeks? One hypothesis is that in recalling the story, the participants focused more on the fact that the story involved undeserved suffering than on who exactly suffered and the details of that suffering. This might have resulted in the increase of general empathic attitudes, and as a consequence greater concern for minorities.[55]

The third hypothesis we wanted to test was whether the lack of impact of "Am I Blue?" on attitudes toward animals in general (as measured by the ATAW scale) might be due to its emphasis on emotional rather than physical suffering. Here too additional statistical analysis of individual items on the scale complicates our finding. While the Physical Abuse version of the story did not have a significant impact on the entire ATAW scale, it did negatively impact readers' agreement with two individual items on that scale: asserting the primacy of human needs over animal needs and the human right to use animals. It seems that adding physical suffering to the emotional suffering of the horse sensitized readers to the needs and rights of animals. Giving additional support to this reading, our data also showed that whereas neither the Original version nor the Animal version made the participants more supportive of the hypothetical meat tax, the Physical Abuse version did.

This suggests that as an argument for animal welfare, the Original version of Walker's story was not sufficiently persuasive for our participants to support a fiscal measure that could potentially affect them personally. Only once physical violence was added were readers more likely to support such a measure. This could be a symptom of the still common assumption that animals do not have emotional lives, despite considerable evidence to the contrary (Bekoff 2008; De Waal 2019; Panksepp 2004). If one does not believe that animals are able to suffer emotionally as humans do, one might simply disregard as unreasonable any arguments or stories that impute such an ability to animals. As a result, one might not see the argument or story as a realistic depiction of the world; even if one does feel transported and moved, it might not lead to attitude or behavior change.[56] This amounts to a negotiated reading (Hall [1973] 1980), but in the case of "Am I Blue?," the message encoded is not a dominant one in the United States.[57] We contend that Walker's story is what Sven Ross (2011) calls a "radical text" (8) in that it challenges dominant beliefs regarding animal emotions—especially those of farm animals—and the permissibility of eating animal flesh (Adams [1980] 2010; Herzog 2011; Joy 2010). Readers who come to such a radical text from a culturally dominant perspective might negotiate its central message in ways that effectively "neutralize" that message (Ross 2011, 8).

What we can take away from this, then, is that although environmentally oriented texts that challenge dominant ideologies might be able to engage a wide range of recipients and transport them into their storyworlds—the transportation index was consistently high for all three story conditions—the ecological messages encoded in them might fail to have an effect on readers who are skeptical about certain textual claims. It depends in part on their own worldviews whether readers believe that what they read is a realistic, accurate depiction of the world and that it needs to be addressed. If this is so, it might have an impact on their attitudes and behavior.

Another counterintuitive result was that the control text garnered higher support than the story conditions for a bill that would make animal abuse a federal crime. In other words, people who read any of the three versions of "Am I Blue?" were actually less inclined to support such a bill than people who read a journalistic story about the discovery of the Higgs boson particle. Shouldn't it have been the opposite? Isn't one of the apparent goals of "Am I Blue?," in whatever version, to demonstrate the suffering of a sentient animal? Yes, but again, that must not mean that all readers decoded the message in the same way. The essay's condemnation of certain practices as abusive or unethical (meat and milk production, mating two horses and then separating them) might have seemed exaggerated or even radical (Ross 2011) to the majority of our readers; they therefore might have become worried, for example, that the bill would be written from an equally radical position, leading to exaggerated criminal sentences. Primed by the story, they might have worried that the bill would classify as abusive or otherwise unethical certain actions they themselves rely on, or even engage in.[58] They might have therefore been more reluctant to support such a bill than those in the control condition, who most likely associated animal abuse with rare acts of violence unrelated to them.

Supporting this conclusion would require further studies, but its possibility suggests that animal rights advocates should be careful when making claims that a large part of the target group will likely perceive as radical. Although activists, scholars, and teachers might imagine that such a strategy will lead to a reconsideration of commonly accepted attitudes and behaviors, it could also backfire, leading to cognitive dissonance and an outright rejection of the argument. The problem is not unlike the one we discussed earlier with reference to common notions about animal emotions, only in this case, recipients might feel so personally offended or threatened by the encoded message that instead of engaging in a negotiated reading, they completely oppose or simply "neutralize" (Ross 2011, 8) that message.

Finally, and independent of the reception of "Am I Blue?," our data yielded one result that might be of interest to ecocritics and that, to our knowledge, has not been reported elsewhere. We found that people who read more fiction held more favorable views toward animal welfare.[59] This is a correlation, but it does not tell us whether it is reading fiction that makes people more animal-welfare friendly or whether fiction is more attractive to people who already care about animals. After all, by its very nature, fiction is a form of entertainment that involves caring for beings unlike oneself. This is the same chicken-and-egg conundrum posed by empirical research findings that people who read more fiction are more empathetic (Koopman and Hakemulder 2015) and have better social cognition (De Mulder et al. 2017; Kidd and Castano 2013).[60] In this context, this result invites additional studies on the effects of fictional texts on pro-animal attitudes and on the relationship between readers' empathetic engagement with such texts and attitude change or policy support.

Conclusion

Our study used quantitative empirical methods to investigate the narrative impact of Alice Walker's "Am I Blue?" on contemporary readers in the United States. In doing so, it sought to replicate the results of an earlier study while also testing new hypotheses about the cultural situatedness of reception and the impact of two text-immanent features. Some of our results were counterintuitive. They did not confirm our hypothesis that the lack of impact of Walker's original story on attitudes toward animal welfare observed in the Polish study was due to the Polish participants' insufficient familiarity with the story's cultural context. Even though they are likely to be familiar with that context, the American participants did not show any improved attitudes toward animal welfare either.

Our other two hypotheses were at least partially confirmed. While the Physical Abuse version did not have more influence on attitudes toward animals measured with the ATAW scale than the Original version, it did have more influence on two particular items from that scale, thereby supporting the notion that the Original version's lack of impact on attitudes toward animal welfare may be due to its focus on the mental suffering of its animal protagonist. This notion was further supported by the fact that only the Physical Abuse variant made our participants more supportive of a hypothetical meat tax. The hypothesis that the original essay's lack of attitudinal impact was due to the parallel it drew between the suffering of human minorities and nonhuman animals was at least partially confirmed by the result showing that only the Animal version had any impact on our participants' agreement with the statement that "The low costs of food production do not justify maintaining animals under poor conditions." Taken together, these results suggest that both the mental suffering hypothesis and the dreaded comparison hypothesis might be right to some extent.

Some other hypotheses and avenues for further research were suggested to us by the results of this study. First, the fact that there was more support among the participants in the control group than in the experimental groups for a bill making animal abuse a federal crime drew our attention to the possibility that the original story's lack of attitudinal impact might have also been due to its attempts at a radical reevaluation of certain commonly accepted practices in the United States, such as factory farming. Second, the results concerning attitudes toward ethnic minorities suggest that the use of the dreaded comparison may have unintended negative consequences when it comes to both attitudes toward animals and attitudes toward people of color, and that those consequences are sensitive to cultural context.

These suggestions remain valid even when we take into account the limitations of our study, both as a freestanding experiment and as a replication of the original experiment in Poland. These limitations include the differences between the two studies. The Polish study was conducted in person and on high school students, whereas the American study was conducted online and on people of more varied ages and educational backgrounds. There were also some significant differences between the questionnaires used in the two studies, including an item that was crucial to one of our main hypotheses. Furthermore, participants in the American study perceived the control text to be less absorbing than any of the experimental texts. This not only makes the results of the American and Polish studies less comparable, as no such thing was observed in the Polish experiment, but it is a potentially confounding factor in the American study itself. In order to fully address these limitations, further studies would be needed. For now, one important takeaway from our study is the insight that more radical texts—that is, texts that challenge culturally dominant ideologies—may fail to have the desired effect on readers who do not already share their radical position because they might find ways to oppose or neutralize the message. Just because a radical text is being encountered by a large number of readers does not mean it will have the intended or expected impact. Given that many radical texts address themselves to readers who embrace dominant cultural ideologies, their authors would do well to consider their narrative strategies carefully if they want to invite more negotiated responses from such readers, and ultimately effect a shift in their attitudes, beliefs, or behavior.

A second and more basic conclusion is that reception is often far more complicated than expected, especially when the stimulus used is as complex as a literary text. Purely text-based ecocritical approaches to Walker's "Am I Blue?" have highlighted the story's narrative complexities, intertexts, and larger cultural context; they can also speculate on the ways it engages readers of various backgrounds. Such fine-grained textual analysis has been a crucial part of our empirical ecocritical approach, but we have also studied the reception of real readers in the past (using historical data) and contemporary real readers (using a controlled experiment). These results suggest that, especially for the kinds of rich, complex texts that ecocritics tend to focus on, the range of impacts on readers' beliefs and attitudes is quite difficult to accurately predict. As such, ecocritics should be wary of making assumptions about the reception of environmentally oriented texts. At the very least, we should ground our claims within the existing empirical scholarship.

Notes

This work was supported by Poland's National Science Center (grant 2012/07/B/HS2/02278) and by the Austrian Science Fund (FWF grant P 31189-G30).

1. In the Polish study, the item "Cultural minorities should be protected" was originally part of the camouflage in the questionnaire, but it revealed a significant effect on

readers of "Am I Blue?" In Poland, the government's protection of the rights of cultural minorities has been a significant concern and political issue (Mucha 2000).

2. Text manipulation is a common practice in fields such as the empirical study of literature, where it is frequently used to "gain deep causal insight into literary response" because "without it, firm conclusions are difficult" (Dixon and Bortolussi 2008, 75).

3. We refer to "Am I Blue?" as a story because that is the most neutral term and can refer to a work of either fiction or nonfiction.

4. Angela Harris (2009) reminds us that the comparison is "difficult to face" because of the historic and contemporary racist practices of animalizing African Americans: "The dreaded comparison [...] ignores the dynamic relationship between people of color and animals given their historic linkage in the white western mind. Animals [...] activate, I think, this urge to disassociate on the part of people of color, based on the intuition that our dignity is always provisional" (27).

5. Hogan (2003) defines situational empathy as "a shift in the structuring perspective that leads us to put ourselves in the place of someone else and thus results in empathy, associating our own feelings with those of the other person" (142). Importantly, for situational empathy to occur, it is not necessary that the empathizer share a lot of categorical traits with the person being empathized with. There must merely be enough experiential overlap to understand the other person's situation.

6. It would also be too simple to assume that Walker is simply anthropomorphizing here, projecting human emotions onto an unfeeling horse. Cognitive ethologists and neuroscientists alike have shown that animals do feel emotions and that we can genuinely feel along with them across species lines (Bekoff 2008; Bradshaw and Watkins 2006; De Waal 2009, 2019; Panksepp and Panksepp 2013).

7. Hall's ([1973] 1980) encoding/decoding model focuses primarily on the study of television, arguing that the events encoded into televisual stories reflect an intended meaning that is in accordance with a society's dominant ideologies. Hall, rejecting textual determinism, proposes that although viewers can exhibit the preferred response—that is to say, decoding the intended, dominant meaning—they can also engage in negotiated or oppositional readings, rejecting the intended meaning partially or entirely.

8. Willis (2017) mentions sociological and ethnographic research as suitable approaches to examine the reception of "contemporary real-life readers" (88). To these, one would have to add psychological research along with methods such as the experiment, the survey, the focus group, and many other methods presented in this book.

9. The decision to make slaughter of horses and sale of horsemeat for human consumption a felony in California was only made in 1998—ten years after the original publication of Walker's story, and four years after the story was banned. Proposition 6, "Criminal Law: Prohibition on Slaughter of Horses and Sale of Horsemeat for Human Consumption," was an initiative statute that appeared on the November 3, 1998, California general election ballot and was subsequently signed into law.

10. We thank cognitive linguist Alexander Onysko for his making us aware of the potential blockage of the metonymic associative link between the individual horse in the story and animals in general.

11. African American literary and cultural criticism, including ecocritical criticism, has been highly sensitive to the fraught discourse around Black Americans and animality, agriculture, and the continued legacies of slavery in present-day racial injustice and discrimination (Ruffin 2008; Outka 2013; Klestil 2023).

12. Prolific (https://www.prolific.co/) is a platform that assists researchers in recruiting paid participants for online surveys and experiments.

13. Our questionnaire did not include a question on participants' ethnicity, but Prolific provided us with information about the ethnic makeup of our sample.

14. For the Animal version of the text, we simply removed all passages referring to slavery and containing other human–animal comparisons. For the Physical Abuse version, we added one paragraph in which Blue was severely beaten by humans. The additional text passage was written by Alexa Weik von Mossner, who is an experienced fiction writer. Both versions are available in the digital repository.

15. People participating in studies and surveys on morally charged topics can provide answers that represent how they would like to be seen rather than what their actual attitudes and beliefs are. Given that our study concerned such a topic, the probability of such reactions from our participants was high.

16. In Green and Brock's (2000) definition, transportation into a narrative world is "a distinct mental process, an integrative melding of attention, imagery, and feelings" (701). The original version of the Transportation Scale developed by Green and Brock proved too long for our purposes, so we decided to shorten it by excluding two of the fifteen items.

17. According to Miall and Kuiken (1999), "Literariness is constituted when stylistic or narrative variations strikingly defamiliarize conventionally understood referents and prompt reinterpretive transformations of a conventional concept or feeling" (122).

18. For ease of reading, we provide numerical data in footnotes. $F(3, 794) = 1.23$, $p = .27$. We used analysis of covariance (ANCOVA).

19. $t(394) = 2.64, p = .009$.

20. $t(394) = -2.29, p = .023$.

21. $t(399) = -2.17, p = .030$.

22. The question was, "Consider a policy that introduces a small tax (< 5 percent) on U.S. meat products, using the revenue generated to ensure the humane treatment of animals in factory farms. To what degree would you support such a policy?" We conducted a one-way analysis of variance with post hoc contrast; $F(3, 796) = 3.71, p = .01$, partial $\eta^2 = .014$, contrast $t(796) = 2.96, p = .003$, partial $\eta^2 = .011$.

23. The question was, "The U.S. Senate and Congress have recently passed the Preventing Animal Cruelty and Torture (PACT) Act. The bill now awaits the president's

signature. The bill would make animal cruelty and torture a federal crime. Violators of the new law would face felony charges, fines, and up to seven years in prison. To what degree would you support such a law?"

24. While we call the combined three versions of "Am I Blue?" the "story condition," it should be remembered that the control context is a narrative too.

25. $F(3, 794) = 3.81, p = .05$, partial $\eta^2 = .005$ (ANCOVA).

26. A total of 17.3 percent answered two correctly, 0.9 percent answered only one correctly, and 0.4 percent answered none correctly.

27. $t(797) = 18.95, p < .001$. This might arguably be seen as a limitation of this experiment. It might be argued that because the control and experimental texts were not equally absorbing, this might account, at least in part, for any observed differences in attitudinal influence between the former and the latter. However, because our main results show there to be no difference in attitudinal influence between the control text and the experimental texts, and because it would not be plausible to attribute those results to the difference in how absorbing the texts are, the latter is not a significant limitation.

28. Criteria for a correct recall coding included mention of the relationship between humans and animals or parallels to human slavery within the story. Responses that only mentioned characters (such as "the story had a horse") did not meet the criteria for correct recall of the theme. There were 50.9 percent of such responses overall.

29. Mean (M) = 4.84, standard deviation (SD) = 1.10.

30. M = 4.83, SD = 1.08.

31. $F(3, 794) = 0.23, p = .88$.

32. $F(3, 794) = 0.39, p = .53$, partial $\eta^2 = .001$.

33. M = 1.79, SD = 0.94.

34. M = 1.43, SD = 0.70.

35. M = 1.79, SD = 0.81 and M = 1.53, SD = 0.75, respectively.

36. $\chi^2(6) = 3.59$.

37. The item "Ethnic minorities should be protected" was used in the U.S. study because one goal was to replicate as much as possible the Polish study. It can be considered a limitation that the notion of "protection" is somewhat vague in the U.S. cultural and political context and might therefore have been interpreted in different ways by participants. Arguably, the term "minorities" is also less salient as a term now in the United States, and "racial minority" might be a more appropriate term than "ethnic minority" in the U.S. context.

38. $F(3, 793) = 0.65, p = .585$, partial $\eta^2 = .002$.

39. $F(3, 721) = 2.5, p = .057$, partial $\eta^2 = .011$.

40. M = 5.91, SD = 1.24.

41. M = 5.61, SD = 1.46, $t(553) = -2.57, p = .01$.

42. M = 6.12, SD = 1.00.

43. $F(1, 794) = 49.38, p < .001$, partial $\eta^2 = .060$.

44. $F(1, 794) = 23.64, p < .001$, partial $\eta^2 = .030$.
45. $F(1, 793) = 5.34, p = .007$, partial $\eta^2 = .007$.
46. $F(1, 794) = 34.95, p < .001$, partial $\eta^2 = .043$.
47. $F(1, 794) = 35.90, p < .001$, partial $\eta^2 = .044$. This result is consistent with the results of other experimental studies, based on data obtained from thousands of participants of different ages and social background, which also show that women have more concern for animals than men (Malecki et al. 2019).
48. $F(3, 794) = 6.02, p = .01$, partial $\eta^2 = .008$.
49. $F(1, 794) = 8.67, p = .003$, partial $\eta^2 = .011$.
50. $F(1, 793) = 18.13, p < .001$, partial $\eta^2 = .023$.
51. $F(1, 793) = 24.71, p < .001$, partial $\eta^2 = .031$.
52. $F(1, 793) = 7.36, p = .007$, partial $\eta^2 = .009$.
53. It is possible that there was no effect on the entire ATAW because the scale is not specific enough in terms of the kinds of animals mentioned in its items. Reading the Animal version of Walker's story might have an impact on readers' attitudes toward animals that live on farms and/or are kept for their meat, but not for other animals more generally, which would also include pets and wild animals. Additional studies would be needed to determine whether this is indeed the case.
54. It is also possible that the notion of "protection" mentioned in the item ("Ethnic minorities should be protected") did not resonate with American readers, or at least not in the intended way. As mentioned earlier, this formulation was kept in order to replicate as closely as possible the questionnaire of the Polish study, but in the American cultural context, the idea that minorities should be "protected," rather than be afforded equal rights and opportunities, might not receive much support, regardless of the story condition.
55. That this increase in empathic attitudes did not lead to the improvement of ATAW might be explained by the anthropocentric bias inherent in story reception.
56. A similar phenomenon is observed by Schneider-Mayerson (2018) in his qualitative study on climate fiction.
57. According to Hall ([1973] 1980), a negotiated reading of a mass media text partially accepts the preferred reading intended by the encoder, but sometimes also resists and modifies it in a way that reflects readers' own social position. The empirical applicability of Hall's model has been tested by researchers such as David Morley (1980), whose results support Hall's ideas about negotiated responses.
58. Note that 86 percent of our participants identified themselves as omnivores who eat meat with nearly every meal (16.1 percent), with most meals (42.5 percent), or occasionally (27.4 percent).
59. $F(1, 793) = 5.34, p = .007$, partial $\eta^2 = .007$.
60. Literary scholars (Booth 1961), philosophers (Nussbaum 1997), and psychologists (Hoffman 2000) have also claimed that reading fiction improves empathy.

References

Adams, Carol J. (1980) 2010. *The Sexual Politics of Meat: A Feminist-Vegetarian Critical Theory.* 20th anniversary edition. London: Bloomsbury Academic.

Appel, Markus, and Tobias Richter. 2007. "Persuasive Effects of Fictional Narratives Increase over Time." *Media Psychology* 10 (1): 113–34.

Barthes, Roland. 1977. *Image–Music–Text.* Translated by Stephen Heath. New York: Hill & Wang.

Bekoff, Mark. 2008. *The Emotional Lives of Animals: A Leading Scientist Explores Animal Joy, Sorrow, and Empathy—And Why They Matter.* Foreword by Jane Goodall. Novato, Calif.: New World Library.

Bekoff, Mark. 2010. *The Animal Manifesto: Six Reasons for Expanding Our Compassion Footprint.* Novato, Calif.: New World Library.

Benzel, Karen. (1994) 1996. Letter to the editor of the *San Francisco Chronicle.* In Walker 1996a, 69.

Booth, Wayne. 1961. *Rhetoric of Fiction.* Chicago: University of Chicago Press.

Bradshaw, Gay A., and Mary Watkins. 2006. "Trans-species Psychology: Theory and Praxis." *Spring Journal* 75: 69–94.

De Mulder, Hannah, Frank Hakemulder, Rianne van den Berghe, and Fayette Klaassen. 2017. "Effects of Exposure to Literary Narrative Fiction: From Book Smart to Street Smart?" *Scientific Study of Literature* 7 (1): 129–69.

De Waal, Frans. 2009. *The Age of Empathy: Nature's Lessons for a Kinder Society.* New York: Three Rivers.

De Waal, Frans. 2019. *Mama's Last Hug: Animal Emotions and What They Tell Us about Ourselves.* New York: Norton.

Deckha, Maneesha. 2008. "Disturbing Images: PETA and the Feminist Ethics of Animal Advocacy." *Ethics and the Environment* 13 (2): 35–76.

Dixon, Peter, and Marisa Bortolussi. 2008. "Textual and Extra-textual Manipulations in the Empirical Study of Literary Response." In *Directions in Empirical Literary Studies: In Honor of Willie van Peer,* edited by Sonia Zyngier, Marisa Bortolussi, Anna Chesnokova, and Jan Auracher, 75–78. New York: John Benjamins.

Eco, Umberto. 1979. *The Role of the Reader: Explorations in the Semiotics of Texts.* Bloomington: Indiana University Press.

Fish, Stanley. 1980. *Is There a Text in This Class? The Authority of Interpretive Communities.* Cambridge, Mass.: Harvard University Press.

Franklin, Robert G., Anthony J. Nelson, Michelle Baker, et al. 2013. "Neural Responses to Perceiving Suffering in Humans and Animals." *Social Neuroscience* 8 (3): 217–27.

Gosling, Samuel D., Peter J. Rentfrow, and William B. Swan. 2003. "A Very Brief Measure of the Big-Five Personality Domains." *Journal of Research in Personality* 37 (6): 504–28.

Green, Melanie C., and Timothy C. Brock. 2000. "The Role of Transportation in the Persuasiveness of Public Narratives." *Journal of Personality and Social Psychology* 79: 701–21.

Grosjean, Bruce. (1994) 1996. Letter to the editor of the *San Francisco Chronicle.* In Walker 1996a, 68.

Hall, Stuart. (1973) 1980. "Encoding/Decoding." In *Culture, Media, Language: Working Papers in Cultural Studies, 1972–1979,* edited by Stuart Hall, Dorothy Hobson, Andrew Lowe, and Paul Willis, 128–39. London: Hutchinson.

Harper, Ami Breeze. 2010. "Whiteness and 'Post-racial' Vegan Praxis." *Journal of Critical Animal Studies* 3: 7–32.

Harris, Angela. 2009. "Should People of Color Support Animal Rights?" *Journal of Animal Law* 5: 15–32.

Herzog, Hal. 2011. *Some We Love, Some We Hate, Some We Eat: Why It's So Hard to Think Straight about Animals.* New York: Harper Perennial.

Hoffman, Martin. 2000. *Empathy and Moral Development: Implications for Caring and Justice.* Cambridge: Cambridge University Press.

Hogan, Patrick Colm. 2003. *The Mind and Its Stories: Narrative Universals and Human Emotion.* Cambridge: Cambridge University Press.

Holt, Patricia. 1996. Introduction to Walker 1996a, 1–17.

Hooker, Deborah Anne. 2005. "Reanimating the Trope of the Talking Book in Alice Walker's 'Strong Horse Tea.'" *Southern Literary Journal* 37 (2): 81–102.

Iser, Wolfgang. 1976. *The Act of Reading: A Theory of Aesthetic Response.* Baltimore: Johns Hopkins University Press.

Jauss, Hans Robert. 1982. *Toward an Aesthetic of Reception.* Translated by Timothy Bahti. Minneapolis: University of Minnesota Press.

Johnson, Dan R. 2012. "Transportation into a Story Increases Empathy, Prosocial Behavior, and Perceptual Bias toward Fearful Expressions." *Personality and Individual Differences* 52 (2): 150–55.

Johnson, Lindgren. 2018. *Race Matters, Animal Matters: Fugitive Humanism in African America, 1840–1930.* New York: Routledge.

Joy, Melanie. 2010. *Why We Love Dogs, Eat Pigs, and Wear Cows: An Introduction to Carnism.* San Francisco: Conari Press.

Keen, Suzanne. 2010. "Narrative Empathy." In *Toward a Cognitive Theory of Narrative Acts,* edited by Frederick Louis Aldama, 61–94. Austin: University of Texas Press.

Kidd, David Comer, and Emanuele Castano. 2013. "Reading Literary Fiction Improves Theory of Mind." *Science* 342 (6156): 377–80.

Klestil, Matthias. 2023. *Environmental Knowledge, Race, and African American Literature.* London: Palgrave Macmillan.

Koopman, Emy, and Frank Hakemulder. 2015. "Effects of Literature on Empathy and Self-Reflection: A Theoretical-Empirical Framework." *Journal of Literary Theory* 9 (1): 79–111.

Kress, Gunther. 1985. *Linguistic Processes in Sociocultural Practice.* Victoria, Australia: Deakin University Press.

Lioi, Anthony. 2008. "An End to Cosmic Loneliness: Alice Walker's Essays as Abolitionist Enchantment." *ISLE: Interdisciplinary Studies in Literature and Environment* 15 (1): 11–37.

Livingstone, Sonia. 1998. "Relationships between Media and Audiences: Prospects for Audience Reception Studies." In *Media, Ritual and Identity: Essays in Honor of Elihu Katz,* edited by Tamar Liebes, James Curran, and Elihu Katz, 237–55. New York: Routledge.

Livingstone, Sonia. 2015. "Active Audiences? The Debate Progresses but Is Far from Resolved." *Communication Theory* 25: 439–46.

Malecki, W. P., Alexa Weik von Mossner, and Małgorzata Dobrowolska. 2020. "Narrating Human and Animal Oppression: Strategic Empathy and Intersectionalism in Alice Walker's 'Am I Blue?'" *ISLE: Interdisciplinary Studies in Literature and Environment* 27 (2): 365–84.

Malecki, Wojciech, Bogusław Pawłowski, Marcin Cieński, and Piotr Sorokowski. 2018. "Can Fiction Make Us Kinder to Other Species? The Impact of Fiction on Pro-animal Attitudes and Behavior." *Poetics* 66: 54–63.

Malecki, Wojciech, Piotr Sorokowski, Bogusław Pawłowski, and Marcin Cieński. 2019. *Human Minds and Animal Stories: How Narratives Make Us Care about Other Species.* New York: Routledge.

Marx, Leo. 1964. *The Machine in the Garden: Technology and the Pastoral Ideal in America.* Oxford: Oxford University Press.

Mazzocco, Philip J., Melanie C. Green, Jo A. Sasota, and Norman W. Jones. 2010. "This Story Is Not for Everyone: Transportability and Narrative Persuasion." *Social Psychological and Personality Science* 1 (4): 361–68.

Miall, David S., and Don Kuiken. 1999. "What Is Literariness? Three Components of Literary Reading." *Discourse Processes* 28: 121–38.

Morley, David. 1980. *The "Nationwide" Audience: Structure and Decoding.* London: BFI.

Mucha, Janusz. 2000. *Kultura dominująca jako kultura obca: Mniejszości kulturowe a grupa dominująca w Polsce.* Warsaw: Oficyna Naukowa.

Nussbaum, Martha. 1997. *Cultivating Humanity: A Classical Defense of Reform in Liberal Education.* Cambridge, Mass.: Harvard University Press.

Outka, Paul. 2013. *Race and Nature from Transcendentalism to the Harlem Renaissance.* London: Palgrave Macmillan.

Panksepp, Jaak. 2004. *Affective Neuroscience: The Foundations of Human and Animal Emotions.* Oxford: Oxford University Press.

Panksepp, Jaak, and Jules B. Panksepp. 2013. "Toward a Cross-species Understanding of Empathy." *Trends in Neuroscience* 36 (8): 489–96.

Peterson, Christopher. 2012. *Bestial Traces: Race, Sexuality, Animality.* New York: Fordham University Press.

Rodrigues, Luis C. 2020. "White Normativity, Animal Advocacy, and PETA's Campaigns." *Ethnicities* 20 (1): 71–92.

Ross, Sven. 2011. "The Encoding/Decoding Model Revisited." Paper presented at the annual meeting of the International Communication Association, Boston, Mass., May 25, 2011.

Ruffin, Kimberly. 2008. *Black on Earth: African American Ecoliterary Traditions.* Athens: University of Georgia Press.

Schneider-Mayerson, Matthew. 2018. "The Influence of Climate Fiction: An Empirical Survey of Readers." *Environmental Humanities* 10 (2): 473–500.

Schneider-Mayerson, Matthew, Abel Gustafson, Anthony Leiserowitz, Matthew H. Goldberg, Seth A. Rosenthal, and Matthew Ballew. 2020. "Environmental Literature as Persuasion: An Experimental Test of the Effects of Reading Climate Fiction." *Environmental Communication,* 1–16.

Sewell, Anna. (1877) 2012. *Black Beauty.* Oxford: Oxford University Press.

Sherrell, Michael. (1994) 1996. Letter to the editor of the *San Francisco Chronicle.* In Walker 1996a, 68.

Singer, Peter. 2009. *Animal Liberation: The Definitive Classic of the Animal Movement.* Updated ed. New York: Ecco Book/Harper Perennial.

Spiegel, Marjorie. 1996. *The Dreaded Comparison: Human and Animal Slavery.* Foreword by Alice Walker. London: Mirror.

Sunstein, Cass R., and Martha C. Nussbaum, eds. 2005. *Animal Rights: Current Debates and New Directions.* Oxford: Oxford University Press.

Vezzali, Loris, Sofia Stathi, and Dino Giovanni. 2012. "Indirect Contact through Book Reading: Improving Adolescents' Attitudes and Behavioral Intentions toward Immigrants." *Psychology in the Schools* 49 (2): 148–62.

Walker, Alice. (1988) 1996. "Am I Blue?" In Walker 1996a, 31–43.

Walker, Alice. 1996a. *Alice Walker Banned.* Edited by Patricia Holt. San Francisco: Aunt Lute.

Walker, Alice. 1996b. Foreword to *The Dreaded Comparison: Human and Animal Slavery,* by Marjorie Spiegel, 13–14. London: Mirror.

Weik von Mossner, Alexa. 2017. *Affective Ecologies: Empathy, Emotion, and Environmental Narrative.* Columbus: Ohio State University Press.

Willis, Ika. 2017. *Reception.* New York: Routledge.

Chapter 7

Screening Waste, Feeling Slow Violence

An Empirical Reception Study of the Environmental Documentary Plastic China

NICOLAI SKIVEREN

It is a pervasive condition of empires that they affect great swathes of the planet without the empire's populace being aware of that impact—indeed, without being aware that many of the affected places even exist.

—Rob Nixon, *Slow Violence and the Environmentalism of the Poor*

In *Slow Violence and the Environmentalism of the Poor,* Rob Nixon (2011) asks, "How can we imaginatively and strategically render visible vast force fields of interconnectedness against the attenuating effects of temporal and geographical distance?" (38). Nixon here poses the question of how to apprehend slow violence, which he defines as a "violence that occurs gradually and out of sight, a violence of delayed destruction that is dispersed across time and space, an attritional violence that is typically not viewed as violence at all" (2). Climate change, plastic pollution, and radioactive fallout are all cases of slow violence, as their effects are spread out across time and space. However, the ephemeral nature of slow violence presents a representational challenge. "How can we convert into image and narrative," Nixon asks, "the disasters that are slow moving and long in the making, disasters that are anonymous and that star nobody, disasters that are attritional and of indifferent interest to the sensation-driven technologies of our image-world?" (3). In this chapter, I present a reception study of the documentary *Plastic China* (2016), which examines the issue of environmental dumping as an example of slow violence. Using qualitative interviewing to investigate the film experience of a group of Danish viewers (N = 14), the chapter sheds light on the following questions: What kinds of film experiences can *Plastic China* create? How might a Danish group of viewers interpret this documentary? And what potentials and challenges might its reception suggest that the documentary film faces in representing slow violence?

To position the study in relation to the existing scholarship in the field, I first offer a brief account of the key studies on empirical ecocriticism, slow violence, and documentary film. Second, I provide a summary of *Plastic China,* along with its formal and narrative characteristics, after which I outline the applied methodology used to examine its reception. Using Stuart Hall's ([1973] 2001) encoding/decoding framework, I then discuss how the reception of this film reflects a variety of dominant-hegemonic, negotiated, and oppositional readings. I suggest adding a new category to Hall's framework, one that is particularly important for contemporary critics and producers of ecomedia to consider: the position of exhaustion. Finally, I conclude the chapter by considering the implications of the study's findings for future studies of environmental media, suggesting that more attention be paid to the complicating variables of longitudinal reception research.

Empirical Study of Environmental Documentary

Surveying the fields of empirical ecocriticism and environmental communication, one finds several post-2000 studies that examine the impact of environmental documentaries. However, as far as I can tell, no one has yet used qualitative interviewing as a means to explore the filmic experiences that environmental documentaries might facilitate. Moreover, most studies within this emergent field have been quantitatively oriented, as scholars have attempted to document the effect of environmental documentaries on quantifiable variables such as attitude, behavior, and intention. Rachel Howell (2014), for instance, uses questionnaires to investigate the transformative potential of the film *The Age of Stupid* on attitude and behavior related to climate change. Adopting a similar method, Jessica M. Nolan (2010, 654) examines *An Inconvenient Truth* (Guggenheim 2006) in terms of its "apparent goals of increasing knowledge and concern about global warming." Grant Jacobsen (2011) investigates the same film but correlates the release of the film with the purchase of carbon offsets. Finally, Henry Janpol and Rachel Dilts (2016) utilize a computer game to trace the impact of the documentary *Explore the Wildlife Kingdom: Dolphins—Tribes of the Sea* (Greisen 2006) on "environmental perception."

At the qualitative end of the spectrum, empirical studies of environmental documentaries are harder to find. Worth mentioning, however, is Matthew Schneider-Mayerson's (2013) ethnographic study of Hollywood disaster movies and its impact on apocalyptic beliefs within the so-called peak oil movement in the United States. Likewise, Adrian Ivakhiv's (2013) reception study of the film *Avatar* (Cameron 2009) explores online fan forums and mainstream media to map the different interpretations and affects mobilized by the film. Finally, in the 2020 *ISLE* cluster on empirical ecocriticism, Pat Brereton and Victoria Gómez (2020) present a study that used focus group interviews to examine media students' experiences of the music video "Dear Future Generations: Sorry" (Williams 2015), focusing on the ability of celebrities to communicate messages about environmental risk.

In this chapter, I explore the representational challenges of slow violence using qualitative interviews to investigate how real flesh-and-blood viewers experience and decode the filmic depiction of poverty and plastic recycling in China. Compared to quantitative studies, the advantage of this method is that it allows viewers to articulate in great detail what characterized their experiences, providing rich descriptions of the different emotional and cognitive responses that watching an environmental documentary might involve. Following Ivakhiv's (2013) reflections on the potential of combining green film criticism with models of cultural circulation, this experimental study explores the heterogeneous landscape of viewer experiences by mapping the multiple interpretations, affects, and sensibilities that can emerge from within the viewer–documentary nexus. In doing so, I demonstrate some of the possibilities that qualitative approaches offer ecocritics interested in the empirical study of environmental documentary and ecomedia in general.

Documenting Slow Violence

As an empirical case study on environmental media, I also wish to extend and respond to some of the analyses offered by other ecocritics who have written about slow violence in documentary film. Following Nixon's aforementioned questions, the representational challenge of slow violence can be defined along three parameters: temporal delay, spatial dislocation, and causal dissociation. Simply put, slow violence is characterized by taking place gradually, elsewhere, and "unmoored from [its] original causes" (Davies 2019, 2). To render the invisible visible, Nixon (2011) considers the work of writer-activists whose imaginative writing, he argues, can "help make the unapparent appear, making it accessible and tangible by humanizing drawn-out threats inaccessible to the immediate sense" (5). A number of ecocritics have recently suggested that environmental documentaries hold similar potential. For instance, Alexa Weik von Mossner (2015) shows how the documentary *There Once Was an Island* (March 2010) gives form to slow violence through its personalized and participatory portrayal of disappearing island communities in the Pacific as a result of anthropogenic climate change. In a related study, Chia-ju Chang (2016) examines the depiction of slow violence in the documentary *E-Wasteland* (Fedele 2012), noting in particular how its observational mode puts trust "in the power of the image and the audience's ability to engage in active, reflective thinking about the condition of a contemporary lifestyle that is saturated with e-products" (100). Finally, Christine L. Marran (2017) explores multiple works by the Japanese director Noriaki Tsuchimoto, arguing that his detailed depiction of victims of mercury poisoning, as seen in films such as *Minamata: Victims and Their World* (Tsuchimoto 1971), provides clear examples of the capacity of the medium of film to make slow violence tangible. What unifies these scholars' considerations of slow violence in documentary film is the contention that the genre affords viewers a chance to inhabit the affective space of seemingly distant others and thereby bridge the spatial and temporal gap so characteristic of slow violence. Moreover, because of their perceived status as nonfiction—their assumedly unique relationship with external realities and truths (Nichols 1994; Winston 1995)—these documentaries can mobilize unique experiential and affective spaces that invite viewers to realize that they are part of the world they see on the screen (Eitzen 2005). This final point is not lost on Weik von Mossner (2017), who argues that our experience of documentary is emotionally salient "not only because there are real-world consequences for the people portrayed in the film but also because, theoretically at least, the viewer herself could be affected" (71).

In this study, I provide empirical evidence in support of and in response to such scholarship, as I examine the reception of the environmental documentary *Plastic China.* As I explain in more detail below, this film is in many ways similar to those examined by Weik von Mossner (2015), Chang (2016), and Marran (2017), as it too features observation, engaging characters, and unhealthy bodies. However, instead of mercury poisoning, electronic waste, or sea-level rise, the Chinese recycling industry, including its relationship to the international waste trade, is the topic of engagement.

Plastic China

It is hard to imagine an example of slow violence more pertinent than environmental dumping, which refers to the practice of so-called developed countries designating so-called developing countries as sites for dumping excess waste. Until 2018, two thirds of all plastic waste in the world was exported to China (Crawford and Warren 2020; Wang et al. 2019), which since 1992 has received an unfathomable 106 million metric tons of plastic waste (Brooks, Wang, and Jambeck 2018).[1] As a global industry, the international waste trade thus makes possible a violence that, following Nixon (2011), is delayed, dispersed, and disconnected, especially in places where inexpensive labor comes at the cost of poor working conditions and environmental safety. With years, even decades, between the moment we throw a piece of plastic in the bin to the moment it causes organ failure in a body on the other side of the planet, the violence of waste is inherently slow, and tracing its origins is virtually impossible. Additionally, environmental dumping is itself a good example of how ecological issues are inherently entangled in social issues such as poverty, stigma, and unemployment (Bauman 2004; Braidotti 2019; Guattari 2000; Morton 2010). In many ways, Nixon's (2011) own work on slow violence is a testimony to this meeting of the social and the ecological, as he seeks to cultivate the interdisciplinary nexus between ecocriticism and postcolonialism. In doing so, Nixon's work not only highlights the "out of sight, out of mind" mentality that underpins processes such as environmental dumping but also poses the question, out of sight and out of mind to whom (Bell 2019; Davies 2019; Liboiron 2018)?

This question is one of the central themes in Chinese filmmaker Jiu-liang Wang's 2016 documentary *Plastic China,* which portrays the lives of two families who work in the Chinese plastic recycling industry. In the opening scene, viewers see a majestic container ship approaching a Chinese harbor. Through a series of tracking shots, viewers follow a container as it makes its way onto the mainland, where its contents are unloaded into the hands of the protagonists, whose job is to sort, process, and sell the imported plastic waste. Early on, viewers are introduced to an eleven-year-old girl, Yi-Jie, whose father cannot afford to send his children to school. Instead, Yi-Jie takes care of her younger siblings in the piles of garbage that make up their home.

Much of the documentary shows the hard work of transforming the imported plastic into tiny plastic pellets, which can be sold for a marginal profit on the Chinese market. The work of recycling, however, is not the only subject of the documentary. The workshop soon turns into a setting for a drama about Yi-Jie and her struggle to navigate the dreams and demands that define her existence. On the one hand, the prospects of going to school, going home, or going to see her grandmother represent the promise of a better future. On the other hand, Yi-Jie's life is restricted by the forces of slow violence, represented by the endless dumping of plastic waste, the poverty of the family, and the family's lack of social security. Ultimately, these restrictions are reflected in the plot, which offers little in terms of narrative

Figure 7.1. Yi-Jie holds her younger sibling. Screenshot from *Plastic China* (Wang 2016).

progression, apart from a few key scenes that provide subtle yet unfulfilled promises of an alternative future for Yi-Jie and her family.[2]

In terms of form, two cinematographic features characterize *Plastic China.* With no voice-over and no direct participation by the camera operator, the documentary uses what Bill Nichols (2001) calls the "observational mode," in which viewers "look in on life as it is lived" (111). Documentaries using this mode "purport to depict everyday life, with minimum intrusion by the filmmaking process, conveying a sense of unmediated access to local ways of experiencing the world" (Wolfe 2014, 145). *Plastic China* thus invites viewers to become a fly on the wall, allowing them an experiential proximity to the lived reality of Yi-Jie and her family.

The other relevant cinematographic feature is the tendency to use close-ups of waste objects that seem to have been strategically chosen to elicit specific responses in the viewer. Generally speaking, the film's mise-en-scène represents a relatively monotonous gray landscape of waste, but occasionally the camera lingers on discarded artifacts, causing them to stand out from the otherwise unidentifiable plastic scraps. By periodically bringing its background to the foreground, *Plastic China* encourages viewers to recognize the raw material that makes up Yi-Jie's life and in turn reflect on their ties to the environment of waste her family is subject to.

In this way, the film invites viewers to occupy the spaces of the families while asking viewers to recognize that their perception of the cinematic environment differs from the subjects that inhabit it: What is for the protagonists a source of income and play (but also of pollution and sickness) is for the audience members something they, at least in the case of the Danish viewers who were involved in the study, can and do throw away on a daily basis. This perspective is echoed paratextually by the filmmakers, who write on the documentary's website that the film is meant to confront viewers with "the truth that the world is flat and issues don't go away by changing time and location. At the end of the day, as a global nation, we are all in this together, and we all play a part in this ever changing world" (CNEX, n.d.). In ecocritical terms, *Plastic China* can thus be read as evoking what Ursula Heise (2008) refers to as "eco-cosmopolitanism," a form of transnational ethics that "envision[s] individuals and groups as part of planetary 'imagined communities' of both human and nonhuman kinds" (61). By showing the socioenvironmental consequences of globalized consumer capitalism, *Plastic China* thus calls attention to the fundamental insight of slow violence, which highlights that what takes place in one setting is often intricately linked to what goes on in another. In doing so, the film attempts to bridge the many gaps—spatial, temporal, and causal—that otherwise tend to distance the Western consumer from the dumping grounds that result from such consumption.

Method and Research Design

To undertake the reception study, I used the method of qualitative research interviewing. This method "attempts to understand the world from the subject's points of view, to unfold the meaning of their experiences, to uncover their lived world prior to scientific explanations" (Brinkmann and Kvale 2009, 1). Epistemologically speaking, the qualitative research interview is rooted in a phenomenological theory of knowledge, which means that the focus is on the lifeworld of its subjects and that the interviews aim to be descriptive (rather than explanatory), open (rather than expectant), and focused on meaning (rather than measurement) (Brinkmann and Kvale 2009; Giorgi 2009).[3]

As an intrinsic case study, the purpose was to describe the subjective film experience of a small group of viewers and not to make claims about representativeness. Participants were recruited using convenience sampling (Ritchie and Lewis 2003, 81),[4] meaning that no selection criteria determined participant eligibility. The only requirement was that the participants reside in Denmark and spoke Danish to ensure consistency and optimal communicative ability.[5] Although background information did not figure as a formal selection criterion, details such as sex/gender, age, occupation, education, and political affiliation were collected. Importantly, this information was not collected to construct generalizations about the participants' experiences; it was used to ensure maximum diversity within the cohort and to provide a more in-depth understanding of individual participants' experiences.

The empirical data were collected by conducting two semistructured one-on-one interviews with each of the fourteen participants.[6] The first interview would take place immediately after a screening and the second a week later, each lasting approximately thirty minutes. The documentary was screened for participants individually. The first interview was designed to flesh out the salient aspects of each participant's film experience. The second interview was iteratively based on the first and included questions about how their experience of the film had unfolded, if at all, with the passing of time. The study's interview guides are provided in the appendix.

To identify the salient clusters of meaning across the twenty-eight interviews, multiple cycles of coding were carried out (Brinkmann and Kvale 2009; Coffey and Atkinson 1996).[7] This resulted in a series of interrelated codes, ranging from descriptions of scenes, characters, and emotions to reflections on abstract themes such as consumerism and agency. The content of the codes was examined by close reading of "the linguistic terms and categories through which respondents construct their worlds and their own understandings of their activities" (Morley 2013, 25). In this way, it was possible to elucidate the participants' experience by paying close attention to the language they used to describe it.

To analyze the participants' testimonies, the study drew on audience reception research.[8] Informing the approach in particular was Hall's encoding/decoding model of communication. At its core, this model emphasizes that the study of media products should not only be concerned with the text itself—that is, how its producers construct specific meanings and messages (encoding)—but also with the moment of consumption in which the subject actively constructs meanings and messages from the text (decoding). Crucially, Hall's ([1973] 2001) model rests on the contention that "encoding and decoding may not be perfectly symmetrical" (510), meaning that "there is no necessary correspondence between encoding and decoding, the former can attempt to 'pre-fer' but cannot prescribe or guarantee the latter, which has its own conditions of existence" (515). This meant that studying the reception of *Plastic China* entailed paying close attention to both tendency and variation, as well as whether the participants' descriptions (decoding) reflected the messages that the filmmakers intended (encoding).

This method, like any other, has its limitations. First, there is the issue of the validity of self-reporting. Researchers might, for instance, unintentionally direct participants toward the answers they desire. To mitigate this risk as much as possible, the participants were informed that the purpose of the study was to describe experience and that there were no right or wrong answers. Moreover, the study used open-ended questions to ensure that the interviewee's description remained focused but unguided. Second, screenings took place at a university campus, meaning that the study excluded the ecological context of media consumption, which can be crucial for understanding the experience of real audiences (Biltereyst and Meers 2018; Stacey 1994).[9] Third, interviews about films are ekphratic, meaning that they involve translating from one medium (audiovisual) to another (verbal). This introduces ambiguity. The same uncertainty applies to what Sara Ahmed (2010) calls "the transparency of self-feeling (that we can say and know how we feel)" (5). In other words, it had to be assumed that some participants would struggle to articulate what their experience was like. Finally, as Schneider-Mayerson (2018) emphasizes in his qualitative study of cli-fi readers, "any reading experience might have effects of which readers are not aware, and that readers can be mistaken about their own experiences" (477). The same is true for film experiences. Our memories—not just of film but in general—can be tainted, forgotten, and sometimes even invented (Hyman and Kleinknecht 1999; Schooler and Eich 2000). Likewise, there may be aspects of our film experience that are not available to consciousness, which makes them difficult to describe.[10]

Despite these limitations, what the methodology lacks in terms of specificity and generalizability, it makes up for in richness of description and in the possibility of producing novel and unexpected findings, making it an ideal tool for an exploratory study of a specific group of viewers' experiences of a specific film.

A Plastic Ocean

What did the cinematic environment of *Plastic China* look like to the eyes of the participants? What language did they use to make sense of this place? We start our journey through the findings of the reception study exploring the varied descriptions of the film's settings. What many participants highlighted first was the widespread presence of plastic waste. "There is plastic everywhere," as one participant, Louise, a social science student, put it.[11] By itself, this observation is unsurprising. But the language with which participants described this environment did reveal some noteworthy affinities concerning the distinct feel of the waste. A key impression was related to size: “Everything they do, they do among mountains of plastic,” explained Peter, a thirty-three-year-old unemployed man.[12] Other metaphors of scale were also used, including “a plastic landscape,” “a plastic ocean,” and in one case even “a plastic universe.”[13] In this way, the participants’ descriptions of the waste-ridden settings were repeatedly conceptualized through metaphors of more-than-human scales, suggesting that the encounter with the film’s cinematic environment was for many participants an encounter with the immense and the insurmountable. For Louise, the “enormous piles of plastic” even seemed “immeasurable,” stressing not only the Sisyphean task required to recycle it, but also a more fundamental difficulty of being able to grasp the sheer amount of waste shown in the documentary.[14]

A related theme that ran through the participants’ descriptions of the environment featured in the film was the danger it was seen to impose on the subjects inhabiting it. What occupied the participants in this regard were scenes and images that indexically signified the health risks that working and living at the recycling plant entail—as seen, for instance, in one sequence where the camera provides a close-up of burning plastic, which is then followed by a tracking shot of the resulting black smoke as it enters the surroundings where the protagonists are busily engaged in manual labor. Another scene, which caught the attention of Michael, a thirty-five-year-old office clerk, was a short monologue by the recycling plant owner, Kun, who explains that lumps have started to form across his body—symptoms that he is reluctant to have examined for fear of the results.[15] In describing scenes such as these, several of the participants expressed feeling apprehensive toward the cinematic environment, as they repeatedly voiced concerns about the risk that the toxic environment might seep into the lungs of the characters, manifesting as cancer and other forms of illness.

Yet out of all the descriptions of the film’s setting as a hazardous place of risk, one scene stood out among all the testimonies. In this scene, we see children at the local river. An establishing shot shows a tiny waterfall, which is followed by a series of medium and close-up shots of the water, revealing that the river is contaminated by waste of all shapes and sizes, ranging from old fishing nets to tiny plastic scraps.

While some participants mourned the waterway’s pollution itself, what seemed to trouble the participants most was the realization that the children

Figure 7.2. The children fish at the local river. Screenshot from *Plastic China* (Wang 2016).

come to the foamy green water to bring home the small dead fish floating lifelessly on the surface of the polluted water. Some participants reported experiencing various bodily reactions such as disgust and nausea in response to this scene, with one participant, Christian, a twenty-one-year-old media studies student, reporting having to look away from the screen.[16] These reactions continued in response to the subsequent scene in which the families, now back at the workshop, eat the fish—along with the harmful microscopic toxins they likely contain. Emblematic of slow violence, the actual impact of what viewers see in these images is not only ephemeral in the sense that they cannot *see* the harmful substances (let alone pinpoint the parties responsible for their existence); it is also likely that their detrimental effects will only come into effect years from now, leaving their underlying causalities impossible to trace empirically. But despite this representational challenge of showing something that cannot be seen, the participants (and their bodies) never doubted that they were bearing witness to something dangerous, even potentially lethal. "You are not only destroying your bodies with physical labor, you are also destroying your insides with fumes, toxins and polluted water," said Kirsten, an unemployed elementary school teacher, highlighting her feelings of apprehension toward the visually imperceptible substances contained by the cinematic environment.[17]

Feeling Stuck: "You Can See Her Future"

When asked what they thought of the different characters featured in the documentary, a central theme that was present in nearly all of the participants' descriptions was the notion of immobilization. "They're stuck in that damn plastic place," said Camilla, a thirty-six-year-old gardener.[18] This sense of confinement carried a distinctly spatial meaning in the minds of several participants. Many recounted their experience of one of the last parts of the film in which Yi-Jie's family sits outside a train station, discouraged because they have been denied tickets to visit their grandmother because they lack IDs and money for tickets. In this scene, the viewer sees Yi-Jie, her brother, and her father in a series of close-ups and medium close-ups as they gaze at passersby with doubtful and contemplative facial expressions, making tangible their disappointment at not being able to leave the train station and their uncertainty about what to do next. The negative emotional quality of the scene is further emphasized by a nondiegetic soundtrack, which features a series of melancholic vocalizations in Mandarin. In the context of the overall narrative, this event represents the low point of the film because it extinguishes the hope that the protagonists might escape their predicament—and, by analogy, the possibility that the story might include a happy ending. Consequently, many participants reported responding emotionally to this scene, sharing the feeling of sadness and disappointment it contains. One participant, Susan, began to weep during the parts of the interview in which she recounted her experience of the scene.[19] "They really wanted to move on. This was the only thing the girl had to look forward to," said Lars, who felt sympathetic toward the children because of the false hope they suffered.[20] For Camilla, the experience of restriction could not be confined to this scene alone: "They were disappointed, but at the same time, I thought that they are probably used to the fact that there isn't a lot of things they can change."[21] In other words, the sensation of being stuck was for her not only felt in moments that made this sensation manifest; rather, it permeated the film and the characters' lives in a much more general sense, as she saw Yi-Jie's entire existence as one of being trapped in the same place, with scant prospect of change.

Supplementing this literal sense of spatial confinement, participants described immobility in a more figurative sense. Characters were considered unable to break through immaterial barriers such as social status, economic poverty, and lack of education. Julie, a student of music therapy, explained seeing the characters' lives as a "vicious circle," lamenting how their poverty would "pass down from one generation to the next."[22] "I thought it was hopeless," Camilla said, "because if his children won't attend school, then they are not going anywhere. They'll be trapped in this situation."[23] Echoing this perspective, Lars even claimed being able to "see her future, because all you have to do is look at her parents."[24]

Incidentally, this figurative conception of immobilization also constituted a source of individual variation, as participants drew on different semiotic codes in their interpretation of the characters' situation. Some used discourses of class and caste (which do not figure explicitly in the film) to describe their confinement: "They won't be able to get out of that caste system," Lars explained. "They could marry in order to advance in the social hierarchy, but how are they going to meet anyone, when they don't go to school?"[25] Christian conceived of the characters through the prism of an urban–rural divide: "There are these people, and then the rest who live in the city [. . .] who all look down upon those who live out there, because they can't do anything."[26] This dichotomy was further developed by Anna, another medical student, who observed differences in the characters' behavior in different settings: "In the factory, the father [Peng] can be strict and he is in charge of things, but in the city he is just a little man. He didn't even know how to get around, where to go, and how everything worked."[27] Finally, Michael's experience of watching the characters of the documentary brought to mind the Danish reality TV show *Luksusfælden* (The luxury trap), which features financial advisers who try to help economically irresponsible Danes save themselves from the pitfalls of debt. He explained how Yi-Jie's family members, like the individuals on the Danish TV show, all "smoke, they have children, they drink, and they have social problems. They fail to prioritize."[28] In this way, the interpretive construals of the social issues depicted in the documentary differed considerably: Whereas some emphasized the influence of social structures related to status and caste, others described the characters' situation as a result of poor life choices.

While most of the participants thus tended to describe the lives of the characters in predominantly negative terms, several participants did find the experience of watching the children's creative play with garbage to be pleasant and inspiring. Throughout the film, there are several sequences featuring handheld low-angle close-ups shots that show the children laughing and smiling as they play. Viewers watch as newspapers are turned into superhero capes and leftover e-waste is transformed into a toy computer.[29] Accompanying these sequences is a subtle nondiegetic piano soundtrack that is neither entirely melancholic nor entirely playful, yet somehow both. The participants responded to this combination of sound and image in at least two ways. First, it allowed many to experience a variation in the otherwise negative affective register of the film, which in turn enabled some of the participants to relate to the story in more positive terms. Tina, a forty-three-year-old elementary school teacher, described how Yi-Jie and her brothers "[remind] me of my own children in the way they are so curious and always find things to play with. [. . .] This made me feel like it was not all that sad, that there was also a certain joy of life, a curiosity, and, I don't know, perhaps some hope?"[30] Yet inasmuch as the shots of playing children led some to feel more connected, others found this aspect of the film to be a source of dissonance. Kirsten, for instance, explained how the children exhibited what she described as a "look of innocence," which for her highlighted the differences in perception between herself as a viewer and documentary's subjects: "Where I see toxic fumes, poor conditions, and contaminated water, they see a truck filled with toys. They sense the positive aspects of their life, whereas the adults tend to focus on the negatives."[31] In other words, the sympathetic attachment that most of the participants formed with the children was for some also characterized by a kind of disconnect that made them reflect on their own positionality as a privileged viewer.

(Dis)connections: "I Could Have Thrown That Ad Away"

A crucial moment in the film that attracted the attention of virtually the entire cohort takes place about halfway through. In this scene, the camera looks over the shoulder of Yi-Jie's brother as he enthusiastically pretends to eat a series of food products depicted in one of the many supermarket ads that litter the workshop. The shot is in medium close-up, allowing the viewer an overview of both the contents of the ad and the brother. He flips a page, exclaims, "Oh, meat!," and begins to imitate eating products right off the page. While the discordant contrast between the ad's abundance of food and the child's poverty was itself disturbing to many participants, what enabled the scene's emotional impact was the moment that the participants realized that the leaflet is from a well-known Danish supermarket.

"That's the game changer," Anna said in response to the scene. "That's when it hits you [. . .] because that's the moment it gets personal. [. . .] Then I am also guilty."[32] When the participants recognized this uncannily familiar and culturally specific object, many immediately felt the distance between themselves and the world of the documentary diminish: "It all came closer. [. . .] It became more real," Lars explained, highlighting how the presence of this single object changed his perception of the authenticity of the film world.[33] Where the film's mountains of waste represented a homogenous gray mass of anonymous lifeless scraps, in scenes like this one, the otherwise passive waste objects seemed to come alive: They "return," as Peter put it metaphorically, "gazing right back at us."

With this perceptual shift from quantity to quality—from pile of rubbish to identifiable object—came a shift in many participants' sense of responsibility. "It's confirmed that the trash is not just coming from some country

Figure 7.3. Danish advertisement. Screenshot from *Plastic China* (Wang 2016).

in Eastern Europe or Asia," said Martin, a self-employed twenty-nine-year-old man, "it could've been mine [. . .] I could've thrown that ad away."[34] As such, we may observe how this brief sequence with the boy and the ad led many participants to reflect on how they—despite geographical and temporal distance—"contribute to this giant waste problem."[35] And yet, despite the causal connection between Denmark and China that the scene inadvertently exposes—or perhaps, indeed, *because* it reveals this connection—some participants felt compelled to voice reservations that served to distance themselves from this otherwise familiar artifact. For instance, Louise explained that she "found the scene quite difficult to relate to [. . .] when there are no specific data." For her, the visual evidence was perceived to be inadequate because it only revealed *that* a connection exists; and not how, why, and to what extent. "It's not that I don't think we have a responsibility," she specified, "but I struggle to understand the amounts [. . .] I don't know exactly what I as a consumer have contributed."[36] While the ad exposed these participants to a disconcerting causal connection, it also left a lot of questions unanswered, which in some cases led to confusion among the participants or, at best, left them longing for more information. Another way that the participants sought to disconnect themselves from the ad was expressed in terms of agency: "We're involved in this, because my trash is also down there. But then again, I don't really see it as my fault, because the problem is that our countries export it," said William, exemplifying the tendency to see the issue of waste disposal as a structural instead of an individual problem.[37] Just as the lack of explicit information allowed participants to maintain a distance to what they had seen, their perceived lack of agency also served to mitigate liability. In other words, where some participants resisted the feeling of responsibility altogether, others felt implicated, but not guilty as such.

Compassion, Powerlessness, and Inaction

When participants were asked about their mood after watching the film, the most common answer was an ambivalent state of compassion and powerlessness. "I strongly feel like it's something we should do something about," many would say, "but, at the same time, I also have a feeling that it's something you can't do anything about."[38] Several of the participants thus found themselves within a register of affective tension dominated by two seemingly contradictory impulses, one driving action and the other negating it. When asked to describe this affective tension, many would engage in processes of rationalization that justified inaction. Julie, for instance, recast environmental dumping as a kind of service: "It gives them a job, and they help sort the waste."[39] In the same vein, Peter argued that "if you remove the plastic, then you also remove their source of income."[40] Others juxtaposed the documentary with other socioecological issues. Michael, for instance, explained how "there are also horrible things happening in South Africa. [. . .] I have more pity for them [. . .] because they have no opportunity. It's just one big black hole." Finally, some contemplated concrete courses of action, but these would seem impossible or inappropriate: "In a sense, I feel like going to China in order to adopt those four children, bringing them home, and looking out for them. But they wouldn't get anything good out of that, nor would I."[41]

This logic of wanting to literally move the children revealed not only how salient the forces of immobilization had been to many participants, but also how their intuitive self-perception reflected asymmetrical power relations between themselves and the subjects of the documentary. At times, this asymmetry was conceptualized in terms of an East–West divide: "As a Westerner, you feel like helping, but how would I help, if I wanted to help this specific family?" Camilla asked, illustrating how being Western in this case manifested itself in an affective amalgam of guilt, duty, and doubt.[42] Yet while this self-imposed cultural identity compelled some to action, albeit entirely hypothetical action, others would refer to similar essentialist presumptions in order to justify their inaction: "I don't think you could make them stop," William said. "You could give them money, but I think they would still be doing this. Then they'd just have more money. It's just part of their culture, I think."[43] In such cases, the notions of global interconnectivity and shared responsibility that the film calls attention to were undermined, as the cultural essentialism that guided such interpretations effectively restricted the issues depicted in the documentary as being a product of Chinese culture alone.

Finally, many participants highlighted that their feeling of powerlessness was simply a result of the magnitude of the problems presented in the documentary. This lack of self-efficacy was reflected in the participants' language:

Where many would refer to "a plastic ocean" and how the subjects were "drowning in chemicals,"[44] they would incidentally also talk about themselves as "a drop in the ocean"[45] and about creating "ripple effects"[46] to enact change around them. In other words, participants would describe their experience through a metaphorical vocabulary of water, which indicated that inasmuch as the issues of waste felt formless, fluid, and intangible, then so did the participants' sense of agency in the face of it. Ultimately, the only action the participants felt capable of pursuing was changing their individual consumer habits, which highlighted the paradox that the only conceivable thing participants saw as capable of mitigating the presence of plastic was the consumer culture that generates it.

Reception of the Observational Documentary Mode

Earlier I described *Plastic China* as an observational documentary. I also argued that because of this aesthetic mode, the film has potential to render slow violence perceptible in the sense that it makes tangible the lived reality of distant others. To some extent, the interviews revealed that the film was able to facilitate such a connection. However, they also showed that the experience of proximity was not the only outcome of this filmic mode, as some participants thought that the documentary lacked adequate information about the issues it presented. *Plastic China* simply left the participants in the study with too many unanswered questions. Susan wondered about health risks: "What kind of plastic is it? And how does it affect them?"[47] Peter found the absence of a guiding voice problematic: "It was very good at showing what the problem was, but I would have liked some suggestions about what you can do differently."[48] This made Kirsten worry about missing the point: "Was there a message? Did they want something? I couldn't really figure it out."[49] For Michael, this even became a source of frustration: "It doesn't give me anything I can use [. . .] and that's what irritates me about the film, because I don't get anything out of watching it. There are no figures."[50] Finally, Martin applauded the fact that "there is no explicit narrative about me having to change my behavior, [. . .] no voice from the outside moralizing for me; it is my inner voice, my conscience, and my reason."[51]

These responses suggest several things. First, they indicate how centrally the expository documentary mode figured in the participants' conception of the documentary genre. Second, they show that deviation from the conventions of this mode (voice-over, argumentation, models and figures) cue affective responses at both ends of the spectrum, ranging from gratification to annoyance. Third, they indicate that, for some participants at least, the power of the image was not enough to satisfy their desire to comprehend the issues it presents. Fourth, they demonstrate varying individual preferences for documentary didacticism: some enjoyed the absence of explicit moralization, but others felt bewildered without it. Finally, although many participants struggled to tolerate the ambiguity of observation, the film nevertheless sparked curiosity about the issue of environmental dumping. Many participants found themselves reflecting on the "journeys of trash,"[52] where it comes from, and "what you can do"[53] to mitigate it. The fact that *Plastic China* generally raises more questions than it answers therefore did not necessarily seem to be categorically counterproductive because it cultivated a responsiveness to a world of relations in which the participants, as viewers and waste-producing consumers, found themselves entangled.

Reverberations

In this section, I present key findings from the second round of interviews, which also exhibited a great deal of individual variation. Though everyone was fully able to summarize the film's narrative, participants gave quite different answers when asked whether, how, and to what extent the film had stayed with them in the week that followed. Where some reported having had specific scenes, images, or characters on their minds, others explained that it was the emotions and moods triggered by the film that occupied them the most.

This is not to say that the participants were continuously preoccupied with the film and their affective memory of it. On the contrary, thoughts and feelings about the film came and went, depending on the situations participants found themselves in. Some explained how Yi-Jie's face had been on their minds in moments when no immediate demands required their attention. In these cases, the film appeared to have taken on a life of its own, as affective and visual memories of the film came flooding back when participants least expected it, such as when lying in bed at night. Others reported that, when faced with consumer choices or when handling household waste, they found themselves thinking of the film and reliving their affective response to it. Many, for instance, claimed that the film had changed their otherwise mundane experience of grocery shopping. Gazing down the aisles of the supermarket, filled with everything from wrapped vegetables to ketchup bottles, some experienced a newfound awareness of the ubiquity of plastics and the frustrating difficulty they faced as consumers in their futile attempts to avoid them. Others claimed that their perception of advertising had changed. Martin, for instance, explained that the juxtaposition of the ad in the film and the ads he would encounter in the supermarket had made them seem "extremely colorful, overly expressive, and altogether awfully celebratory."[54] Consequently, this led to a heightened awareness of the "high level of consumption and manipulation" that he saw as characterizing contemporary consumer culture. But where the film produced—in some participants and at certain times—a heightened awareness of the long-term socioecological consequences of their everyday choices, it made Susan doubt the ethical importance of otherwise well-regarded environmental practices, such as recycling:

> When I sorted my waste at home, I suddenly thought to myself, do I actually want to do this? [...] If that's what I contribute to, then it almost seems more reasonable to burn it here [...] where we can handle it properly. I was actually quite sad to discover that I felt this way, because I still believe it's a good idea to sort and recycle. Just not in that way. But I can't oversee the whole process. I can only make the initial decision and await the consequences.[55]

In Susan's case, the experience of *Plastic China* led to uncertainty about whether the practice of recycling might be counterproductive to the efforts of mitigating plastic pollution—along with the social issues that the disposal of waste can be seen to exacerbate on a global scale. While the practical implications of such a reaction should of course be taken seriously as itself a potential source of other forms of slow violence (if, for instance, it would lead Susan to simply stop recycling altogether, thus causing her own waste to become sources of pollution), it is worth noting that despite her doubts, she now experienced a distinct perception of the geographically vast flows of waste stretching across the planet and the violence these flows exert on a host of human and nonhuman life-forms. In sum, it made the slow violence of the global waste trade almost literally tangible, as she was now able to relate this ephemeral force to something as concrete as a piece of plastic.

Finally, in addition to these specific moments, the second interviews identified a more general experience of fatigue and exhaustion among the participants. "You are constantly bombarded with things you should and shouldn't do," Tina explained.[56] In other words, the experience of *Plastic China* was for many an experience of one audiovisual text amid a plethora of other media texts. "It's like, just another thing that's wrong with the world,"[57] explained Martin, who echoed Susan's contention that "it soon becomes yesterday's news."[58] Reflecting on this experience of exhaustion, Tina specified that "it's not that I don't have sympathy, but you end up becoming immune to this kind of agony and despair." As a result of this fatigue, many would explain actively seeking to distance themselves from the concerns evoked by the documentary. Louise, for instance, declared, "I need my life to be something else than relating to other people's misery. It really sucks to say, and I know it isn't very cosmopolitan *[verdensborgeragtigt]*, but that's how I feel."[59] The longitudinal findings thus revealed, among other things, the considerable influence that other media products with similar themes had on the participants' capacity and willingness to hold on to the emotional responses and cognitive insights of the documentary.

Discussion: Decoding Slow Violence

In sum, the reception study of *Plastic China* revealed both tendencies and variations. To use the terminology of Hall's encoding/decoding model, we might say that participants adopted and shifted between different semiotic codes and ideological positions in their efforts to make sense of the film and their experience. In this final section, I attempt to map these codes and positions onto the empirical findings to discuss some of the potentials and challenges that a documentary like *Plastic China* presents as a form of ecomedia.

Some descriptions reflected what Hall ([1973] 2001) refers to as the "dominant-hegemonic" position (515). This refers to cases in which the participant's interpretation of the film was aligned with the intentions of the filmmakers. This symmetry was particularly evident in the descriptions of the cinematic environment, which echoed the filmmakers' intention to produce a documentary that shows how "the work of recycling plastic waste with their bare hands takes a toll not only on their health, but also their own dilemma of poverty, disease, pollution and death" (CNEX, n.d.). It was clear that the participants involved in the study were highly perceptive of what Stacy Alaimo (2010) has described as transcorporeal vulnerability, which highlights the porousness of the body and the trans-actions that—for better or worse—take place between us and the environment. The film used character engagement to highlight this message, and it appeared to do so successfully. Moreover, the recurrent mentions of scenes featuring burning plastic and black smoke suggest that this visual motif was particularly effective in this regard. As such, the interviews lend evidence to support the claim that "the emotions we feel in response to a cinematic landscape are in fact rarely disconnected from our concern for the protagonist's fate" (Weik von Mossner 2017, 62).

Other descriptions reflected what Hall ([1973] 2001) calls the "negotiated position" (516). This refers to cases in which participants acknowledged the messages of the film but restricted its appeal by modifying it to fit their situational perspective. The descriptions of the scene featuring the Danish ad were an evident example of this. By itself, this scene suggests that the consumer capitalism of the Western world is an undeniable part of the slow violence depicted in the film. However, when describing their understanding of this connection, the participants voiced concerns about their lack of agency with regards to this very relationship. These negotiated responses suggest at least two things. First, they indicate that a lack of voice-over does not necessarily entail a lack of moralization; the images of recognizable waste matter were sufficient in terms of eliciting moral reflection and, in many cases, feelings of guilt. Second, they suggest that even though the striking image of the Danish ad in many ways exemplifies what Nixon (2011) refers to as an "iconic symbol" with "dramatic urgency" (10), such symbols are not always sufficient to evoke feelings of responsibility, even for those who recognize their meaning. Though all the participants were struck by this culturally specific piece of storied matter (Iovino and Oppermann 2014), the scene appeared to do more to reinforce their preexisting feelings of powerlessness than it did to evoke reflections on the necessity of an ecocosmopolitan ethics (Heise 2008).

This lack of individual and collective efficacy may also have been influenced by the sublime character of the wastescape that makes up the backdrop for most of the documentary. The way the participants described the waste in *Plastic China* often resembled Timothy Morton's (2013) hyperobject—that is, an entity whose presence is largely incomprehensible because it is so "massively distributed in time and space relative to humans" (1). Like global warming, plastic here seems to represent yet another example of how we, as humans, have "manufactured materials that are already beyond the normal scope of our comprehension" (131). Ultimately, this perception of plastic waste as a form of technological sublime or Mortonian hyperobject appeared to supplement the structural nature of slow violence in ways that only increased the participants' perceived lack of agency.

Another aspect that reflected efforts of negotiation on part of the participants could be traced in the documentary's attempt to engage viewers emotionally. Here the repeated use of close-ups of characters, along with the emotive score, would often lead to sympathetic attachment and emotional contagion, but occasionally participants would also experience such features as overly sentimental and overtly manipulative. Such responses suggest that inasmuch as appeals to emotion can be an effective means for engaging viewers in narratives about environmental risk (which the stories about going to the supermarket appeared to confirm), they can also lead to the exact opposite: disengagement and lack of interest. Individual preference for documentary didacticism and observational aesthetics suggests similar things. Some viewers may find the invitation to become a fly on the wall liberating, but others experience this mode as restrictive, even frustrating. A fly, after all, cannot do anything about what it observes. Consequently, the study suggests that the potential that an observational documentary such as *Plastic China* shows in terms of making available the lived experience of slow violence comes with the risk that viewers who find the aesthetics of observation too demanding, too ambivalent, or simply too tedious may ultimately lose interest.

Finally, some viewers' descriptions resembled what Hall ([1973] 2001) refers to as the "oppositional position," in which the participant recognizes the message of the text but decodes it "in a *globally* contrary way" (517). This was particularly evident among the participants who saw the events depicted in the film as evidence of Chinese culture rather than a by-product of the globalized system of waste disposal. Contrary to the ecocosmopolitanism articulated by the filmmakers, these participants interpreted the poverty of Yi-Jie and her family as the result of poor life choices and of cultural dispositions. While it is true that "the full-blown ascent of Chinese authoritarian capitalism" (Nixon 2011, 43) plays a crucial role in the state of affairs depicted in the documentary, these oppositional readings nevertheless actively undermined the structural—and in this case global—nature of slow violence. By ignoring the narrative frame of the international waste trade, encoded in the opening sequence with the arrival of the container ship, these participants favored the reductive semantics of cultural essentialism over globalization and interconnection in their reading of the film. Ultimately, this suggests that even though the strength of *Plastic China*'s observational mode lies precisely in its ability to represent socioecological calamities in ways that invite reflection rather than reduction, the fact that its relatively unguided images are open to interpretation also creates the risk that viewers will misconstrue the global nature of the networks of exploitation that it attempts to depict.

Finally, considering the frequency with which the participants would relate their experience of the documentary to the oversaturated media environment they navigate on a day-to-day basis, it seems appropriate to supplement Hall's three spectator positions with a fourth position that recognizes the widespread experience of fatigue that the participants described. I dub this the position of exhaustion. Like the oppositional reading, this position is characterized by a resistance to the text; however, in contrast to the oppositional reading, this resistance does not originate in a semiotic disagreement with the messages encoded in the film but rather from the affective experience of feeling overwhelmed by the normative demands it poses. Similar responses have been noted by other scholars, such as Janpol and Dilts (2016), who express concern about the risk of what they call an "inoculation effect [. . .] whereby people become desensitized to an issue due to repeated exposure" (91). On the one hand, this effect suggests that insofar as documentary filmmakers wish to engage viewers in narratives of environmental risk by appealing, in one way or another, to their ethical sensibility, doing so inevitably involves the risk of overwhelming and paralyzing the viewer. On the other hand, the experiences of exhaustion identified in this study are also indicative of a challenge that points beyond the formal variables of the individual text and toward the "structures of feeling" (Williams 2015) that characterize this historical moment of media consumption. Rosi Braidotti (2019), for instance, addresses the profound ways in which "we are caught in contradictory pulls and spins that call for constant negotiations in terms of time, boundaries and degrees of involvement with, and disengagements from, the same technological apparatus that frames our social relations" (55). In short, we become exhausted because we are subjected to the constant stress of a range of issues, including anthropogenic climate change, deforestation, pandemics, and global inequality. In a similar vein, Simon Estok (2016) has suggested that "eco-media are embedded in a period in which our continuous partial attention runs hand in hand with our compassion fatigue" (x).

The findings of this case study point in the same direction. Consequently, future empirical research ought to explore in more detail how phenomena such as exhaustion, fatigue, and emotional numbing influence our perception of environmental issues while also remaining attentive to how such predispositions may be perpetuated by the media culture that viewers are required to navigate on a day-to-day basis.

Conclusion

The documentary *Plastic China* shows that being subject to the slow violence of environmental dumping means being stuck in the margins of society, outside the center, and outside the urban, but never outside the immobilizing restraints (and hazards) of class, capitalism, pollution, and environmental change. That such messages are encoded into a cultural text is, however, no guarantee that viewers and readers will recognize, let alone accept, such messages. By contrast, the reception study presented in this chapter shows that the landscape of viewer experience was inherently heterogeneous. Some reported feeling responsible for what they had seen, some emphasized the significance of structure, and some rejected the documentary's message about environmental accountability altogether. As such, one of the main results of the study is evidence in support of the claim that we as viewers experience film differently, whether it be in terms of the meaning we ascribe to the various scenes, settings, and characters or the emotions we feel when watching such films. Within the small audience of this study's sample, a wide range of emotions was identified, including sadness, anger, frustration, fear, guilt, shame, embarrassment, and disgust as well as more ambiguous states such as anxiety, powerlessness, hopelessness, doubt, cynicism, and indifference—not to mention more positive reactions such as compassion, sympathy, happiness, inspiration, and curiosity. When ecocritics strive for a more complete understanding of how moving images move us, it is therefore essential that we engage with the rich affective pluralism of a real flesh-and-blood audience.

Finally, the study raises questions about the complicated longitudinal impact of ecomedia. Existing empirical research suggests that the impact of environmental narratives on people's attitudes decrease significantly within one or two months (Malecki et al. 2019; Nolan 2010). Likewise, the study of *Plastic China* found that the affective impact of the film diminished significantly after a week, and sometimes within hours of its screening. However, although such evidence of diminishing impact is compelling, I cannot help but wonder whether we ought to complicate the linearity with which we conceptualize the effect of environmental narrative. What about cases in which works of art suddenly resonate in us with renewed intensity long after we have encountered them (Chapman, Lickel, and Markowitz 2017; Felski 2020)? Indeed, this empirical survey of *Plastic China* suggests that the impact of the documentary was influenced just as much by the situational context as it was by the passing of time; above all, it was viewers' encounters with objects, people, conversations, and atmospheres that prompted the film to reverberate in their minds and bodies. Discussing the challenges of longitudinal research designs, Janpol and Dilts (2016) argue that "intervening variables that may occur over time would have the potential to confound results, making causal attributions somewhat more difficult" (96). While I agree with this observation, it is nevertheless worth noting that in the context of this study, it was precisely the intervening variables that led the film to reemerge in the days that followed. This study therefore suggests that as empirical ecocritics, we should be tentative when using a rhetoric of transformation because the perception of ecological issues emerges not only in the aesthetic experience and its immediate aftermath. Rather, our visions, feelings, and impulses come and go, depending on the circumstantial, micropolitical situations that we engage in on a day-to-day basis (Hawkins 2005;

Scherer 2007). As such, the irony is that if the destruction of slow violence is attritional, dispersed, and ephemeral, then so is our awareness of it, which also appears to fluctuate between our embodied habits, our momentary reflections, and the affective spaces we occupy.

Appendix

The guides below, here translated from Danish into English, served to structure the two interviews with each participant. The guides were memorized by the interviewer and were not physically present in the interview situations; nor were the guides followed systematically, as the conversational format usually made many of the questions unnecessary.

Bullets (•) refer to follow-up questions that were used in cases where the original question did not result in any answer or description. More often, however, the interviewer formulated follow-up questions on the basis of participants' statements. For instance, this was done by asking participants to clarify ("Can you say some more about that?") or by recasting the participants' own descriptions as questions ("You felt sorry for her?").

Items marked by an asterisk (*) refer to information previously given by the participant.

Interview Guide 1: Immediate Impressions

OPENER/COMPREHENSION CHECK

Thank you again for taking the time to do this interview.

For this interview, what I am interested in is above all your experience of watching the film. Now, this also means that there are no right or wrong answers, because you have simply experienced what you experienced, and there is nothing more right or wrong in that.

So let's start out with a very broad question: Could you try to describe for me some of what you have just seen?

Could you give me a summary of the film?

FILMIC EXPERIENCE

- Was there anything in particular that made an impression on you?
- How did you experience the different characters in the film?
- What did you think of the young girl?
- What did you think of the two fathers?

- What do you think it would be like to be [name of character]?
- Pretend that I haven't seen the film, and then describe for me in as much detail as possible your experience of the place in which the film takes place.
- What do you think it would be like to be there? To live there? Why?
- Does the place remind you of something you've seen or felt before? It could be on film or in real life.
- Take a moment and try to pay attention to how you feel right now. How would you then describe the mood you are in?

REFLECTIONS + STILLS

- We have now talked a bit about your immediate experience of the film. If we try to move on from your experience of the film and toward your thoughts about the film as a film, then what did you think of it?
- Would you watch it again? Why (not)?
- Would you recommend it to others? Why (not)?
- Would you have watched it on your own accord? Why (not)?
- Did you think it was boring/entertaining? Why?
- Was there anything in the film that came as a surprise to you?

Following this question, the interviewer brings in three stills (from the ad scene, the fishing scene, and a workshop scene) and asks:

- What went through your head when you saw these scenes?
- What is it like to watch them again?
- How does it make you feel?

CONCLUDING REMARKS

This is all for now, unless you have any questions for me, the film, or what we have talked about.

Thank you for your participation. As you know, we would like to do a second interview a week from now. You don't have to prepare for it in any way. All we need from you is half an hour of your time to follow up on some of the things we have talked about today.

See you next week!

Interview Guide 2: A Week Later

OPENER/MEMORY CHECK

Welcome back. Thank you for taking the time to do this second interview. We really appreciate it.

For this interview, I have prepared a few questions that we can use—just like last time—to guide our conversation. And like last time, I am first and foremost still interested in learning about your experience. This also means that, just like our last interview, there are no right or wrong answers. There are three things I am going to ask you about. First, I'm going to ask you what you remember from the film you saw last week. Second, I have a few follow-up questions about some of things we talked about last time. And finally, at the end, I have some broader questions about waste.

But first, could you describe for me what you remember in particular from the film you saw last week?

- Where does the film take place?
- Who is in the film?
- What happens to them?
- How did it make you feel?
- Could you say a few words about how you remember the place in the film? It could also just be an atmosphere.
- If you close your eyes and think of the place in the film, how does it present itself to you?

AFFECTIVE CONTINUITY

- Last week, you mentioned that the film made you feel [X*]. Do you still feel this way?
- When would you say these thoughts/feelings/concerns stopped?
- Did the feeling/thoughts/concerns change in any way?

The above questions would be repeated two or three times, depending on how many feelings/thoughts the participant had described in the first interview.

- Since our last interview, has there been any situations where you thought of the film that you watched last week?

- Can you describe for me in as much detail as possible the situation when you came to think of the film?
- Have you talked with anyone about the film since you saw it?
- What came up in the conversation?
- Who did you talk to?
- Is there a reason you haven't talk to anyone about it?

WASTE

If you take a moment and go through the last week in your mind, has there then been any situations where you have encountered some waste that caught your attention? If so, please describe for me in as much detail the situation and what went through your mind at the time.

- It could be something you have seen somewhere.
- It could be something someone else was holding/handling.
- It could be something you were handling yourself.

CONCLUDING REMARKS

Is there anything else you would like to add?

Thank you for your participation.

Notes

I would like to express my gratitude to the editors of this volume for their invaluable guidance: Matthew Schneider-Mayerson, Alexa Weik von Mossner, Wojciech Malecki, and Frank Hakemulder. I would also like to thank those who have provided feedback throughout the chapter's various stages: Svend Erik Larsen, Peter Mortensen, Tobias Skiveren, Magnus Andersen, Morten Gustenhoff, Soo Ryu, Stephanie Volder, Josefine Brink Siem, Mai Ørskov, and Anna Solovyeva. The chapter would not have been the same without your critical input.

1. In this context, it is worth noting that "since January 1, 2018, China's import ban on waste plastics has been put into force, which has had a far-reaching effect on global plastic production and solid waste management. Southeast Asian countries like Malaysia have replaced China as the leading importer of plastic wastes from the U.S." (Wang et al. 2019, 72).

2. For the purposes of the reception study, the film was screened in its original language (Chinese) with Danish subtitles. A slightly shorter, fifty-three-minute version of the film had to be used because no Danish subtitles exist for the original version. The

running time of the original version is eighty-two minutes. The main difference between the two versions is that a number of plot-redundant shots have been omitted and that the original version presents, in the opening shots, an intertitle that reads: "China is the leading importer of plastic wastes from Japan, Korea, Europe, and USA." The absence of the intertitle as well as Danish subtitles might have affected the viewers' experiences in a number of ways, though it is beyond my scope to explore these in detail.

3. The fact that the phenomenological interview emphasizes meaning does not mean that interviews exclude affective and emotional elements. On the contrary, some scholars describe the interview practice as a "situated affective encounter" (Ayata et al. 2019), stressing that affect, emotion, and atmosphere in fact influence the interview process in significant ways. According to Gabriel and Ulus (2015), these emotional elements can be observed in, at least, three ways: Interviewees may express emotions directly ("I feel . . ."), indirectly (by using allegories and metaphors), and through body language. To include such observations in the data set, the interviews were recorded on video, and body language was transcribed whenever affectively intense moments occurred (Knudsen and Stage 2015).

4. Recruitment postings were distributed via social media and in public spaces such as libraries, cafés, supermarkets, and schools around the city of Aarhus. The recruitment efforts resulted in eighty-four interested applicants, fourteen of whom were selected. None had prior relations with the researcher.

5. The fact that interviews were conducted in Danish meant that excerpts had to be translated into English for this publication. Following the recommendations of van Nes et al. (2010), the study worked in the original language as long as possible and as much as possible "to avoid potential limitations in the analysis" (315).

6. The choice to conduct one-on-one interviews instead of group interviews was based on the assumption that the interview topic could be sensitive to some participants, and that such sensitivity might inhibit some interviewees from sharing their genuine views and experiences, had the interview simulated a social situation involving multiple individuals—see, for instance, Rose (2016) on Gray (1992) and Morley (1980).

7. To carry out the coding in practice, I used NVivo12 Pro software.

8. For an overview of the field of audience reception studies and its development, see Alasuutari (1999), Jensen (2012), and Hill (2018).

9. The concept of ecological validity refers to the question of "whether the findings of a study can be generalized to naturalistic situations" (Andrade 2018, 499). In the context of media consumption, naturalistic situations can entail many different sociocultural contexts, such as private households, public cinemas, school classrooms, and film festivals. Media reception studies that choose to pay particular attention to the question of ecological validity often emphasize the act of consuming media as a social event and is often motivated to discern "the role of film in people's daily lives"

(Biltereyst and Meers 2018, 35). Although it would have been interesting to consider in more detail the naturalistic context of the consumption of environmental documentaries, its social dynamics, and so on, the intention of this study lies elsewhere.

10. Much contemporary scholarship in film studies, both in the cognitive tradition and elsewhere, has emphasized the salience of the nonconscious (affective) aspects of film experience (Laine 2011; Plantinga 2009; Rutherford 2011). At stake in many of these studies is the controversial distinction between affect (unconscious) and emotion (conscious), which has generated much debate, not only in film studies but in a handful of other fields, including literary criticism (Hogan 2018), social science (Ahmed 2014), and philosophy (Massumi 1995, 2002). Although I agree that policing this distinction is productive for our theoretical understanding of film experience, dwelling on this distinction in depth is beyond my scope here.

11. Louise: female, twenty-four, social science student, center right, higher education (medium). The participants have been assigned a pseudonym and are specified by sex, age, occupation, political identity (out of the following options: left, center left, center, center right, right), and educational background. These indicators are provided throughout to give readers an idea of who the individual participants were. Crucially, the intention of doing so is not to suggest that the descriptions of the participants and their associated categories may be used as the basis for generalization. Furthermore, as noted by Schneider-Mayerson (2018), "there is a danger that these generic categories mask nuanced differences, or even reify them" (491). Despite this risk, I believe the indicators to be helpful for the reader in navigating through the testimonies of the cohort.

12. Peter: male, thirty-three, unemployed, left, higher education (long).

13. Kirsten: female, thirty-one, unemployed elementary school teacher, center right, higher education (medium). Michael: male, thirty-five, office clerk, center right, education level unknown. Christian: male, twenty-one, media studies student, left, higher education (medium).

14. Louise: female, twenty-four, social science student, center right, higher education (medium).

15. Michael: male, thirty-five, office clerk, center right, education level unknown.

16. Christian: male, twenty-one, media studies student, left, higher education (medium).

17. Kirsten: female, thirty-one, unemployed elementary school teacher, center right, higher education (medium).

18. Camilla: female, thirty-six, gardener, left, elementary school.

19. Susan: female, thirty-six, unemployed, center, higher education (long).

20. Lars: male, twenty-six, unemployed, center left, higher education (medium).

21. Camilla: female, thirty-six, gardener, left, elementary school.

22. Julie: female, twenty-four, music therapy student, left, higher education (medium).

23. Camilla: female, thirty-six, gardener, left, elementary school.

24. Lars: male, twenty-six, unemployed, center left, higher education (medium).

25. Lars: male, twenty-six, unemployed, center left, higher education (medium).

26. Christian: male, twenty-one, media studies student, left, higher education (medium).

27. Anna: female, twenty-seven, medical student, center left, high school.

28. Michael: male, thirty-five, office clerk, center right, education level unknown.

29. Even though the camera in many of these scenes is in close proximity to the subjects, the children largely ignore its presence throughout. In part, it must be assumed that this is the result of editing. However, it is also worth noting that Wang lived closely alongside the two families for a year and half during the documentary's shooting, which enabled the director to film his subjects without making his presence particularly conspicuous (Zhao 2017).

30. Tina: female, forty-three, elementary school teacher, left, higher education (medium).

31. Kirsten: female, thirty-one, unemployed elementary school teacher, center right, higher education (medium).

32. Anna: female, twenty-seven, medical student, center left, high school.

33. Lars: male, twenty-six, unemployed, center left, higher education (medium).

34. Martin: male, twenty-nine, self-employed, center right, higher education (long).

35. Peter: male, thirty-three, unemployed, left, higher education (long).

36. Louise: female, twenty-four, social science student, center right, higher education (medium).

37. William: male, twenty-one, medical student, center left, high school.

38. William: male, twenty-one, medical student, center left, high school.

39. Julie: female, twenty-four, music therapy student, left, higher education (medium).

40. Peter: male, thirty-three, unemployed, left, higher education (long).

41. Kirsten: female, thirty-one, unemployed elementary school teacher, center right, higher education (medium).

42. Camilla: female, thirty-six, gardener, left, elementary school.

43. William: male, twenty-one, medical student, center left, high school.

44. Anna: female, twenty-seven, medical student, center left, high school.

45. Anna: female, twenty-seven, medical student, center left, high school.

46. Kirsten: female, thirty-one, unemployed elementary school teacher, center right, higher education (medium).

47. Susan: female, thirty-six, unemployed, center, higher education (long).

48. Peter: male, thirty-three, unemployed, left, higher education (long).

49. Kirsten: female, thirty-one, unemployed elementary school teacher, center right, higher education (medium).

50. Michael: male, thirty-five, office clerk, center right, education level unknown.

51. Martin: male, twenty-nine, self-employed, center right, higher education (long).

52. Oscar: male, twenty-eight, self-employed, center, higher education (long).

53. Louise: female, twenty-four, social science student, center right, higher education (medium).

54. Martin: male, twenty-nine, self-employed, center right, higher education (long).

55. Susan: female, thirty-six, unemployed, center, higher education (long).

56. Tina: female, forty-three, elementary school teacher, left, higher education (medium).

57. Martin: male, twenty-nine, self-employed, center right, higher education (long).

58. Susan: female, thirty-six, unemployed, center, higher education (long).

59. Louise: female, twenty-four, social science student, center right, higher education (medium).

References

Ahmed, Sara. 2010. *The Promise of Happiness.* Durham, N.C.: Duke University Press.

Ahmed, Sara. 2014. *The Cultural Politics of Emotion.* Edinburgh: Edinburgh University Press.

Alaimo, Stacy. 2010. *Bodily Natures: Science, Environment, and the Material Self.* Bloomington: Indiana University Press.

Alasuutari, Pertti. 1999. "Introduction: Three Phases of Reception Studies." In *Rethinking the Media Audience: The New Agenda,* edited by Pertti Alasuutari, 2–21. London: Sage.

Andrade, Chittaranjan. 2018. "Internal, External, and Ecological Validity in Research Design, Conduct, and Evaluation." *Indian Journal of Psychological Medicine* 40 (5): 498–99.

Ayata, Bilgin, Cilja Harders, Derya Özkaya, and Dina Wahba. 2019. "Interviews as Situated Affective Encounters—A Relational and Processual Approach for Empirical Research on Affect, Emotion and Politics." In *Analyzing Affective Societies: Methods and Methodologies,* edited by Antje Kahl, 63–77. London: Routledge.

Bauman, Zygmunt. 2004. *Wasted Lives: Modernity and Its Outcasts.* Oxford: Polity.

Bell, Lucy. 2019. "Place, People, and Processes in Waste Theory: A Global South Critique." *Cultural Studies* 33 (1): 98–121.

Biltereyst, Daniel, and Phillipe Meers. 2018. "Film, Cinema, and Reception Studies: Revisiting Research on Audience's Filmic and Cinematic Experiences." In *Reception Studies and Audiovisual Translation,* edited by Elena Di Giovanni and Yves Gambier, 21–42. Amsterdam: John Benjamins.

Braidotti, Rosi. 2019. *Posthuman Knowledge.* Cambridge: Polity.

Brereton, Pat, and Victoria Gómez. 2020. "Media Students, Climate Change, and YouTube Celebrities: Readings of 'Dear Future Generations: Sorry' Video Clip." *ISLE: Interdisciplinary Studies in Literature and Environment* 27 (2): 385–405.

Brinkmann, Svend, and Steinar Kvale. 2009. *InterViews: Learning the Craft of Qualitative Research Interviewing.* London: Sage.
Brooks, Amy, Shunli Wang, and Jenna Jambeck. 2018. "The Chinese Import Ban and Its Impact on Global Plastic Waste Trade." *ScienceAdvances* 4 (6): 1–7.
Cameron, James, dir. 2009. *Avatar.* 20th Century-Fox.
Chang, Chia-ju. 2016. "Wasted Humans and Garbage Animals: Deadly Transcorporeality and Documentary Activism." In *Ecodocumentaries: Critical Essays,* edited by Alex K. Rayson and Susan Deborah Selvaraj, 95–114. London: Palgrave Macmillan.
Chapman, Daniel A., Brian Lickel, and Ezra M. Markowitz. 2017. "Reassessing Emotion in Climate Change Communication." *Nature Climate Change* 7: 850–52.
CNEX. n.d. "Storyline." Accessed October 17, 2022. https://www.cnex.tw/plasticchina.
Coffey, Amanda, and Paul A. Atkinson. 1996. *Making Sense of Qualitative Data: Complementary Research Strategies.* Thousand Oaks, Calif.: Sage.
Crawford, Alan, and Hayley Warren. 2020. "China Upended the Politics of Plastic and the World Is Still Reeling." Bloomberg, January 21, 2020. https://www.bloomberg.com/graphics/2020-world-plastic-waste/.
Davies, Thom. 2019. "Slow Violence and Toxic Geographies: 'Out of Sight' to Whom?" *Environment and Planning C: Politics and Space* 40 (2): 409–27.
Eitzen, Dirk. 2005. "Documentary's Peculiar Appeals." In *Moving Image Theory,* edited by Joseph Anderson and Barbara Fisher Anderson, 183–99. Carbondale: Southern Illinois University Press.
Estok, Simon. 2016. "Foreword: Packaging Concerns." In *Ecodocumentaries: Critical Essays,* edited by Alex K. Rayson and Susan Deborah Selvaraj, vii–xiv. London: Palgrave Macmillan.
Fedele, David, dir. *E-Wasteland.* 2012. Short film. http://www.e-wastelandfilm.com/.
Felski, Rita. 2020. *Hooked: Art and Attachment.* Chicago: University of Chicago Press.
Gabriel, Yiannis, and Eda Ulus. 2015. "It's All in the Plot: Narrative Explorations of Work-Related Emotions." In *Methods of Exploring Emotions,* edited by Helena Flam and Jochen Kleres, 36–45. New York: Routledge.
Giorgi, Amadeo. 2009. *The Descriptive Phenomenological Method in Psychology: A Modified Husserlian Approach.* Pittsburgh, Pa.: Duquesne University Press.
Gray, Ann. 1992. *Video Playtime: The Gendering of a Leisure Technology.* London: Routledge.
Greisen, Steven, dir. 2006. *Explore the Wildlife Kingdom: Dolphins—Tribes of the Sea.* Reel Productions.
Guattari, Félix. 2000. *The Three Ecologies.* Translated by Ian Pindar and Paul Sutton. London: Athlone.
Guggenheim, Davis, dir. 2006. *An Inconvenient Truth.* Participant Productions.
Hall, Stuart. (1973) 2001. "Encoding, Decoding." In *The Cultural Studies Reader,* edited by Simon During, 507–17. London: Routledge.

Hawkins, Gay. 2005. *The Ethics of Waste: How We Relate to Rubbish.* Lanham: Rowman & Littlefield.

Heise, Ursula. 2008. *Sense of Place and Sense of Planet: The Environmental Imagination of the Global.* Oxford: Oxford University Press.

Hill, Annette. 2018. "Media Audiences and Reception Studies." In *Reception Studies and Audiovisual Translation,* edited by Elena Di Giovanni and Yves Gambier, 3–20. Amsterdam: John Benjamins.

Hogan, Patrick. 2018. *Literature and Emotion.* New York: Routledge.

Howell, Rachel. 2014. "Investigating the Long-Term Impacts of Climate Change Communications on Individuals' Attitudes and Behavior." *Environment and Behavior* 46 (1): 70–101.

Hyman, Ira E., Jr., and Erica E. Kleinknecht. 1999. "False Childhood Memories: Research, Theory, and Applications." In *Trauma and Memory,* edited by Linda Williams and Victoria Banyard, 175–88. London: Sage.

Iovino, Serenella, and Serpil Oppermann. 2014. *Material Ecocriticism.* Bloomington: Indiana University Press.

Ivakhiv, Adrian. 2013. *Ecologies of the Moving Image: Cinema, Affect, Nature.* Waterloo, Ontario: Wilfrid Laurier University Press.

Jacobsen, Grant. 2011. "The Al Gore Effect: *An Inconvenient Truth* and Voluntary Carbon Offsets." *Journal of Environmental Economics and Management* 61: 67–78.

Janpol, Henry L., and Rachel Dilts. 2016. "Does Viewing Documentary Films Affect Environmental Perceptions and Behaviors?" *Applied Environmental Education and Communication* 15 (1): 90–98.

Jensen, Klaus Bruhn. 2012. "Media Reception: Qualitative Traditions." In *The Handbook of Media and Communication Research: Qualitative and Quantitative Methodologies,* edited by Klaus Bruhn Jensen, 171–85. New York: Routledge.

Knudsen, Britta, and Carsten Stage. 2015. *Affective Methodologies: Developing Cultural Research Strategies for the Study of Affect.* Basingstoke: Palgrave Macmillan.

Laine, Tarja. 2011. *Feeling Cinema: Emotional Dynamics in Film Studies.* New York: Continuum.

Liboiron, Max. 2018. "Waste Colonialism." Discard Studies, January 1, 2018. https://discardstudies.com/2018/11/01/waste-colonialism/.

Malecki, Wojciech, Piotr Sorokowski, Bogusław Pawlowski, and Marcin Cieński. 2019. *Human Minds and Animal Stories: How Narratives Make Us Care about Other Species.* New York: Routledge.

March, Briar, dir. 2010. *There Once Was an Island: Te Henua e Nnoho.* On the Level Production.

Marran, Christine L. 2017. *Ecology without Culture: Aesthetics for a Toxic World.* Minneapolis: University of Minnesota Press.

Massumi, Brian. 1995. "The Autonomy of Affect." *Cultural Critique* 31: 83–109.

Massumi, Brian. 2002. *Parables for the Virtual: Movement, Affect, Sensation.* Durham, N.C.: Duke University Press.

Morley, David. 1980. *The Nationwide Audience: Structure and Decoding.* London: British Film Institute.

Morley, David. 2013. "Changing Paradigms in Audience Studies." In *Remote Control: Television, Audiences, and Cultural Power,* edited by Ellen Seiter, Hans Borchers, Gabriele Kreutzner, and Eva-Maria Warth, 16–43. London: Routledge.

Morton, Timothy. 2010. *The Ecological Thought.* Cambridge, Mass.: Harvard University Press.

Morton, Timothy. 2013. *Hyperobjects.* Minneapolis: University of Minnesota Press.

Nichols, Bill. 1994. *Blurred Boundaries: Questions of Meaning in Contemporary Culture.* Indianapolis: Indiana University Press.

Nichols, Bill. 2001. *Introduction to Documentary.* Bloomington: Indiana University Press.

Nixon, Rob. 2011. *Slow Violence and the Environmentalism of the Poor.* Cambridge, Mass.: Harvard University Press.

Nolan, Jessica M. 2010. "*An Inconvenient Truth* Increases Knowledge, Concern, and Willingness to Reduce Greenhouse Gases." *Environment and Behavior* 42 (5): 643–58.

Plantinga, Carl. 2009. *Moving Viewers: American Film and The Spectator's Experience.* Berkeley: University of California Press.

Ritchie, Jane, and Jane Lewis. 2003. *Qualitative Research Practice: A Guide for Social Science Students and Researchers.* London: Sage.

Rose, Gillian. 2016. *Visual Methodologies: An Introduction to Researching with Visual Materials.* London: Sage.

Rutherford, Anne. 2011. *What Makes a Film Tick? Cinematic Affect, Materiality, and Mimetic Innervation.* New York: Peter Lang.

Scherer, Matthew. 2007. "Micropolitics." In *Encyclopedia of Governance,* edited by Mark Bevir, 563–64. Thousand Oaks: Sage.

Schneider-Mayerson, Matthew. 2013. "Disaster Movies and the 'Peak Oil' Movement: Does Popular Culture Encourage Eco-apocalyptic Beliefs in the United States?" *Journal for the Study of Religion, Nature, and Culture* 7 (3): 289–314.

Schneider-Mayerson, Matthew. 2018. "The Influence of Climate Fiction: An Empirical Survey of Readers." *Environmental Humanities* 10 (2): 473–500.

Schooler, Jonathan W., and Eric Eich. 2000. "Memory for Emotional Events." In *The Oxford Handbook of Memory,* edited by Endel Tulving, 379–92. Oxford: Oxford University Press.

Stacey, Jackie. 1994. *Star Gazing: Hollywood Cinema and Female Spectatorship.* London: Routledge.

Tsuchimoto, Noriaki, dir. 1971. *Minamata: The Victims and Their World.* Higashi Productions.

van Nes, Fenna, Tineke Abma, Hans Jonsson, and Dorly Deeg. 2010. "Language Differences in Qualitative Research: Is Meaning Lost in Translation?" *European Journal of Ageing* 7 (4): 313–16.

Wang, Jiu-liang, dir. 2016. *Plastic China.* CNEX.

Wang, Wanli, Nickolas J. Themelis, Kai Sun, et al. 2019. "Current Influence of China's Ban on Plastic Waste Imports." *Waste Disposal and Sustainable Energy* 1 (1): 67–78.

Weik von Mossner, Alexa. 2015. "Slow Violence on the Beach: Documenting Disappearance in *There Once Was an Island.*" In *The Beach in Anglophone Literatures and Cultures: Reading Littoral Space,* edited by Ursula Kluwick and Virginia Richter, 175–91. London: Routledge.

Weik von Mossner, Alexa. 2017. *Affective Ecologies: Empathy, Emotion, and Environmental Narrative.* Columbus: Ohio State University Press.

Williams, Raymond. 2015. "Structures of Feeling." In *Structures of Feeling: Affectivity and The Study of Culture,* edited by Devika Sharma and Frederik Tygstrup, 20–28. Boston: De Gruyter.

Williams, Richard (Prince Ea). 2015. "Dear Future Generations: Sorry." YouTube, April 20, 2015. https://www.youtube.com/watch?v=eRLJscAlk1M.

Winston, Brian. 1995. *Claiming the Real: The Documentary Film Revisited.* London: British Film Institute.

Wolfe, Charles. 2014. "Documentary Theory." In *The Routledge Encyclopedia of Film Theory,* edited by Edward Branigan and Warren Buckland, 144–50. New York: Routledge.

Zhao, Kiki. 2017. "China's Environmental Woes, in Films that Go Viral, Then Vanish." *New York Times,* April 28, 2017. https://www.nytimes.com/2017/04/28/world/asia/chinas-environmental-woes-in-films-that-go-viral-then-vanish.html.

Chapter 8

All the World's a Warming Stage

Applied Theater, Climate Change, and the Art of Community-Based Assessments

SARA WARNER AND JEREMY JIMENEZ

We are all social actors in the drama of climate change, with Americans playing leading roles in the production of greenhouse gas emissions. What might the performing arts contribute to activism, public opinion, and policy debates about the climate crisis? How can theater stage an intervention into this emergency situation? What kinds of empirical data can we glean from performance-based research? These are the questions guiding a three-year experiment conducted by a group of academics, artists, and local residents who created community-based plays about the impact of climate change in the Finger Lakes region of New York. Our plays, *Climates of Change* (2017) and *The Next Storm* (2019), highlight the potential for applied theater to serve as both a method of science communication and mode of knowledge production.[1]

The Art of Science Communication

With the existence of a global climate crisis now indisputable and its consequences concomitantly pervasive, many people look to scientists to secure a sustainable future. Because demagogues can manipulate, obscure, and completely fabricate facts, it is imperative that citizens understand the causes and effects of climate change in order to make informed and empowered decisions. To achieve this, we must make climate science engaging and relevant so that people care enough about the crisis to take meaningful action. One of the most effective ways to make scientific data intelligible is to deliver it as a compelling story (Dahlstrom 2014). Climate scientist Kate Marvel (2017) recognizes this as she narrates the future consequences of warming oceans for her son:

> To be a climate scientist is to be an active participant in a slow-motion horror story. These are scary tales to tell children around the campfire. . . . The culprit is the teller, both victim and villain. . . . We continue to burn fossil fuels and the gases they make continue to trap heat. . . . The heat is mixed deep into the ocean, a long slow slog to equilibrium. There is no way to stop it. What do I tell my son? A monster awaits in the deep, and someday it will come for you. We know this. We put it there.

Although stories have long encouraged prosocial behavior (Gottschall 2012; Johnson 2012), storytelling is not necessarily easy, especially in science communication. As one of our ensemble members explains:

> I think that science communication requires a story. . . . There is a lack of people in this world. . . . Carl Sagan was one . . . who can understand how to read data, understand how to think about science, and understand how to create and tell a story. And I think that it's really important to be able to turn that science into a story that I could tell somebody on the street and that could affect them, not up here [points to head], but here in their heart.

Because emotion rather than empirical evidence often informs people's climate change views (Ejelöv et al. 2018; Lehman et al. 2019), the affect-producing machine of theater can nurture such learning, providing innovative ways to respond to the science of global warming (Hurley 2000). Applied theater—embodied storytelling designed to provoke or shape responses to a specific social problem—offers visceral ways of communicating that highlight the need for emotion and creativity in addressing climate change (Ackroyd 2000; Thompson 2009). Equally important, applied drama recognizes and fosters forms of knowing often excluded from positivist paradigms that disqualify personal, affective, and communal forms of evidence. By reconfiguring what we accept as empirical evidence, applied theater represents a broadening of who counts as creators, inviting everyone to participate in knowledge generation (Brown et al. 2017; Denizen 2009; Prentki and Preston 2013). Our community-based plays begin with the idea that those who experience a problem must participate in addressing it.

Playmakers and audiences alike attest to theater's ability to entertain and educate the public, foster empathy, encourage creative and critical thinking, and transform individuals and communities. These are grand claims, to be sure, and there's considerable debate about what constitutes proof of performance's efficacy. On the one hand, performance offers a surfeit of evidence; it is simply of a variety—personal, ephemeral, and anecdotal—that the scientific community routinely dismisses. As such, artists do not wield the power to define the kinds of materials that count as evidence; nor can they easily validate the criteria, methods, and standards used to evaluate that evidence (Denizen 2009).

On the other hand, applied drama should be able to demonstrate its value in terms that a variety of stakeholders can understand. Metrics too often focus on the extrinsic value of theater, including economic benefits that accrue to communities through the arts (Brown and Ratzkin 2012; McCarthy et al. 2004; Reason and Rowe 2017). This instrumentalist view of performance is imbued with, and further reifies, market-based paradigms (often critiqued by applied theater advocates). Moreover, these assessments fail to consider the real benefits of live theater: the range of emotional and intellectual transformative experiences that affect participants and audiences engaged in the arts. It is this imprint, in the moment of performance, that can redistribute energy into social action and other areas of extrinsic value—in our case, empathic identification, behavioral modification, and community engagement around environmental issues. The question is, how do we translate the easily perceived yet difficult-to-define experience of live performance into something measurable, assuming one could capture, in any kind of holistic way, the entirety of an aesthetic experience?

In an attempt to bridge the methodological gap between performance research and the scientific method in assessing the impact of our climate change project, we use what Dwight Conquergood (2002, 151) calls "rare hybridity," a "co-mingling of analytic and artistic ways of knowing" that highlights the crossroads of the arts and sciences: creativity. Our aim was not to prioritize empirical design in making art; rather, we sought to demonstrate that aesthetic and empirical goals, while divergent, need not be entirely divorced in an applied theater setting. By drawing on participant impact surveys, interviews, and other assessment strategies embedded throughout the process, we discuss how applied theater can offer an alternative method for understanding what counts as empirical evidence while simultaneously engaging participants in the work of creating new knowledge.

Performing Ecoempirical Research

While many plays deal with environmental concerns, the "applied theatre sector has been relatively slow to pick up the ecological baton" (Bottoms 2010, 121). It is difficult to balance the provision of scientific data within an aesthetic structure, and many practitioners acknowledge a lack of scientific knowledge or a disinterest in scientific approaches, preferring to focus on human stories (Forgasz 2013). Audiences can be critical of, if not alienated by, work that is overly didactic and data driven. Last, it can be hard to harmonize socioecological grand narratives and personal experiential encounters in order to determine when and how to apply fictional frames to already dramatic real-life situations. This is particularly true in community-based experiments with participants who have little or no theater experience (Heddon and Mackey 2012).

Empirical studies documenting applied theater's effectiveness to influence environmental behavior are relatively sparse but have been slowly growing in the last decade. Audience surveys capture immediate individual responses; they do not address sustained behavioral transformation and long-term effects on communities (Davis and Tarrant 2014). Most after-show questionnaires are plagued by low response rates and lack of depth (Marshall 2005; Roose et al. 2002). More elaborate forms of assessment, such as longitudinal interviews, require significant investments of time and resources that most arts-based projects lack. Another key point is that audience members generally have not consented to participate in research studies (although it is possible to obtain such informed consent); therefore, we opted not to include illustrative audience surveys as contributing evidence of our project's efficacy.[2]

University-based projects like ours must acknowledge selection bias; our audiences and participants do not represent the general population, and this poses a threat to external validity. Any attempted curation of a representative sample, correcting for the liberal bias of Cornell University and the broader Ithaca community, would violate the inclusive mission of our community-based project. In other words, even if we actively recruited conservative-leaning participants for our show, they would likely be notably outnumbered by their progressive-leaning peers, and we wouldn't wish to exclude these latter participants just to generate a more balanced, more broadly representative sample. Further, theatrical collectives themselves can engender social desirability biases, insofar as participants might feel compelled to respond to survey and interview prompts in ways that will not jeopardize their feeling of belonging to a community with shared values.[3] We were able to mitigate this, at least to some extent, by asking our participants preproduction survey questions with no perceived community consensus and by asking broad questions that encouraged a multiplicity of responses that would not be discordant with other answers. Although we could have used other methods to reduce social desirability bias, as others have done (Nederhof 1985), this wasn't among our priorities.

Another issue difficult to tease out is the possible role that our impact studies had on the ensemble, especially given that we used multiple rounds of assessments. In addition to survey fatigue, there was the danger that any assessment that asks participants about the various ways an experience has affected them can actually prime the participant to notice, or even seek out, such experiences thereafter, thus potentially inflating later reported impact "gains."[4] One way we minimized this possibility was keeping our questions quite broad, and we generally refrained from providing examples that could inadvertently trigger such responses in future assessments.

The methodological differences between performance-based and scientific research are most clearly manifest in project aims and the different role repetition plays. In contrast to the scientific method, applied drama invites participants to immerse themselves in creative experiences, trusting that the processes will have impacts that cannot easily be predicted, defined in advance, or duplicated. As a general rule, artists privilege impulses over clearly delineated aims and welcome surprises that take them in unpredictable directions. As Michael Etherton and Tim Prentki (2006) observe, "Impact manifests itself in many forms, including the material, the physiological, the psychological, the social and the cultural" (140). Plays, if they are to be compelling and engaging (as opposed to didactic and prescriptive), should promiscuously court multiple, even conflicting, interpretations. This, hence, poses difficulty in establishing outcomes that might persuade social scientists of their empirical validity. Furthermore, in the sciences, empirical data are typically gathered by multiple investigators who independently replicate experiments.[5] Theatrical events, although iterative, can never be precisely duplicated. Performances vary night after night, and each audience is unique. Were our process to be repeated by another group, the results would likely diverge because the cast, audience, and location would be different, as would the exigencies of another community ensemble (NEF 2005). In fact, repetition with a difference is not only axiomatic to performance, it is also what constitutes theater's ability to inspire, model, and foster change. It is difference within repetition that interrupts our habitual ways of experiencing and perceiving—making the familiar strange and the strange familiar—that allows us to consider how things might be otherwise.

These limitations should not lead to the conclusion that applied theater is not a venue that could benefit from collecting and analyzing empirical data; rather, it requires humility in making generalizable claims while simultaneously proffering alternative standards for evaluation. In this vein, we devised a variety of assessment strategies to attempt to bridge this methodological gap. Over the course of our three project iterations, we surveyed participants using Likert scales, compiled written reflection exercises, recorded improvised skits by participants in both the university course and community sessions, conducted interviews lasting between twenty and thirty minutes with open-ended questions before, during, and after the project's completion, and took copious notes on discussions with participants during rehearsal as well as audience members during our postshow talkback sessions. Instead of maintaining consistency in data collection for each project, we utilized design thinking (Brown 1992) to guide our assessments; that is, we altered our surveys and interview questions on the basis of participant feedback in order to facilitate greater participant comfort in expressing what they learned and found most valuable throughout their engagement. We also attempted to be more systematic in our data collection through each iteration; thus, with each phase, we focused more on how we could triangulate our data to assess the extent to which qualitative and quantitative feedback might be aligned.

Scenes from the Anthropocene: *Climates of Change*

In 2007, Cornell theater professor Sara Warner collaborated with Godfrey L. Simmons Jr., founding artistic director of Civic Ensemble Theater, on a multiyear community-based course where students and residents of Tompkins County worked together to create plays about local effects of climate change.[6] Simmons and Warner (who served as the grant principle investigator, a member of the writing collective, and the producer) worked closely with Toby Ault, a Cornell climate scientist, and Sarah K. Chalmers, also of Civic Ensemble. The class—composed of thirteen students and seven community members—met for three hours every Friday to discuss readings, rehearse theater techniques, and devise the play from interviews and story circles with members of the public.[7] We invited anyone interested in climate change or community-based theater to join our ensemble, regardless of previous performance experience.

Our first play, *Climates of Change,* follows Zola Richards, a recent college graduate and newly minted chemical engineer from a small town in New York, where her family is one of the few farmers of color in the region. She returns to find that her homestead faces foreclosure after several poor growing seasons caused by extreme weather. Zola can save the farm if she accepts a lucrative job with a big oil firm, but at the expense of compromising her principles and her opposition to fossil fuel extraction and consumption. The play ends ambiguously, prompting audiences to imagine her future actions. Designed to foster both visceral empathy and critical distance, the finale encourages audiences to think about the ways their privilege, complacency, and complicity fuel climate change and environmental injustice.

We cast Zola with a student who initially self-identified as a climate skeptic. A devoutly religious person, she believed any changes in the climate were part of God's plan. As a first-generation student of color, she thought there were more pressing problems than climate change, including institutional racism, gun violence, and police brutality—views shared by other participants. The student playing Zola found the play imperative in demonstrating that combating environmental racism is an integral aspect of climate activism. Indeed, several people reported in their open responses that the most rewarding aspect of the process was our focus on climate justice. A Caribbean American student wrote that her favorite assignments were those that discussed "how poverty plays into climate change narratives because it seems like an obvious point yet it was one that [she] had not previously considered and one [she] doubts many others did either." A white female student highlighted readings that focused on communities of color "because they opened [her] eyes to social justice aspects of climate change that [she] had not previously considered." We also compared these qualitative comments with these respondents' Likert survey responses. The student playing Zola revealed that her knowledge of climate science increased from a 1 (little to no prior knowledge) to a 3 (a basic understanding) over the course of the project. The other two students also reported increases, from 3 to 4 (advanced knowledge) and 2 (minimal knowledge) to 3, respectively.

We assumed that students might have different motivations for joining the project than community members, but the facilitators were surprised to learn that climate change was not the driving force for either demographic. Eleven of the thirteen students wanted to work on a theater project and would have enrolled in the course regardless of the topic, and only two of the local residents identified themselves as climate activists. The majority of community members had worked on previous applied theater projects (not related to climate change) with our grant partner Civic Ensemble. Intake questionnaires and multiple cycles of evaluation (administered weekly in the form of individual and group reflections and postproject surveys) revealed two things. First, the initial motivation for participation for both demographics, students and community members, was not specifically climate change—our applied theater project's extrinsic goal. Rather, people joined in search of theater's intrinsic values: emotional resonance, aesthetic growth, intellectual stimulation, and social bonding (Brown and Ratzkin 2012). Second, motivations changed as the project developed. As participants learned more about the science of climate change, they rated taking immediate climate action in their personal lives and communicating with others about the dangers of climate change as increasingly important.

The class benefited from what David Diamond (2007) calls "intentional feedback loops," or embodied activities that promote behavioral change by fostering "a shared system of beliefs, explanations, and values" (16). Feedback loops empower participants to co-construct the theory of change guiding the applied theater experiment. Facilitators modified the devising and rehearsal processes in response. For example, participants repeatedly complained that climate change is "depressing," so we rewrote scenes to present the crisis as absurd rather than tragic. Productions with comedic elements frequently yield higher social outcomes, as laughing together foments and strengthens relationships (Brown and Ratzkin 2012). We created a character named Climate, the surly and petulant teenage son of Mother Nature, who, amped up on caffeinated beverages, wreaks havoc and mayhem on the world for sport. To enhance the effect, we made Climate invisible to most characters. We also incorporated a story line about the affective dimensions of climate change in humorous scenes personifying the five stages of climate grief (Running 2007).

Applied theater, much like applied physics, involves applying disciplinary techniques to real-world problems, with the aims of developing new or improved methods of addressing challenges (Ackroyd 2000; Freebody et al. 2018; Prentki and Preston 2013; Snyder-Young 2013; Thompson 2012). Our project used performance to stimulate emotional reactions in a safe space and encourage audiences to imagine alternative, nondystopic climate futures. In hopes of inspiring dialectical thinking, we developed the play in the style of a Living Newspaper, an interactive mode of applied theater. Originating in Russia and Germany at the turn of the twentieth century, Living Newspapers are fact-based dramatizations of current events that eschew conventions of commercial theater in favor of experimental, agitprop techniques such as audience participation, slide projections, and actors who break character to comment on the action. The form gained prominence in the United States with the Federal Theatre Project, part of President Roosevelt's New Deal during the Great Depression in the 1930s (Brown 1989). Because our plays considered the viability of a Green New Deal, we thought it fitting to revive this historical mode of performance.

The Living Newspaper format enabled us to incorporate interactive debates on climate science (with scenes of Zola's college class on "Energy"), which positioned audiences as students. Other scenes transformed spectators into attendees at a town hall meeting on windmill technology and as

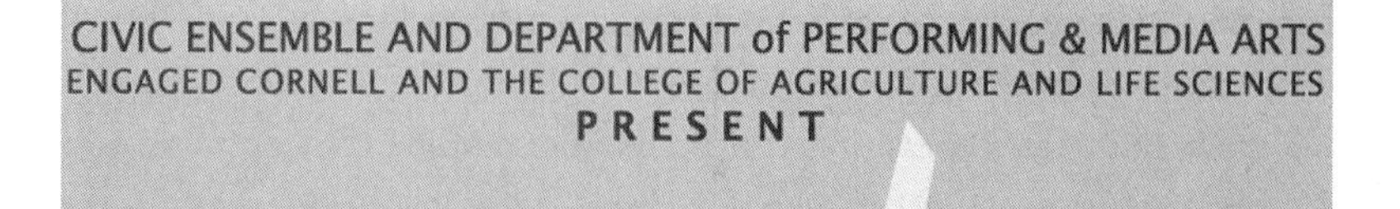

Figure 8.1. Poster for *Climates of Change: A Living Newspaper Play* (2017).

market patrons debating a plastic bag ban. These techniques that blur distinctions between spectators and actors foment audience engagement and foster collaborative problem solving.

Approximately 650 people attended the five free performances, but attendance alone is a poor indicator of a project's success. Better informal metrics are the number of people in the lobby discussing the performance with other patrons after the curtain closes and the variety of opinions voiced during after-show conversations (Schneider-Mayerson 2018). We hosted curated talkbacks after each performance, which drew sizable crowds[8] and provided an important opportunity for what Brown and Ratzkin (2012) term moments of curatorial insight, facilitated opportunities to help audiences make meaning from their aesthetic experiences. Our sessions, hosted by ensemble members trained in community dialogue, aimed to increase social connectivity, especially among people working at the intersection of arts and climate activism. These dialogues encourage critical thinking, a measure that previous research has shown correlates with positive impacts, including heightened levels of social cognition and changes in attitude (Brown and Ratzkin 2012).

Our postshow conversations began in pairs, moved to small groups, and then to a community-wide forum. In these sessions, audience comments tended to privilege theater's intrinsic values (most notably emotional resonance and intellectual stimulation) more than they referenced the extrinsic mission of sparking action to mitigate climate change. As one respondent proclaimed, "It has been an exceedingly rewarding experience for me at the intellectual level as well as the level of affective engagement with issues and with myself." We also perceived our audiences to be visibly and audibly moved by our performances, which was further evidenced by postshow compliments to production team members and participants alike.

From an impact perspective, being affected is what matters in theater, but this evidence is difficult to measure. Attempts to quantify performance's efficacy are generally plagued by analytical complications (Brown and Ratzkin 2012; Denizen 2009; McCarthy et al. 2004; Tompson 2009). Thus, our study has considerable limitations with regard to classic empirical validity, as objective, scientific proof of theater's impacts is difficult to ascertain. Moreover, gathering such evidence can involve tactics antithetical to the intrinsic and extrinsic values of applied theater. Because change is not an immediate or linear process, our assessments encouraged people to "think evaluatively" rather than simply evaluate a set of readings or embodied activities (Snyder-Young 2013). Nonetheless, we saw value in gathering empirical data not only for our assessment purposes but also because it would benefit some of our stakeholders, especially our climate science partner. This information seemed crucial for the project's sustainability, which requires external sources of support—ideally large, multiyear science grants. We therefore used a greater variety of impact assessments in hopes of more rigorously investigating what "succeeded" and "failed," and why—in our second applied theater project on climate change, *The Next Storm.*

The Waters Will Rise, and So Will We: *The Next Storm*

Our collective began work in 2019 on a second community-based play, *The Next Storm.* We welcomed some new members to Civic Ensemble Theater: Sage Clemenco and Julia Taylor, as well as new community members, including Jeremy Jimenez, associate professor from neighboring SUNY Cortland. Jimenez first heard about *The Next Storm* while participating in a Civic Ensemble story circle on racial justice; he enthusiastically embraced this opportunity for active participation, sensing that this play could be an ideal environment (a local community) and method (the arts) to contribute to more effective climate change communication and advocacy.[9]

Several differences characterized the second play that impeded our ability to compare the two productions. First, our second grant was significantly smaller, so we could not travel to different audiences. We staged the show at a single location as part of Cornell University's main-stage performance season. The two-week run totaled five performances. We benefited from the expertise of talented designers and their arsenal of technological wizardry, but the emphasis on professional production values pressured, and at times overshadowed, the community-based spirit of our applied theater project. Audiences paid a nominal fee for tickets, but we offered free admission through Civic Ensemble to offset financial barriers to attendance. To allow ourselves more time to develop the work for the mainstage, we extended the class from one semester to two. To increase community participation, we held twinned devising sessions, one on campus on Friday afternoons and

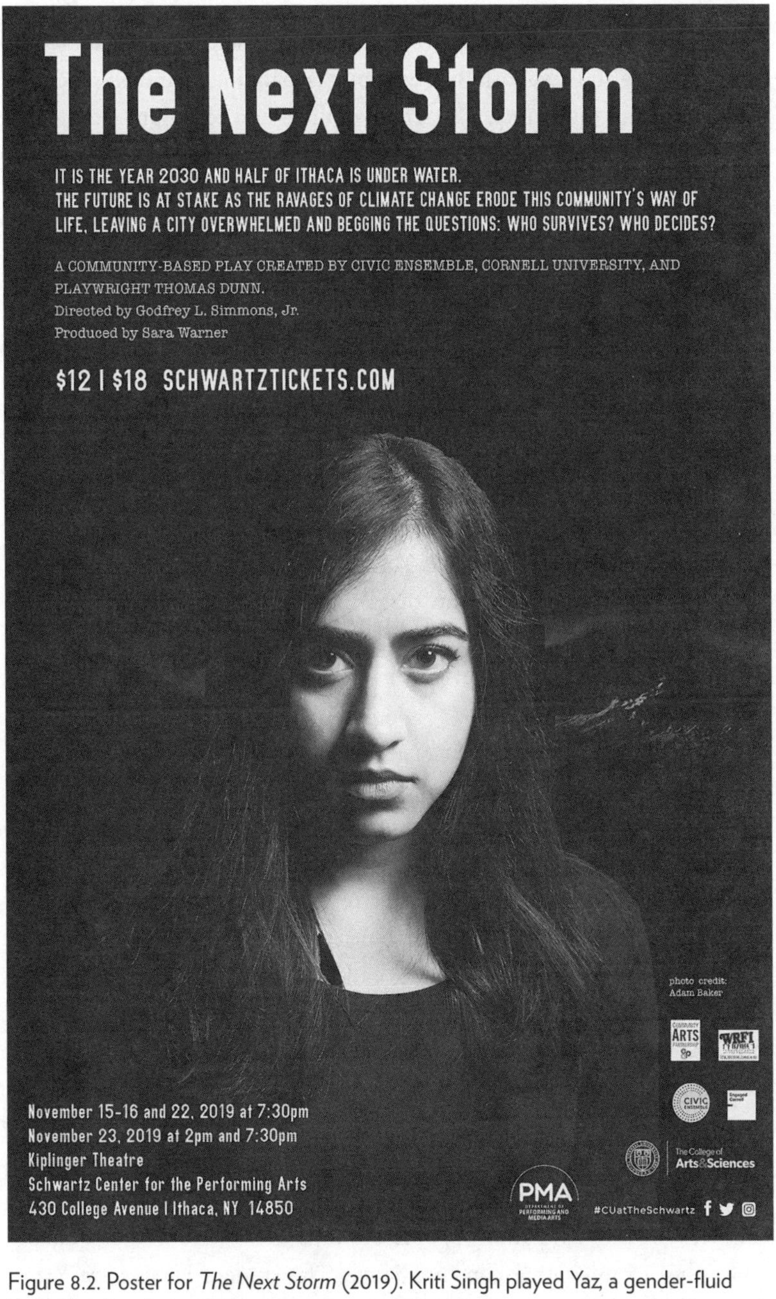

The Next Storm
IT IS THE YEAR 2030 AND HALF OF ITHACA IS UNDER WATER.
THE FUTURE IS AT STAKE AS THE RAVAGES OF CLIMATE CHANGE ERODE THIS COMMUNITY'S WAY OF LIFE, LEAVING A CITY OVERWHELMED AND BEGGING THE QUESTIONS: WHO SURVIVES? WHO DECIDES?
A COMMUNITY-BASED PLAY CREATED BY CIVIC ENSEMBLE, CORNELL UNIVERSITY, AND PLAYWRIGHT THOMAS DUNN.
Directed by Godfrey L. Simmons, Jr.
Produced by Sara Warner
$12 | $18 SCHWARTZTICKETS.COM
photo credit: Adam Baker
COMMUNITY ARTS PARTNERSHIP
WRFI
CIVIC ENSEMBLE
The College of Arts&Sciences
November 15-16 and 22, 2019 at 7:30pm
November 23, 2019 at 2pm and 7:30pm
Kiplinger Theatre
Schwartz Center for the Performing Arts
430 College Avenue | Ithaca, NY 14850
PMA
DEPARTMENT OF PERFORMING AND MEDIA ARTS
#CUatTheSchwartz

Figure 8.2. Poster for *The Next Storm* (2019). Kriti Singh played Yaz, a gender-fluid high school student and climate activist.

one on Thursday evenings at an alternative high school that emphasizes student choice and responsibility, with pupils planning their own schedules and helping run the school. The school doubles as a community center for Ithaca's racially and economically diverse West End.

A key difference between *The Next Storm* and *Climates of Change* was the script. In this second iteration, we synthesized our existing archive of interviews and story circles with data from the two 2019 sessions and commissioned a local dramatist and climate journalist, Thom Dunn, to help us devise our play. By setting the action ten years in the future, in 2030, Dunn created a speculative, cautionary world in which global warming had carried on more or less unabated. Caring about a future that may not affect us personally requires a level of empathy and engagement beyond mere factual understanding. One of the ways theater can contribute to climate action is focusing people's hearts and minds on the future. As psychologist Elke Weber notes, "Engaging people through the arts—not necessarily in ways that are alarmist and fear-provoking, but in ways that are thought-provoking—can remind them of their long-term goals. The arts can comment on abstract things—like the sustainability of planet Earth—so much better than statistics can" (Bilodeau 2013).

The Next Storm takes place in a partially flooded Ithaca, a city with a history of epic weather events. Residents debate constructing a hydroelectric dam and floatovoltaic energy grid to fuel an affordable housing complex for flood victims, the majority of whom live in a historically Black neighborhood.[10] Tensions mount when citizens realize the developer (an environmentalist turned libertarian) plans to block access to a beloved waterfall and demolish prized single-family homes in a white, affluent section of town. The conflict strains relationships and intensifies a generational rift when the students revolt against their well-meaning parents and city officials, whom they accuse of apathy and neglect. The final scene, inspired by local youth activists, involves a group of high schoolers disrupting the hydroelectric dam's ground-breaking ceremony. In a direct address to the audience (positioned as citizens at the celebration), the protestors ask spectators to stand in solidarity with young people and their demand for more equitable approaches to mitigating climate change. The student leaders are Yaz, a gender-fluid visual artist of color, and their best friend, Terry, a charismatic ciswhite male, who struggles to understand his social privilege and share the spotlight with the more introverted and economically disadvantaged Yaz. Yaz expresses millennials' collective feelings of frustration against the baby boomer generation, who are privileged to eschew radical action on climate change because they will likely be gone before it seriously affects them. It's "deeply unfair," Yaz charges, "that this [younger] generation [has] to deal with these major issues and their impacts when they had nothing to do with [creating them]."

Trying to mediate this generational divide are Eve (a Black elder, played by guest artist Rhodessa Jones), Alex (a lesbian activist), and Carson, a grieving climate scientist (played by Jimenez), who lost his fiancée in a hurricane relief mission.[11] The inclusion of Carson, an endearingly awkward character who attends a climate support group at a Unitarian church and talks through a sock puppet named Climate Dog when words fail him, provided moments of comic relief that went beyond traditional factual climate presentations by creating emotional +connections between scientists and publics. When participants and audiences identify with flawed yet lovable characters like Carson, there is greater incentive to learn, modify one's behavior, and encourage better habits in others. Jimenez bonded with Carson's character, which allowed him to unite his fascination with science and his passion for social justice.[12] As a result of his participation in the play, Jimenez became more motivated to incorporate topics into his classroom, as well as to more wholeheartedly pursue a research agenda focused on environmental justice; individually, he has since become a certified master composter and volunteer at community events, committed to rewilding his yard with native plants, and joined his town's environmental advisory council to advise how development projects can be more ecologically friendly. Supportive connections and participatory engagement in meaningful activities encourage people to invest their energies in personal growth and social transformation (Cahill 2018). Applied theater outputs better accomplish their extrinsic goals when they envision the capacity for change rather than pedantically expressing what should happen; we thus refrained from making Carson, our climate scientist, an undisputed authority figure.

Engaging a controversial topic like climate change, actively making room for diverse (even dissenting) viewpoints, and massaging this material into a story shared with the public requires a fair amount of consensus through compromise. This unruly aspect of applied theater is its source of inspiration, innovation, and knowledge creation (Thompson 2012). Script devising is an exercise in participatory democracy. While Dunn's script retained some aspects of a Living Newspaper—including instructional classroom scenes, audience participation at town hall events, and an ending designed to rally audiences to action—*The Next Storm* deviated from this fact-based form. In particular, the future setting of the play presents both a depressingly apocalyptic vision of the world and an optimistic reminder that life could be otherwise. Despite all the cynicism pointed at the social and moral failures sardonically dramatized in the script, the conclusion—youth protest—is designed so people leave the theater believing that the world of the future will belong to those with the resolve to face its challenges. This perspective reflects the shared vision of the ensemble, who insisted on a hopeful ending.

The Next Storm played to larger audiences than our first play—over 800 across five shows.[13] Because the show was part of Cornell's theatrical season, the rehearsal process was quite rigorous, with the cast assembling five nights a week for four hours, plus weekends, as opening night drew near. The intensity of this time commitment limited the number and range of community members who could participate; we experienced a sizable turnover between the fall and spring sessions. Only five people (two students and three community members) took part in both semesters. In the spring, a new group (ten students and four community members) joined the five continuing ensemble members to rehearse and mount the production. Collaboration is an easy goal and a difficult practice.

Like *Climates of Change, The Next Storm* actively eschewed the moralistic style of much environmentally engaged art. A play alone cannot teach audiences how to save the planet; nor is its best aim to provide a checklist of things people can do to ameliorate environmental impacts. Climate change is a global problem requiring more than just individual sacrifices. Rather than generating facts and figures, theater promotes deep affective and cognitive engagement with a problem in order to imagine how the world might be otherwise (Freebody et al. 2018). Our plays attempt to address what Deirdre Heddon and Sally Mackey (2012) note as missing in ecocritical approaches to applied theater, while heeding the words of Nicole Seymour (2018), who cautions against instrumentalist approaches to climate activism, as these replicate the "prescriptive tendencies of mainstream environmentalism" (28). We must not let our yearning for prompt changes distract us, Seymour advises, "from the real job of criticism": to explore "problems and make things messy rather than neatly resolving them" (28). Our play, while hopeful, privileged uncertainty—the uncertainty of forecasting the future, of scientific fact, of performance practices, and of civic ensembles—to engage us in the kind of critical thinking necessary to interrogate and interrupt social norms that may inhibit personal and political change. Exploring this affective–cognitive engagement will be a primary concern of the data analysis that follows.

Performing Ecocriticism: Survey Data

We were eager to assess the extent to which our applied theater project could serve educational purposes, given how few of our participants had taken classes in environmental issues or climate change, and how few students or community members identified as climate activists. As the project evolved, we added nuance to our assessment strategies and collected more empirical data for the second play. In addition to weekly reflections (individual written responses and collective discussions), we administered intake questionnaires and Likert scale surveys at midterm and the conclusion of class for both semesters. The response rate for the written assessments and interviews was 100 percent.[14]

Our five-point Likert scale asked participants to rate their knowledge about climate change both before and after the entire show experience, with 5 indicating participants were very knowledgeable about climate science and 1 indicating students had little to no prior knowledge.[15] All participants indicated an increased understanding of climate science through participation in the devising process.[16] Although our sample sizes are too small to draw robust conclusions, it is possible that as facilitators increased their own content knowledge with each iteration, they were able to provide greater learning opportunities for the ensemble (Shulman 1986).

As sources of knowledge acquisition, student participants rated interactions with community members particularly high (and vice versa).[17] We also asked participants to rate which aspects of the project were most useful in advancing their knowledge of climate change, surveying both traditional classroom materials as well as techniques specific to performance-based research. The options were: class readings; lectures by professors; participating in story circles and interviews; community workshops; independent research projects; weekly journal assignments; guest visits (artists, activists, and scientists rated individually); devising the script; rehearsals; acting in productions; and after-show conversations.

Community workshops included content creation through story circles and interviews (2017), followed by script devising sessions (2019). These almost universally positive rankings for performance-based methods stand in contrast with mixed rankings of other, more traditional sources of knowledge acquisition, namely lectures, readings, independent research, and journals (homework). Assessments measured immediate impacts, not long-term effects, and we have no way of knowing what ripple effects this experience will have on participants' lives. Impact assessments, although important in gauging theater's efficacy, seldom provide evidence of sustained transformation (Etherton and Prentki 2006). For us, more valuable insights into participants' learning processes emerged from instructor field notes and semistructured interviews that promoted interpersonal engagement. We turn to these next.

Empirical Insights: Participant Interviews

We conducted thirty-minute interviews with each member of the cast of *The Next Storm* during the final two weeks of our performances, with 87.5 percent of cast members (fourteen of sixteen) being willing and able to participate. The questions (listed in the appendix) assess participants' climate knowledge before and after the show, what they learned from the process, whether their work increased or decreased their anxiety about climate change, and what lifestyle changes they made, if any, as a result of this work.

While reviewing transcribed interviews (with Temi software), Warner and Jimenez utilized grounded theory (Charmaz 2005) to independently analyze what themes emerged from the data. We each generated our own codes, shared our codes with each other, and then discussed how to combine and organize our codes into primary themes and secondary subthemes. We identified the following:

- Assigned primary blame/responsibility to persons/groups for climate change.
- Experienced climate change induced fear/anxiety.
- Experienced perspective shifts (from participation).
- Assigned high value of applied theater for social change.
- Became a climate change advocate (in their social networks).
- Endorsed policy recommendations (mitigate climate change/advance environmental justice).
- Placed high value on cocreating new artistic work (as opposed to performing existing play).
- Experienced personal growth from embodying their role in the play.
- Reflected on their personal privileges.
- Raised the issue of hope.
- Emphasized the value of applied theater for personal development.

We then recoded our data and organized direct interview quotes according to these themes. Given the richness inherent in our participant interviews

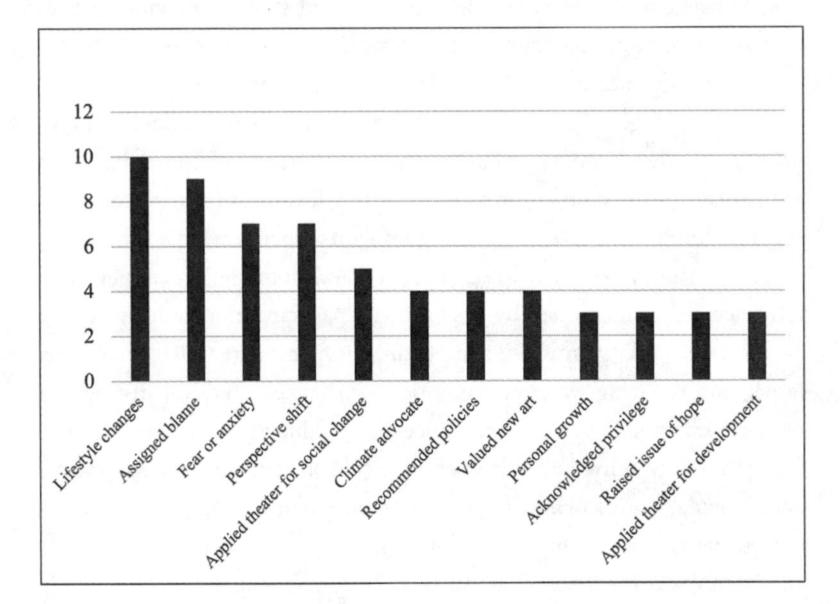

Figure 8.3. Themes from interviews.

after *The Next Storm,* the remainder of our analysis explores insights gleaned from this assessment.

In general, our interviews corroborated and elaborated on the survey results. The most common response involved the importance of interaction with community members in the learning process. One student described community engagement as "beautiful and transformative even when painful," which speaks to theater's efficacy in facilitating difficult dialogues. The person who performed the role of Yaz stated that she "really just enjoyed the community-based element . . . of getting to interact with people outside of [her] bubble of . . . people [her own] age with similar backgrounds," highlighting that she thinks this facet "added a lot to the show and to our ability to show different perspectives." Echoing this sentiment, another student said that this "community-based play allowed me to better understand . . . the intergenerational divide. . . . Being on a college campus full of young people [and] being with people that have young children or someone who is an elder in the community allowed me to encounter various perspectives that would be hard to grasp from news feeds or regular interviews."

Our participant interviews also corroborated studies showing that although research participants claimed a significant increase in climate science knowledge acquisition, they made few significant pro-environmental behavioral changes (Vermeir and Verbeke 2006). When asked what modifications they've made, our fourteen cast members indicated participation in the play convinced them to focus on eating less meat (28.6 percent), recycling more (28.6 percent), shutting off lights when leaving a room (21.4 percent), using less plastic (14.2 percent), and flying less (14.2 percent). Other changes, mentioned by only a single participant, include switching to a green electricity provider, composting, using more recyclable materials in art, and reducing water consumption. Participants did not distinguish between high individual impact choices (i.e., reducing meat consumption, flying) and low individual impact choices (i.e., shorter showers, turning off lights), and, besides flying, more often mentioned low-impact choices as the ecohabits they planned to adopt (Schneider-Mayerson 2018; Wynes and Nicholas 2017). These patterns seem to corroborate research that finds many people will commit to making minor green lifestyle changes in order to compensate for unsustainable behaviors they are keen to continue while neglecting to make more substantive changes (Sörqvist and Langeborg 2019).

One participant reflected on how difficult it is to commit to long-term habit changes. "Since doing this play," reflected one actor, "I've definitely thought maybe I should eat less meat . . . to adjust my personal impact, but not . . . lead a protest [which his character does in the play] or anything like that." Several members of the ensemble shared this ambivalence. "I normally keep the heat up," a student admitted. "Because of this play, I thought, let me lower the heat and put on some more layers [of clothing] and be a little better to the environment. But maybe an hour later, I turned it all the way back up," she confessed. "So in terms of the environment, I don't think I've changed my habits. But I think I have more conversations, like my mom was talking to me about buying plastic water bottles, and I tried to convince her not to." We might attribute students' ambivalence, which can sometimes border on hypocrisy, to their age, circumstances, or relative privilege as students of Cornell University. Community members voiced similar responses, though they often cited specific, more compelling, reasons for resisting substantive change: the lack of public transportation in rural areas, jobs requiring air travel, hectic lifestyles involving take-out meals, and, for some, a sense that as oppressed people of color, they have sacrificed enough. These participants think that those who have profited the most from ecologically harmful practices should be the ones to make major concessions.

Although we secretly hoped for converts to climate activism (and we can claim two!), we knew raising participant awareness was a much more likely outcome. Indeed, we found notable evidence of transformation in both their thinking and communicating about environmental issues. Stating that she did not dwell on climate change "until [she] started taking this class," one student described how the environment "went to the forefront of what I was thinking about in my everyday [life] and influencing my lifestyle. . . . Over the summer, a person came canvassing in my neighborhood back home and they asked, 'What is your number one issue that you were thinking about in the upcoming elections?' And I said, 'Climate change.' And I don't think I would've said that beforehand." The student discussed this response with her parents. Her African American mom identified police brutality and her white stepfather "keeping American jobs in the United States" as more important issues than climate change. This led the student to wonder:

> How does climate change influence all of those things? . . . If we're all gonna die in ten years, then race relations really don't matter that much, and . . . jobs might be obsolete because no one will be able to pay for the work. So I think this was a very helpful conversation, and I felt like I made progress with two people who I care very much about and I've learned a lot from. And I think starting small felt like a win, and there are small wins that I do every day.

It was encouraging that this student's assessment of the daunting immensity of the climate crisis ("we're all gonna die in ten years") didn't lead her to despair, as she found the motivation to pursue the "small wins," especially given that her doomsday clock projection of a decade ("ten years") mischaracterized the 2018 U.N. report's summary findings.[18]

In terms of cognitive impacts, almost all participants reported being open to learning more about climate change. One community member, who had avoided the topic "because the picture is so grim," indicated that the play's futuristic setting in the year 2030 made her "more inclined to learn more details now, because of raising [her] son who's only twelve, and . . . who knows what the next twelve years of his life will bring and beyond." Being a mother, she added, "You can't *not* be concerned" about the collapse of ecosystems around the world and the resulting potential impacts on our future well-being. Categorizing this awareness as both inspiring and exhausting, she confessed it made her realize the impact that even a single object—"paper cups on airplanes"—can have on the planet. The trash "is here and it's happening all over this country, all over this world. And imagine in one single day how many of these are being thrown away or being released into the environment. . . . I have many multiple moments like this throughout the day, of [being] overwhelmed." While such insight might be engendered by a documentary or news report, she, along with many other participants, discussed how they had "no time" to learn more about climate change on their own. In this way, community participants more drawn to theater than science can be immersed in an environment that encourages knowledge acquisition.

As the director of a study abroad program at a local college, this participant deals extensively with student travel. Airplanes were a recurrent topic of conversation among the cast, many of whom (ourselves included) are frequent flyers. Our play helped educate the ensemble about greenhouse gas emissions and their personal carbon footprints. As one student who takes several transatlantic flights per year opined, "I never understood what the whole carbon emissions was about . . . how people traded carbon . . . what is it . . . credits? I sort of have an understanding of it now and . . . you know, we all want to travel all over the world, but you feel guilty, and you think . . . should I be flying home in December? . . . Is it really necessary?"

In addition to these important cognitive shifts, many participants reported being much more likely to debate environmental issues with friends and family. One participant said the play prompted her to regularly discuss environmental issues with her roommate; the two enrolled in a climate change class the following semester, which better equipped them for further advocacy. A freshman chorus member, who is "very good friends" with another actor in the play, exhibited the most profound transformation in awareness. At the beginning of the process, he recalls, "we didn't care about . . . climate change, but working through the project both of us have realized that it's a serious issue. . . . Now, when we meet with our other friends, we tell them about the play, about climate change, and that they need to talk about these things. It's not something other people will solve. We are the ones who have to solve it." Yet this participant didn't report taking any significant actions as a result of his newfound awareness. This gap between the participant's recognition of the severity of the climate crisis and their willingness to make radical and immediate changes to their lifestyles plagued not only our cast but also public perception more broadly (Wang et al. 2019).

Applied theater seeks to make ecological issues less psychologically distant by fostering emotional connections and empathic identifications that encourage pro-environmental behaviors, including advocacy. One student reported becoming something of an eco–police officer who pesters friends with comments like this: "Why didn't you recycle that? That's really bad . . . or, why did you do this specific thing, and . . . I feel like I'm crazy. It's almost as if they don't acknowledge it at all. . . . My friends will say, 'What are you talking about? . . . What's up with you and the environment?' And I'm like,

'What's up with you and *not* the environment?'" While advocacy sometimes took this form of somewhat panicked militancy, most participants expressed more muted sentiments. More typical was a general awakening, as evidenced by one student for whom climate change is "getting a little more alarming," due in large part to "having done . . . [extensive] research for the play." In addition to "starting to think about [global warming] outside of the play," he's "also speaking to people a lot about it as well."

As theater raised awareness, it made most ensemble members anxious about the future; some found they could mitigate this anxiety by pushing for collective action. "Before this play, I didn't feel the urgency as much; it sort of was on the back burner," shared one actor. "But now, I think about it every day. This is scary and this is real and it's not ten years from now—it's now, and we need all hands on deck." Research has found, however, that merely equipping people with knowledge concerning the severity of the crisis is not enough to motivate lifestyle changes and political advocacy (Leiserowitz, Maibach, and Roser-Renouf 2009). In fact, several interviews indicate that increased environmental awareness actually made them less inclined to pursue climate-friendly actions. As one actor lamented, "It takes some time to accept that what we've been told . . . we need to do . . . is just not enough." Another participant felt similarly disempowered from their newly acquired awareness: "I would like to feel empowered [and believe] that things could change, like . . . the climate protesters [in our play] work to stop the falls project . . . but then on a global scale, it's really hard for us to convince Washington to stop funding [fossil fuel] projects, as well as other countries and companies overseas. It doesn't really feel like we have much say in any of that." Indeed, most cast members reported feeling simultaneously empowered by the knowledge gained from working on the play and disempowered to actually make a difference. As one actor noted:

> I'm more empowered because I know that there's different ways to handle the problem now, and more that I could personally do. But at the same time, I think . . . there is a constant disempowering that occurs when you think that here we're making a piece of theater that will touch people, but there are wildfires happening right now, making people move. So it feels like I'm doing what I can right now, but is the scale big enough to make the changes necessary? And is that my responsibility, in any case?

This preoccupation with personal responsibility, shared by most participants, is the bedrock on which climate awareness, action, and resilience are built. Applied theater projects are not designed to instrumentally teach the science of climate change, or any other topic for that matter, but to provide an embodied means of exploration, questioning, and collaborative problem solving (Heras and Tàbara 2014). In many ways, our plays were rehearsals for the difficulty and discomfort of making decisions in the midst of uncertainty. As one ensemble member noted:

> Either we take action or just sit back and let things happen to us. I don't think anyone wants that to be the case . . . so this play reminds us that this is the time to step up. Before, climate change intimidated me. But those discussions that we had in the beginning . . . really opened my eyes. . . . You can see the genuine fear that it puts in people. But I don't think it has made me paralyzed. And I don't think anyone here feels that way. I think we're doing this because we think there's still room for hope, that it's not too late.

This comment encapsulates common sentiments expressed throughout our entire project. The next storm is coming, but the future has not been decided—not yet. There are still windows of opportunity to mitigate the future impacts of our climate crisis. Information and empathy are essential in promoting climates of change. Through compelling stories and characters, we offer alternative perspectives and visions of what could be. Drawing on legacies of intersectional social movements, we join activists such as Madonna Thunder Hawk (2022), Elizabeth Yeampierre (2018), and Rebecca Solnit (2021) in maintaining that hope is the contingent foundation for transformational change and collective action on the scale required to address the climate crisis.

Applied theater offers visceral and impactful ways of generating and communicating scientific data that highlight the need for creativity and innovation in addressing climate change. Equally important, it recognizes personal, affective, and communal forms of knowing as valid and valuable research perspectives. *Climates of Change* and *The Next Storm* helped individuals discover, test, reassess, and revise their perspectives through community interaction. In the devising process, members collected, analyzed, and shaped knowledge, which provided participants with an informed basis for recommending change, abetted by an enhanced sense of solidarity and mutual trust among members. By exploring fictional realities in the safety of the performance space, participants in *Climates of Change* and *The Next Storm* show that our future is not written in stone; we can rehearse and enact alternative paths forward.

Appendix

We conducted interviews, approximately thirty minutes in length, with participants of *The Next Storm,* during the final two weeks of our performances, with 87.5 percent of cast members (fourteen of sixteen) being able and willing to take part. The following is a list of our interview questions:

1. If you had to summarize the main message—or messages—of this play that you expect an audience member to receive, what would that be?
2. How concerned about climate change were you before taking part in this play compared to now?
3. What, specifically, did you learn about climate change from working on the play?
4. To what extent has this play increased or decreased your anxiety about climate change?
5. To what extent has this play empowered or disempowered you? And to what extent did this have to do with the character you played?
6. To what extent have you changed any of your daily habits since beginning this play, and how confident are you that you'll retain these changes three to six months after this play is finished?
7. To what extent did the play inspire you to talk with friends, family, or colleagues about climate change?
8. What does climate justice mean to you and to what extent is this a play about climate justice?
9. Consider your privilege and marginalization. What are the various ways in which you've relatively privileged compared to others, and what are ways in which you're relatively marginalized? Does this affect the way you think about climate change?
10. To what extent do you think this play is an accurate description of Ithaca 2030? In what ways do you think it is, and in what ways was it not?

Notes

1. More information is available at our multimedia archive (https://climateplays.com/).

2. Our initial survey was designed by Lissette Lorenz, a Cornell graduate student in science and technology studies with a history of creating socially engaged community-based performance.

3. This could have been further exacerbated by the fact that, in the interest of triangulating survey and interview data, participant surveys were not always anonymous; in retrospect, it would have been better to assign identifier numbers to surveys and interviews so we could triangulate our data while better mitigating social desirability bias.

4. This could have been better addressed with dummy questions—that is, questions that distract the participants from realizing the main purpose of the research, which might lead them to answer in ways they perceive would be satisfactory to the researchers.

5. This would be difficult for our project because we had a rather small sample of performance nights to analyze. Although having a larger sample size would have helped to mitigate the variability inherent in our limited data set of five public performances, our show—like nearly all nonprofit productions—lacked the resources to produce enough performances to meet such empirical standards. Instead of increasing the number of show performances, however, we could have increased the number of participants (i.e., making our participant sample more robust by minimizing potential outlier variability). However, what might have been gained by a larger number of participants likely would have been offset by the intangible loss of personal intimacy in building a sense of community, which is an essential component of applied community theater experiments.

6. Our projects were funded by grants from the Engaged Cornell initiative to promote public scholarship.

7. We used Roadside Theater's model of story circles (https://roadside.org/program/story-circles) to generate dialogues about climate change and climate justice that became content for our plays.

8. We forgot to take a formal count on some evenings and cannot provide specific numbers. We got swept up in the artistic experience, a testament to the intrinsic impact of the live theater experience.

9. Similarly, Jimenez thinks that applied theater will be an essential asset for smaller communities to survive, even thrive, in a postcarbon future.

10. Centering people of color in our conversation was "huge," in the words of one student. Acknowledging that the "stuff Greta [Thunberg] is doing is amazing," this student highlighted the "many young people of color who were speaking this way before and who don't get this space, and who don't get invited to the TV shows," adding, "This makes me really mad." We discussed local indigenous concerns through an

exploration of the Clinton–Sullivan massacre of 1779, which took place in our community. In addition to centering BIPOC concerns, we positioned women, queer, and gender-nonconforming voices in prominent roles.

11. Rhodessa Jones, a venerated applied theater practitioner, is cofounder of the San Francisco–based Medea Project: Theater for Incarcerated Women. Jones appeared as a guest artist in *The Next Storm* in conjunction with her appointment as a Frank H. T. Rhodes Class of '56 visiting professor at Cornell.

12. Early in his academic career, Jeremy pivoted away from his initial chemical engineering undergraduate major to social science fields that seemed better suited to social advocacy. Throughout his doctorate studies, however, he had the opportunity to learn in depth about climate change from many Stanford professors and fellow graduate students—a topic he became increasingly interested in and concerned with (some friends might say "obsessed"). As he came to understand how climate change and other environmental issues disproportionately and adversely affect marginalized groups (such as women, people of color, and residents of the Global South), he increasingly has been keen to make environmental justice issues among his top academic priorities.

13. We do not have a precise head count partly because of our free ticket policy for community members.

14. Survey results include participants who stayed throughout an entire phase, excluding the perspectives of community members who periodically attended public sessions on a drop-in basis.

15. We used evaluations from story circles in fall 2017 and play-devising sessions in spring 2019 and fall 2019.

16. While the averaged Likert rating gains were somewhat modest, the overall learning gains increased over time, both concerning the before-and-after scores within each group as well as total averaged ratings after each of our three community session iterations. In the fall 2017 story circles, the participants' averages were 3.22 before and 3.89 after (SD = .629 before and .567 after). In the spring meeting of our second play, the participants' averages were 2.83 before and 4 after (SD = 1.34 before and .816 after). In our fall meetings of our second play, the participants' averages were 3.73 before and 4.23 after (SD = .422 before and .512 after).

17. We found an average Likert score of 4.417 (SD = .837) for the spring 2019 meetings and 4.73 (SD = .442) for the fall 2019 meetings.

18. The report indicated that in twelve years, it likely would be too late to prevent particularly debilitating climate-related impacts, perhaps endangering "hundreds of millions of people" (Watts 2018) but not "killing everyone."

References

Ackroyd, Judith. 2000. "Applied Theatre: Problems and Possibilities." *Applied Theatre Researcher* 1: article 1. https://www.intellectbooks.com/asset/755/atr-1.1-ackroyd.pdf.

Bilodeau, Chantal. 2013. "How Can Scientists Raise Awareness about Climate Change? Try Theatre." *American Theatre,* May 1, 2013. https://www.americantheatre.org/2013/05/01/how-can-scientists-raise-awareness-about-climate-change-try-theatre/.

Bottoms, Stephen. 2010. Review of *Theatre Ecology: Environments and Performance Events,* by Baz Kershaw. *Research in Drama Education: The Journal of Applied Theatre and Performance* 15 (1): 121–26.

Brown, Ann L. 1992. "Design Experiments: Theoretical and Methodological Challenges in Creating Complex Interventions in Classroom Settings." *Journal of the Learning Sciences* 2 (2): 141–78.

Brown, Dan, and Rebecca Ratzkin. 2012. "Measuring the Impact of Live Theatre." In *Counting New Beans: Intrinsic Impact and the Value of Art,* edited by Clayton Lord, 65–164. San Francisco: Theatre Bay Area.

Brown, Katrina, Natalia Eernstman, Alexander R. Huke, and Nick Reding. 2017. "The Drama of Resilience: Learning, Doing, and Sharing for Sustainability." *Ecology and Society* 22 (2): 8.

Brown, Lorraine. 1989. *"Liberty Deferred" and Other Living Newspapers of the 1930s Federal Theatre Project.* Fairfax, Va.: George Mason University Press.

Cahill, Helen. 2018. "Evaluation and the Theory of Change." In *Applied Theatre: Understanding Change,* edited by Kelly Freebody, Michael Balfour, Michael Finneran, and Michael Anderson, 173–86. New York: Springer.

Charmaz, Kathy. 2005. "Grounded Theory in the 21st Century: Applications for Advancing Social Justice Studies." In *The Sage Handbook of Qualitative Research,* edited by Norman K. Denzin, Yvonne Lincoln, and Yvonna S. Lincoln, 507–35. Thousand Oaks, Calif.: Sage.

Conquergood, Dwight. 2002. "Performance Studies: Interventions and Radical Research." *TDR* 46 (2): 145–52.

Dahlstrom, Michael F. 2014. "Using Narratives and Storytelling to Communicate Science with Nonexpert Audiences." *Proceedings of the National Academies of Science of the United States of America* 111 (supplement 4): 13614–20.

Davis, Susan, and Michael Tarrant. 2014. "Environmentalism, Stories, and Science: Exploring Applied Theatre Processes for Sustainability Education." *Research in Drama Education: The Journal of Applied Theatre and Performance* 19 (2): 1–5.

Denizen, Norman K. 2009. "The Elephant in the Living Room, or Extending the Conversation about the Politics of Evidence." *Qualitative Research* 9 (2): 139–60.

Diamond, David. 2007. *Theatre for Living: The Art and Science of Community-Based Dialogue.* London: Trafford.

Ejelöv, Emma, André Hansla, Magnus Bergquist, and Andreas Nilsson. 2018. "Regulating Emotional Responses to Climate Change—A Construal Level Perspective." *Frontiers in Psychology* 9: 629.

Etherton, Michael, and Tim Prentki. 2006. "Drama for Change? Prove It! Impact Assessment in Applied Theatre." *Research in Drama Education: The Journal of Applied Theatre and Performance* 11 (2): 139–55.

Forgasz, Rachel Regina. 2013. "Response to the Themed Issue: Environmentalism." *Research in Drama Education: The Journal of Applied Theatre and Performance* 18 (3): 324–28.

Freebody, Kelly, Michael Finneran, Michael Balfour, and Michael Anderson. 2018. "What Is Applied Theatre Good For? Exploring the Notions of Success, Intent, and Impact." In *Applied Theatre: Understanding Change,* edited by Kelly Freebody, Michael Finneran, Michael Balfour, and Michael Anderson, 1–17. New York: Springer.

Gottschall, Jonathan. 2012. *The Storytelling Animal: How Stories Make Us Human.* New York: Houghton Mifflin Harcourt.

Heddon, Deirdre, and Sally Mackey. 2012. "Environmentalism, Performance and Applications: Uncertainties and Emancipations." *Research in Drama Education: The Journal of Applied Theatre and Performance* 17 (2): 163–92.

Heras, María, and J. David Tàbara. 2014. "Let's Play Transformations! Performative Methods for Sustainability." *Sustainability Science* 9 (3): 379–98.

Hurley, Erin. 2000. *Theatre and Feeling.* New York: Palgrave Macmillan.

Johnson, Dan R. 2012. "Transportation into a Story Increases Empathy, Prosocial Behavior, and Perceptual Bias toward Fearful Expressions." *Personality and Individual Differences* 52 (2): 150–55.

Lehman, Betsy, Jessica Thompson, Shawn Davis, and Joshua M. Carlson. 2019. "Affective Images of Climate Change." *Frontiers in Psychology* 10: 960.

Leiserowitz, Anthony, Edward Maibach, and Connie Roser-Renouf. 2009. *Global Warming's Six Americas, 2009.* Yale Program on Climate Change Communication, May 20, 2009. https://climatecommunication.yale.edu/publications/global-warmings-six-americas-2009/.

Marshall, Gill. 2005. "The Purpose, Design and Administration of a Questionnaire for Data Collection." *Radiography* 11 (2): 131–36.

Marvel, Kate. 2017. "We Should Never Have Called It Earth." On Being (blog), August 1, 2017. https://onbeing.org/blog/kate-marvel-we-should-never-have-called-it-earth/.

McCarthy, Kevin F., Elizabeth H. Ondaatje, Laura Zakaras, and Arthur Brooks. 2004. *Gifts of the Muse: Reframing the Debate about the Benefits of the Arts.* New York: Rand Corporation.

Nederhof, Anton J. 1985. "Methods of Coping with Social Desirability Bias: A Review." *European Journal of Social Psychology* 15 (3): 263–80.

NEF (New Economics Foundation). 2005. "Capturing the Audience Experience: A Handbook for the Theatre." London: ITC, SOLT, and TMA.

Prentki, Tim, and Sheila Preston. 2013. *The Applied Theatre Reader.* New York: Routledge.

Reason, Matthew, and Nick Rowe. 2017. *Applied Practice: Evidence and Impact in Theatre, Music, and Art.* New York: Methuen.

Roose, Henk, Daniëlle De Lange, Filip Agneessens, and Hans Waege. 2002. "Theatre Audience on Stage: Three Experiments Analysing the Effects of Survey Design Features on Survey Response in Audience Research." *Marketing Bulletin* 13 (1): 1–11.

Running, Steve. 2007. "The Five Stages of Climate Grief." *Numerical Terradynamic Simulation Group Publications* 173: 1–2. https://scholarworks.umt.edu/ntsg_pubs/173/.

Schneider-Mayerson, Matthew. 2018. "The Influence of Climate Fiction: An Empirical Survey of Readers." *Environmental Humanities* 10 (2): 473–500.

Seymour, Nicole. 2018. *Bad Environmentalism: Irony and Irreverence in the Ecological Age*. Minneapolis: University of Minnesota Press.

Shulman, Lee S. 1986. "Those Who Understand: Knowledge Growth in Teaching." *Educational Researcher* 15 (2): 4–14.

Snyder-Young, Dani. 2013. *The Theatre of Good Intentions: Challenges and Hopes for Theatre and Social Change*. New York: Palgrave.

Solnit, Rebecca. 2021. "Ten Ways to Confront Climate Change without Losing Hope." *Guardian,* November 18, 2021. https://www.theguardian.com/environment/2021/nov/18/ten-ways-confront-climate-crisis-without-losing-hope-rebecca-solnit-reconstruction-after-covid.

Sörqvist, Patrik, and Linda Langeborg. 2019. "Why People Harm the Environment Although They Try to Treat It Well: An Evolutionary-Cognitive Perspective on Climate Compensation." *Frontiers in Psychology* 10: 348.

Thompson, James. 2009. *Performance Affects: Applied Theatre and the End of Effect*. London: Palgrave.

Thompson, James. 2012. *Applied Theatre: Bewilderment and Beyond*. New York: Peter Lang.

Thunder Hawk, Madonna. 2022. "Indigenous Women Leaders Podcast: An Interview with Lakota Elder Madonna Thunder Hawk." Conducted by Mary Kim Titla. *Women's eNews,* January 20, 2022. https://womensenews.org/2022/01/indigenous-women-leaders-podcast-an-interview-with-lakota-elder-madonna-thunder-hawk/.

Vermeir, Iris, and Wim Verbeke. 2006. "Sustainable Food Consumption: Exploring the Consumer 'Attitude–Behavioral Intention' Gap." *Journal of Agricultural and Environmental Ethics* 19: 169–94.

Wang, Susie, Mark J. Hurlstone, Zoe Leviston, Iain Walker, and Carmen Lawrence. 2019. "Climate Change from a Distance: An Analysis of Construal Level and Psychological Distance from Climate Change." *Frontiers in Psychology* 10: 230.

Watts, Jonathan. 2018. "We Have 12 Years to Limit Climate Change Catastrophe, Warns U.N." *Guardian,* October 8, 2018. https://www.theguardian.com/environment/2018/oct/08/global-warming-must-not-exceed-15c-warns-landmark-un-report.

Wynes, Seth, and Kimberly A. Nicholas. 2017. "The Climate Mitigation Gap: Education and Government Recommendations Miss the Most Effective Individual Actions." *Environmental Research Letters* 12 (7): 074024.

Yeampierre, Elizabeth. 2018. "Capitalism Is Going to Kill the Planet: An Interview with Elizabeth Yeampierre, Executive Director of UPROSE Brooklyn, on Climate Justice and Hurricane Recovery in Puerto Rico and the Diaspora." Conducted by Ricardo Gabriel. *NACLA Report on the Americas* 50 (2): 170–76.

Chapter 9

Tracing the Language of Ecocriticism

Insights from an Automated Text Analysis of ISLE: Interdisciplinary Studies in Literature and Environment

SCOTT SLOVIC AND DAVID M. MARKOWITZ

The discipline of ecocriticism (or ecological literary studies) can be traced back to the earliest commentaries on natural themes in human texts, such as conversations about paintings of animals on the walls of European caves or discussions about how ancient songs invoked rain to bring crops to life. In *A Century of Early Ecocriticism,* David Mazel (2001) argues that some of the earliest examples of academic ecocritical studies of literature began in the 1860s with commentaries on works focused on environmental topics, such as the writings of John and William Bartram and Henry David Thoreau. He follows a trajectory of "early ecocriticism" from Henry T. Tuckerman's *America and Her Commentators: With a Critical Sketch of Travel in the United States* in 1864 to the publication of Leo Marx's landmark monograph, *The Machine in the Garden,* in 1964. The term "ecocriticism" was first used in 1978 when William Rueckert published "Literature and Ecology: An Experiment in Ecocriticism" in the *Iowa Review,* suggesting that ecological patterns, such as energy cycles, might exist in the very language of literary texts. Nearly two decades later, in 1996, Cheryll Glotfelty and Harold Fromm sought to institutionalize the field by publishing *The Ecocriticism Reader: Landmarks in Literary Ecology.* In her introduction to the collection, Glotfelty (1996) defines ecocriticism as "the study of the relationship between literature and the physical environment" (xviii). She suggests that ecocritics seek to explore such questions as, "How is nature represented in this sonnet?" and "In what ways and to what effect is the environmental crisis seeping into contemporary literature and popular culture?" (xviii–xix).

By the time Glotfelty and Fromm published *The Ecocriticism Reader* in 1996, Patrick D. Murphy had already created what would come to be the central scholarly journal in the field of ecocriticism, aspiring to help "ecocritics . . . demonstrate the validity and value of their critical activities," (2003, vii). Murphy launched *ISLE: Interdisciplinary Studies in Literature and Environment* in 1992, in time for the title of the journal to be used as the basis for the name of the nascent Association for the Study of Literature and Environment (ASLE), which was founded in October 1992. The first issues of *ISLE* began to appear in 1993, and the biannual publication of the journal (Winter and Summer) became consistent when ASLE adopted *ISLE* as its official journal in July 1995, appointing the organization's founding president, Scott Slovic, as the journal's new editor in chief.[1]

Although Slovic's approach as an editor and organizer in the field of ecocriticism and environmental writing has long been associated with a big-tent openness toward a wide variety of approaches and concerns (Slovic 2000), his research has been engaged with empirical studies of the efficacy of environmental communication since his doctoral work on nature writing and environmental awareness in the late 1980s (Slovic 1992). This early work, which relies on theories of awareness from psychologists ranging from William James to Stephen and Rachel Kaplan, can be considered a kind of second-order empirical ecocriticism in that it applies empirically developed psychological theories to the assessment of the potential impact of environmental texts on readers rather than conducting new (first order) studies of textual impacts. In 1991, Scott Slovic and his father, psychologist Paul Slovic, designed a study that used four questionnaires in a sophomore literature survey course in an attempt to determine students' attitudes toward nature before and after exposure to an assortment of environmental texts during the semester. Although the results of the study remain unpublished, this approach to actively gathering new data about audience responses to textual stimuli was a prototype for some of the current empirical studies, which use methodologies from the social sciences to design prompts to trigger audience responses and survey instruments to gather data about those responses. The current study, however, represents a different approach to empirical ecocriticism: a text analysis of a large body of environmental writing (both scholarly and literary) that appeared in the journal *ISLE*.

The text analysis approach, sometimes referred to as corpus linguistics research, involves the analysis of large amounts of text with the aid of computers (Boyd 2017). Computational linguistics dates back to the 1950s and has become an important subfield within linguistics. As Gena R. Bennett (2010) explains in her overview of corpus linguistics, a "corpus" is a large body of "naturally occurring examples of language stored electronically" (2). Essential to this "corpus approach" is the focus on "authentic language" that occurs in a "real-life situation," such as literary works, class lectures, newspapers, TV shows—or, in this case, material published in a scholarly journal (7). Computer analysis of a particular corpus can quickly discern patterns associated with lexical (vocabulary) or grammatical (syntactic or stylistic) aspects of the collected language. The computer analysis enables researchers to identify what patterns exist in a given corpus, but it does not explain why such patterns occur. Such patterns are still interesting and important because they serve as indicators of social, psychological, or institutional processes that might be difficult to observe at a macro level—for example, how a field of scholarship changes over time. Computational linguistics and digital humanities research can use largely similar automated means to achieve different ends. Computational linguistics research, including the results reported in this chapter, often consider words to be indicators of authors' attention (Boyd and Schwartz 2021). Researchers therefore seek to identify one's psychological focus through language patterns and infer one's internal state. Analyses of language, images, or other artifacts in the digital humanities are not typically psychological in nature and tend to reveal more descriptive patterns of how such cultural products are used. Therefore, computational linguistics traditionally focuses on language as a lens into psychological processes, whereas digital humanities research, by default, does not.

For this study, we selected a large sample of language (N = 713 texts) associated with a specific scholarly discipline (ecocriticism) and subjected it to a particular computational tool, Linguistic Inquiry and Word Count, or LIWC (Pennebaker et al. 2015b), to evaluate the writing style patterns of articles within the corpus. We also contextualized the linguistic patterns identified by the computer analysis by placing them within the history of ecocriticism. In this way, we can offer a snapshot of the discipline and some of its key trajectories through automated text analyses. While some corpora are generalized and may contain millions of words, our particular focus here is on a specialized corpus of one journal within a particular scholarly discipline.

In the next section, we provide a relevant historical context for understanding the development of modern ecocriticism from roughly 1980 to the present. The fifteen-year time period (2004–18) covered by our empirical analysis of the language published in *ISLE,* as will be explained below, encompasses a particularly dramatic period of transition in the discipline of ecocriticism and provides a rather wide lens for viewing linguistic shifts, which is why we have focused on these journal issues in our study.

Fives Waves of Ecocriticism

Ecocriticism has evolved since the creation of ASLE and *ISLE,* moving through successive phases that became known as waves as a result of Lawrence Buell's (2005) description of first- and second-wave ecocriticism in *The Future of Environmental Criticism.* While Glotfelty (1996) emphasizes the study of natural imagery in poetry and broad responses to themes of environmental crisis as represented in cultural texts (e.g., exploring how ecological science influences literary studies), T. V. Reed (2002) seeks to refocus ecocritical attention on big issues, asking questions like "How can literature and criticism further efforts of the environmental justice movement to bring attention to ways in which environmental degradation and hazards unequally affect poor people and people of color?" and "How can ecocriticism encourage justice and sustainable development in the so-called Third World?" (149). Joni Adamson and Scott Slovic (2009) identify the existence of three discernable waves of ecocriticism. In his editor's notes, Slovic comments on a new fourth wave (Slovic 2012), later mentioning a possible fifth wave (2019). The key waves of ecocriticism can be described as follows:

First wave: c. 1980–present. An emphasis is placed on nonhuman nature (including wilderness) and on American and British literature, particularly nonfiction nature writing.

Second wave: c. 1995–present. The field expands to encompass studies of multiple literary genres, multicultural perspectives, and urban and suburban contexts in addition to rural and wild locations. Environmental justice ecocriticism begins to emerge.

Third wave: c. 2000–present. Comparative approaches across national and ethnic cultures begin to appear in ecocriticism. This wave also recognizes the increasing emphasis on tensions between local and global approaches to the concept of place, various studies of gender and environment (ecomasculinity, green queer theory), various theories of animality (including animal subjectivity/agency and posthumanism), and strong critiques of ecocriticism from scholars working within the field, such as Dana Phillips (2003) and Michael P. Cohen (2004).

Fourth wave: c. 2008–present. The key feature of the fourth wave is the rise of material ecocriticism, the result of the publication of Stacy Alaimo's "Trans-corporeal Feminisms and the Ethical Space of Nature" in the 2008 book *Material Feminisms.* The 2014 collection *Material Ecocriticism,* edited by Serenella Iovino and Serpil Oppermann, emphasizes the concept of "storied matter," suggesting that all material phenomena have a certain degree of "agency," by which they tell their "story" through their very existence in the world. The fourth wave also features a distinct turn toward pragmatic applications of ecocritical scholarship, including efforts to apply ecocritical analysis to "basic human behaviors and lifestyle choices, such as eating and locomotion and clothing and dwelling" (Slovic 2012, 619).

Fifth wave: c. 2011–present. Beginning with the publication of Rob Nixon's *Slow Violence and the Environmentalism of the Poor* in 2011, there has come to be an increasing focus, as stated in *ISLE,* "on information management, the psychology of information processing, and on efficacy of various communication strategies and these concerns appear to work in tandem with the efforts of ecocritics to reach out beyond their traditional academic audiences by writing op-eds and blog entries, speaking at public meetings, publishing creative writing in addition to scholarship, and using other creative outlets" (Slovic 2019, 514; see also Slovic 2015).

While these waves of ecocriticism have approximate beginning dates, none has a conclusion, as there continue to be energetic practitioners of each of these modes of ecocritical analysis well beyond the beginning of the later phases. This does not make the earlier styles of ecocriticism passé or irrelevant. Rather, it suggests what Buell (2005) refers to as a "palimpsest" (17), or a layering of the various waves of ecocritical ideas. Each wave brings a new layer, with different emphases, vocabularies, and theoretical lenses brought to bear on the understanding of the complex and fraught relationship between human beings and the planet we inhabit. Slovic (2016) has suggested that the rapid development of the field has created a dizzying challenge for intellectual historians who wish to understand how the field is changing year by year. One might argue that the recent emergence of empirical ecocriticism as a distinctive submovement within the field of ecocriticism toward the end of the second decade of the twenty-first century reflects both the pragmatic tendencies of fourth-wave ecocriticism and the engagement with the psychology of information processing that seems to be a hallmark of the fifth wave.

The Current Investigation: A Case Study of *ISLE*

The history of ecocriticism can also be traced by examining the language of scholarship published in journals. Such publications are cultural artifacts that help document how people discuss the relationship between humans and the natural world (Carmichael 2010). In this chapter, we use texts from *ISLE* to reveal key changes in the discipline's focus, to highlight trends in theory and application, and to better understand patterns in the field. While other empirical ecocritical approaches tend to focus on gathering data about audience responses to particular types of environmental texts, especially literary texts and films, this study demonstrates a different way to use empirical data to question or verify suspected trends in environmental communication. Rather than evaluating the impact that textual prompts might have on readers and viewers, we have sought to understand fundamental shifts in literary and scholarly discourse that appear in *ISLE,* a major journal in the environmental humanities. As one of the first attempts to quantitatively trace the history of ecocriticism through journal publications, we are interested in macrolevel perspectives of the field and do not track explicit questions, theories, or textual selections in *ISLE* during a fifteen-year period. This exploratory approach uses validated measures that indicate social and psychological processes to investigate how text can reveal information about writers in the field (their focus, attention) and ecocriticism itself. These measures are called summary variables because they provide high-level information about writing style (how authors are communicating) instead of content (what authors are communicating about).

Automated text analyses evaluate language found in its natural context, such as the environmental writing that appears in the pages of a journal like *ISLE*—editor's notes, ecocritical articles, nonfiction nature writing, and ecopoetry. As described below, we scrutinized the linguistic characteristics of 713 specific texts using an automated text analysis program typically used in the social sciences to evaluate macrolevel trends in the ecocritical literature.

The value of automated text analysis is that it creates snapshots of the linguistic features of large collections of texts. Also, once the assessment criteria have been established and encoded into the analytical process, the evaluation process is more systematic than would likely be the case if human researchers were evaluating texts one by one. As explained by James W. Pennebaker and colleagues (2015a), in their foundational description of LIWC, their approach was developed in the 1990s as a means to examine the emotional, cognitive, and structural elements in specific samples of verbal and written speech. Here we have focused on the structural properties of discourse published in *ISLE* during the fifteen-year span of the texts used for the project. We suggest that corpus linguistics research of this kind can provide an analytical tool to supplement and bolster more traditional ecocritical approaches, which tend to rely on the more subjective interpretive efforts of the individual scholar, who reads selected examples of relevant texts and seeks to assess the stylistic qualities of those texts on the basis of the scholar's personal and critical lens or perspective.[2]

We argue that content is a characteristic that, in an academic journal, can vary by volume, issue, or even journal. Therefore, we focus on word types that are largely domain independent, which include function words (articles, prepositions) and structural properties (words per sentence, readability) of the text. As detailed in the Method section, we evaluate the rate of verbosity (word count), analytic thinking (the complexity of the writer's thinking style as indicated by function words), jargon (the level of specialized terminology in each publication), and concreteness. Our approach uses these summary variables to gain an understanding of how ecocritical writing is reflected in *ISLE* and may change over time.

Method

We collected all publications archived by *ISLE* from 2004 to 2018 across nine publication types, choosing a fifteen-year period to ensure that we covered a broad time frame. This initially included book reviews, scholarly articles, poetry, nonfiction, editor's notes, interviews, fiction, memorials, and farewells. Some articles in older issues could not be extracted from the publisher, and we also excluded several publication types to create our focal data set. First, we excluded publication types with fewer than twenty texts over a fifteen-year history of *ISLE*. Second, we excluded book reviews because they report on the work of other scholars. These texts would likely provide a signal about the reviewer's writing style instead of the original author's language patterns. This deviated from our empirical goal, so we did not consider book reviews.

After these exclusions, we retained 713 publications from *ISLE* as PDFs (article n = 356, poetry n = 216, nonfiction n = 91, editor's note n = 50), and these data were submitted to automated text analysis to evaluate writing style patterns over time. The unit of analysis in our corpus of texts was the individual *ISLE* article. Metadata such as year and article type were extracted automatically from the article's HTML.

Automated Text Analysis

The text data were quantified by LIWC (Pennebaker et al. 2015a). LIWC is a tool that contains an internal dictionary of words ranging from emotion terms (happy, sad), to parts of speech (articles, prepositions), to summary variables that describe social and psychological characteristics of text (total word count, rate of analytic thinking as reflected by function words such as articles or pronouns). LIWC counts words by identifying whether a word is found in its internal dictionary and then measuring its frequency as a percentage of the total word count. For example, a sentence from Aldo Leopold's (1949) "Thinking Like a Mountain," "My own conviction on this score dates from the day I saw a wolf die," contains fifteen words and increments a host of LIWC dimensions, ranging from self-references (e.g., my, I; 13.33 percent of the total word count) to death words (e.g., die, 6.67 percent of the total word count).

Measures

We evaluated four summary variables that describe the structure and style of the text that other scholars have also evaluated (Tausczik and Pennebaker 2010): word count, analytic thinking (Pennebaker et al. 2014), jargon (Markowitz and Hancock 2016), and concreteness (Larrimore et al. 2011). Word count and analytic thinking are LIWC composite variables, while jargon and concreteness are variables created with LIWC dimensions that have been validated in prior work. These measures allowed us to understand how the publications were written and draw inferences about *ISLE* writers.

Word count. We measured the number of words per text as a raw frequency. Word count is a measure of verbosity and complexity (Tausczik and Pennebaker 2010). That is, longer texts are often more difficult to comprehend than shorter texts (Flesch 1948). On average, *ISLE* publications were 5,108.23 words,[3] and we evaluated a total of 3,642,168 words across all publications.

Analytic thinking. Function words (articles, prepositions, pronouns) are small words that form the connective tissue of a sentence between other words (nouns, verbs). Beyond providing structure, prior work suggests that function words play a crucial role in understanding a host of social and psychological dynamics (Pennebaker 2011). For example, rates of first-person singular pronouns (I, me, my) correlate with personality dimensions such as agreeableness and neuroticism (Ireland and Mehl 2014), first-person plural pronouns indicate social status (leaders tend to use more "we" words than nonleaders; Kacewicz et al. 2014), and people who match on function words tend to have more favorable and satisfactory relationships (Gonzales, Hancock, and Pennebaker 2010).

Research also suggests that function words can indicate how writers think across a single coherent dimension. Pennebaker and colleagues (2014) evaluated the writing style of admissions essays to a large American university and observed that high-achieving students (those who had higher GPAs) used greater rates of articles and prepositions than storytelling words (pronouns, conjunctions, adverbs, negations). They found that articles and prepositions can be indicators of formal and hierarchical thinking, as authors make references to direct objects in their environment and refer to concrete nouns ("Only *the* mountain has lived long enough to listen objectively to *the* howl of *a* wolf") instead of more general and less direct references when they are absent ("Only mountains have lived long enough to listen objectively to wolves howling").

We measured the rate of analytic thinking in *ISLE* articles using the analytic thinking index from LIWC. High rates of analytic thinking indicate a formal and complex thinking style; low rates of analytic thinking reflect a narrative, or simpler thinking by the communicator. This index ranges from 0 (high rate of narrative thinking, low rate of analytic thinking) to 100 (low rate of narrative thinking, high rate of analytic thinking). At the language level, high scores tend to contain high rates of articles and prepositions and low rates of personal pronouns, conjunctions, adverbs, negations, and auxiliary verbs. Low scores tend to contain the opposite pattern and indicate a thinking style that reflects a narrative (e.g., high rates of storytelling words such as pronouns, negations, and auxiliary verbs). Together, this measure allowed us to understand how cognitive thinking style may change over time and across publications.

Jargon. The rate of jargon captures specialized terminology, or words that are uncommon in the English language. Prior work suggests that the rate of jargon can indicate a variety of social and psychological phenomena, including deception (Markowitz and Hancock 2016) and receipt of grant money (Markowitz 2019). Jargon was first calculated by measuring the percentage of words captured by the LIWC dictionary because it represents a collection of the most common words in everyday English (Markowitz and Hancock 2016; Tausczik and Pennebaker 2010). Then we applied the following formula: 100 – Dictionary. Jargon therefore represents the percentage of words that are not incremented by LIWC and are uncommon in everyday English.

Concreteness. Our final summary measure considers the rate of concreteness through function words. Three word types—articles (a, the), prepositions (above, below), and quantifiers (more, less)—are often indicators of concrete relative to abstract speech (Larrimore et al. 2011; Markowitz and

Hancock 2016). Articles and prepositions make direct references to objects in an environment, and quantifiers express degrees of difference between objects (Larrimore et al. 2011). Elaboration using high rates of articles, prepositions, and quantifiers typically suggests a more concrete writing style than an abstract writing style.

Prior work suggests that articles, prepositions, and quantifiers also reveal a host of social and psychological dynamics. For example, Laura Larrimore and colleagues (2011) have found that concreteness indicators positively predict whether people would receive an online, peer-to-peer loan from a stranger. Markowitz and Hancock (2016) report that fraudulent scientists write more abstractly (e.g., less concretely) than genuine scientists. Consistent with prior work (Larrimore et al. 2011; Markowitz and Hancock 2016), we created a concreteness index by adding the standardized rates (*z* scores) of articles, prepositions, and quantifiers from LIWC. Text with a high score on this measure is more concretely written than text with a low score. Although there are many ways to measure concreteness and abstraction (Pollock 2018), we decided that using function words was beneficial because they are context independent and are used at high rates in nearly all English settings (Chung and Pennebaker 2007).

Results

We first visually inspected the rates of our four summary variables over time by collapsing across publication types. As Figure 9.1 displays, *ISLE* publications have generally become longer since 2004; the writing content has become more specialized and word choices less common, as indicated by rates of jargon; and the style has also become relatively more abstract, as indicated by a general decreased rate of concreteness (fewer articles, prepositions, and quantifiers), though concreteness appears to have more of a curvilinear relationship over time. There is also a relative increase in analytic thinking over time, though differences are small on a year-to-year basis. These patterns are generally supported by the significant bivariate correlations represented in Table 9.1.

Are the prior patterns consistent for each publication type? The correlation coefficients separated by publication type describe similar patterns. In general, more recent publications are characterized by higher rates of jargon

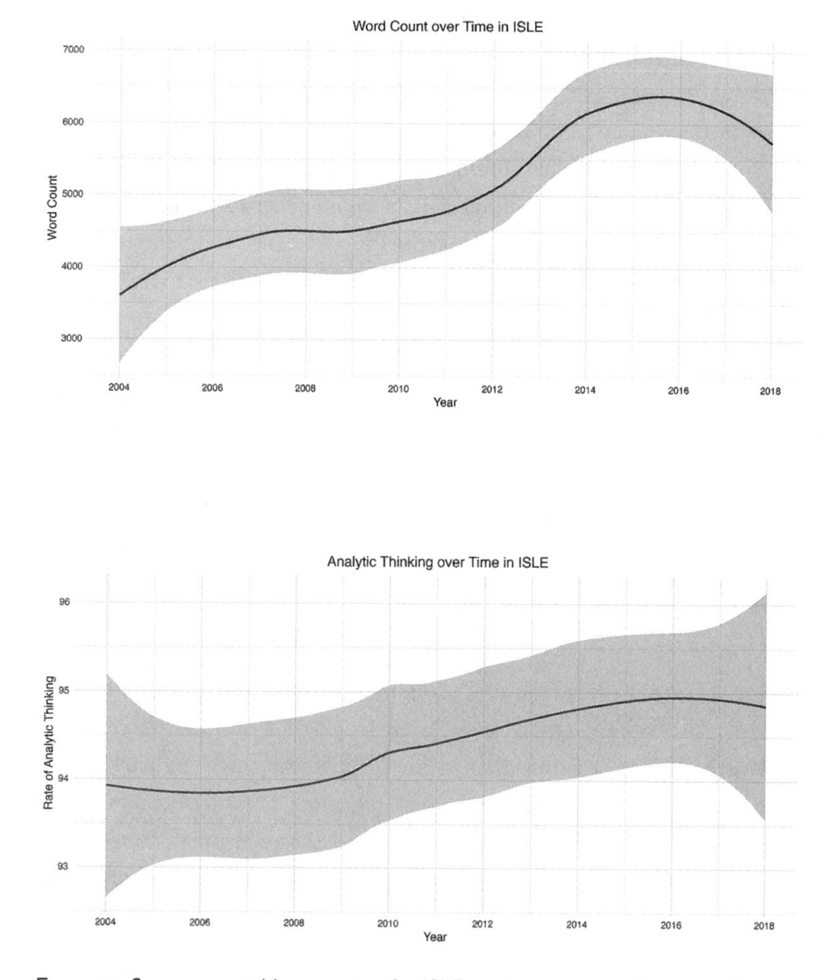

Figure 9.1. Summary variables over time for *ISLE* articles, poetry, nonfiction, and editor's notes (2004–18). Regression line was estimated using the LOESS (locally estimated scatterplot smoothing) or local regression method.

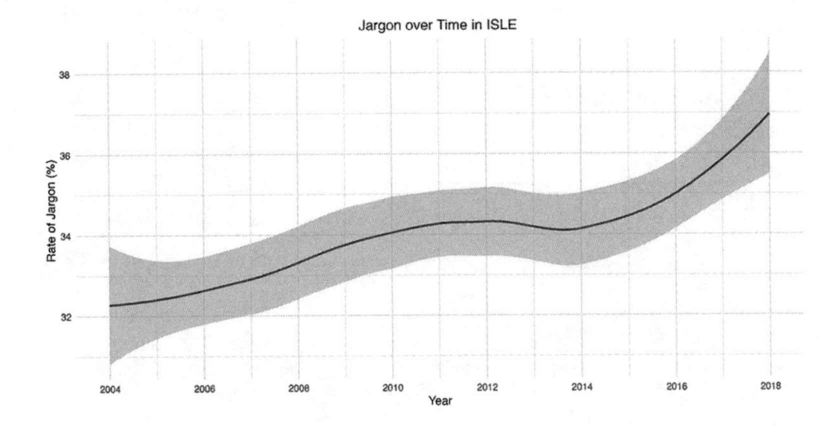
Jargon over Time in ISLE
Rate of Jargon (%)
38
36
34
32
2004
2006
2008
2010
2012
2014
2016
2018
Year

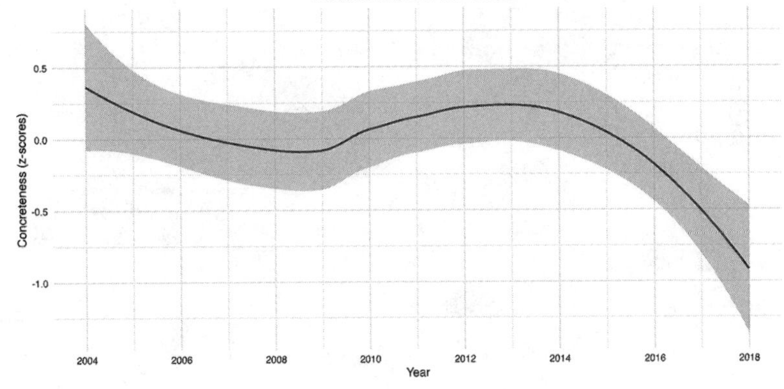
Concreteness over Time in ISLE
Concreteness (z-scores)
0.5
0.0
-0.5
-1.0
2004
2006
2008
2010
2012
2014
2016
2018
Year

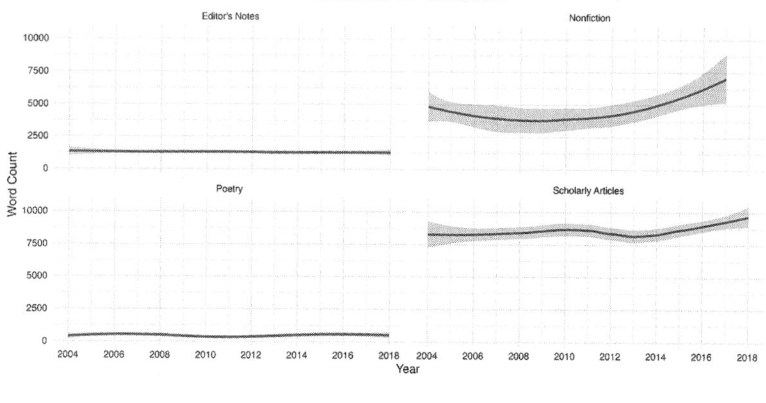

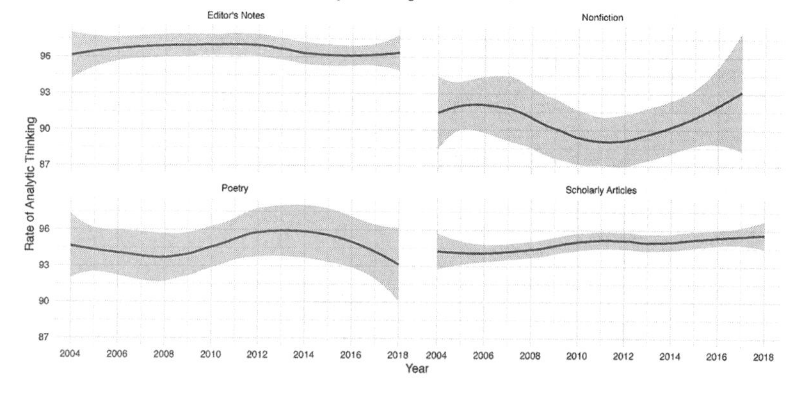

Figure 9.2. Summary variables over time by publication type. Regression line estimated using the LOESS (locally estimated scatterplot smoothing) or local regression method.

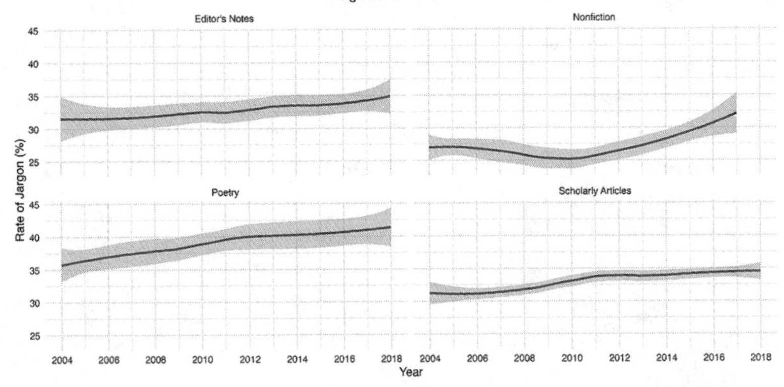
Jargon over Time in ISLE
Editor's Notes
Nonfiction
Poetry
Scholarly Articles
Rate of Jargon (%)
45
40
35
30
25
2004
2006
2008
2010
2012
2014
2016
2018
Year

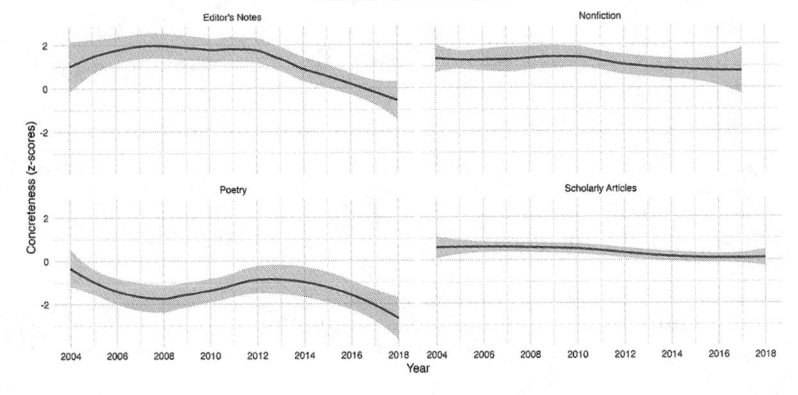
Concreteness over Time in ISLE
Editor's Notes
Nonfiction
Poetry
Scholarly Articles
Concreteness (z-scores)
2
0
-2
2004
2006
2008
2010
2012
2014
2016
2018
Year

Table 9.1. Correlation Matrix of Summary Variables

Characteristic	*Year*	*Word Count*	*Analytic Thinking*	*Jargon*
Overall effect (N = 713)				
Year	—			
Word count	.191**	—		
Analytic thinking	.073†	.054	—	
Jargon	.173**	−.228**	.426**	—
Concreteness	−.097**	.238**	.183**	−.592**
Articles (n = 356)				
Year	—			
Word count	.135*	—		
Analytic thinking	.140**	.128*	—	
Jargon	.275**	.141**	.481**	—
Concreteness	−.165**	−.115*	.168**	−.527**
Editor's notes (n = 50)				
Year	—			
Word count	−.024	—		
Analytic thinking	−.106	−.378**	—	
Jargon	.311*	−.404**	.523**	—
Concreteness	−.518**	.065	.300*	−.470**
Nonfiction (n = 91)				
Year	—			
Word count	.152	—		
Analytic thinking	−.074	.214*	—	
Jargon	.188†	.195†	.583**	—
Concreteness	−.138	.141	.439**	−.166
Poetry (n = 216)				
Year	—			
Word count	.042	—		
Analytic thinking	.001	−.023	—	
Jargon	.225**	.027	.380**	—
Concreteness	−.158*	−.073	.263**	−.484**

Note. Pearson correlations are two tailed. *p* values describe the probability of receiving a false-positive result. Conventions suggest that $p < .05$ is statistically significant, the relationship between variables is statistically meaningful, and it has a less than 5 percent chance of being falsely positive.
*$p < .05$; **$p < .01$; †$p < .075$.

and less concrete writing (Table 9.1). Figure 9.2 visually describes these data; the trends suggest a relatively stable word count for editor's notes and poetry over time. Around 2013, scholarly articles and nonfiction start to increase their word count. Concreteness generally decreases for all publication types over time, with some minor instances of stability or increase, as indicated by the strong negative correlation coefficients between year and concreteness.

Discussion

As Buell (2005) notes, "It would be a mistake . . . to suppose that ecocriticism has unfolded in a tidy, sequential manner, with a new dispensation displacing the old" (138). Despite various efforts to chart the historical development of the field, as exemplified in the opening section of this chapter, it is difficult to discern precise trends other than the subtle topical shifts as new phases emerge. However, the broad trend toward the use of abstract language and jargon in the field is perhaps noticeable in the squabble over the role of theory in ecocriticism, which erupted in the pages of *ISLE* in 2009 and 2010. In the Spring 2009 issue, Simon C. Estok published a much-discussed article on "ecophobia," arguing that antagonism toward nonhuman nature (a bias akin to racism, misogyny, and homophobia in social contexts) is pervasive in human culture and deserves ecocritical attention. As Louisa Mackenzie and Stephanie Posthumus (2013) note, Estok's work "is broadly about the failure of ecocriticism to be both sufficiently activist and coherently theoretical, and the potential of the concept of ecophobia to address both" (760). In the wake of Estok's article, S. K. Robisch (2009) took Estok and the broader discipline to task for overvaluing abstract, jargon-filled academic discourse and failing to attend to environmental and political realities. Robisch threatens:

> Every poststructuralist or culture maven who denies the existence of wilderness or nature should be dropped ten days into the Frank Church Wilderness with a knife and made to find his or her way out. Then I'll be glad to hold some food and water behind my back and have a long conversation about theory and epistemology until the survivor acknowledges the representational value of words like "giardia" and "mullein" and "grizzly bear" and "shelter." Hypothermia's a great way to theorize ecophobia. (705)

The exchange between Estok and Robisch sent shudders through the discipline of ecocriticism—at least among those scholars who followed the field through *ISLE*'s pages. A year later, the journal published a Special Forum on Ecocriticism and Theory to reassure the community of scholars that the journal was dedicated to the notion that "ecocriticism is large, and it contains multitudes" (Slovic 2010a, 639). The special forum itself included fourteen brief statements on ecocriticism and theory, covering such broad issues as "intellectual history and practice; trans-disciplinary perspectives; and notions of internationalization and cultural specificity" (Slovic 2010b, 759).

We have emphasized this turbulent and energetic period in the history of *ISLE* because the current study reveals a notable shift in the language of the scholarly articles, reviews, and creative pieces published in the journal, closely tracking with the theory-related conversation from 2009 to 2010. In particular, work that appeared in the journal, according to the automated text analysis, began to include considerably more jargon around the years 2011 to 2013, a trend that seems to continue to this day. This is the result of no conscious shift in editorial design. Submissions continue to be sent to external readers for blind review and are still subjected to rigorous standards of clarity and relevance. However, there appears to be a notable tilt toward more technical discourse in the pages of *ISLE* in the wake of the 2009–10 discussion of the pros and cons of using theoretical language in environmental humanities writing. To a certain degree, the conspicuous shifts noted in this study are extension of longer trends, dating back to the beginning of this study.

Ironically, just as the dramatic trend toward less concrete language and more jargon was emerging in the field, so too was the discipline beginning to focus on how the humanities, in new materialist philosophy and material ecocriticism, grapple with the "intra-action" between the body of the planet and the human body, and with the sociopolitical roles of writer-activists (and scholar-activists), as discussed in Nixon (2011). The concerns Robisch (2009) expressed in his critique of Estok's theorizing, particularly the worry that theory would result in a detachment from environmental knowledge and meaningful social action, seem not to be borne out in the actual content or language of the journal, or in the field more broadly. Ecocriticism and environmental literature have become acutely attuned to the physical world and to various forms of public engagement, as indicated by the sketch provided above of the waves of ecocriticism, particularly the fourth and fifth.

The emphases on materiality, practical application, and public engagement in many examples of contemporary ecocriticism seem to contradict the linguistic trend toward jargon and abstraction. What seems to be happening, though, is that ecocriticism is evolving toward a greater degree of disciplinary confidence and maturity, resulting in an increasingly distinctive language of its own, which manifests as jargon when analyzed through LIWC. To offer a few examples of this, we will point to some of the topics and terminology from the issues of *ISLE* published in 2019. Although the data collection for this study ended in 2018, the trends clearly seem to be continuing into the present.

The Winter 2019 issue (*ISLE* 26.1) included studies of "trans-scalar" challenges in ecological thinking, "ecological humanist mosaics," and the complexities of personal identity, Freudian rhetoric, black ecotheology, Latourian concepts of "the collective," "insurgency and distributed agency," anthropomorphism in contemporary Chinese literature, "waste and reclamation," and piracy and environmental justice in recent films. The Spring 2019 issue (*ISLE* 26.2) included work focusing on "landscape and the politics of self-relinquishment," "paleo-narratives and white atavism," the "rhetoric of commonplaces," and a set of articles exploring Estok's notion of ecophobia ten years after the original publication of his article. In addition to a special cluster of articles on international approaches to "Human–Animal Entanglements in the Fairy Tale," the Summer 2019 issue (*ISLE* 26.3) includes studies of "the animal novel as biopolitical critique," "seasonal and environmental discourse," "gendered homecomings," and "environmental damage, resource conflict, and literary strategies in the Niger Delta." While each of these articles seeks to unpack the implications of specific cultural texts, it is clear that authors are applying a wide array of theoretical approaches to their ecocritical analyses. Therefore, it may not be surprising that we discern increasing levels of jargon and abstraction in the work published by the journal.

Contributions, Limitations, and Future Directions

This chapter offers an empirical mirror for writers in the field of ecocriticism by showing trends in how articles are written and how the field has developed as reflected through language patterns. First, our study uses language patterns as a prime data source to use in empirical ecocriticism research. Language patterns observed in published research articles are one of many avenues to explore in empirical ecocriticism, given the amount and diversity of writing produced in the field.

Second, the analysis introduces an automated approach to analyzing text. Treating text as data from a quantitative perspective is common in the social sciences to reveal social and psychological characteristics of authors (Pennebaker 2011). We hope that this piece provides an opportunity for critically and qualitatively oriented scholars to consider other approaches for their work; the same approach, for example, can be used on an entire body of fiction to evaluate the styles of writers. Automated text analyses applying validated categories can support suspicions about trends in a field or body of literature (e.g., How does a prolific author's writing change over time?), while also facilitating new avenues of research that might be difficult to evaluate with human coding alone. More sophisticated approaches, such as machine learning and deep learning, can combine human coding and large-scale text analyses to complement the approach taken in the current investigation.

Our investigation only considered a single journal in its evaluation of ecocriticism through language patterns. We also excluded book reviews to ensure that we captured the authors' writing style for each publication type with the highest fidelity possible. Book reviewers, who often report on the work of someone else and quote extensively, are therefore one layer removed from original text's author. Because some of our measures (analytic thinking, concreteness) included word types that are influenced by individual differences (articles, prepositions; Pennebaker 2011), we excluded book reviews to ensure that publication type comparisons could occur without questioning whether the writing style of a piece reflects that of the author or of another individual. However, we suggest that future studies should expand the sample to other outlets and publication types. The trends represented in our chapter are also correlational and do not represent cause and effect. Future research might run experiments (Malecki, Pawłowski, and Sorokowski 2016) to test how people perceive the writing style of articles that have high rates of jargon or abstraction, for example, to draw conclusions about ecocritical readership.

It is also unclear if the trends in *ISLE* are specific to the journal, if they occur in the entire field of ecocriticism, or if they reflect natural patterns of writing for relatively young disciplines. Ecocritical writing has been documented since the 1800s (Mazel 2001), but the journal under investigation has existed for only three decades. Future work should evaluate whether the *ISLE* data reflect recent developments in writing style that are common in other academic disciplines, or if the journal's authorship has unique conventions.

Conclusion: Automated Text Analysis as a Mode of Empirical Ecocriticism

The disciplines of ecocriticism and environmental literature, as represented by fifteen recent years' worth of published material from a central journal in the field, display a vibrant, socially engaged, increasingly international and multicultural range of voices and commitments. Although scholars have frequently used LIWC and other automated text analysis tools to analyze primary literary texts as strategies for determining authorship and charting broad trends in literary history, the use of this technique for validating a perceived trend in a specific scholarly field is less common. As described above, one of the most contentious issues in the history of ecocriticism during the past decade has been the role of critical theory, broadly conceived, in a discipline that many practitioners think should be less academic and more practical and socially engaged. Ironically, while the LIWC data show a field that has clearly embraced more abstract and technical language, the history of ecocriticism described earlier in this article—from the first to the fifth waves—suggests that exploring theories of ecohorror (Summer 2014), transportation technologies (Winter 2017), migrant ecologies (Spring 2017), nature and liminality (Summer 2017), vegan ecocriticism (Fall 2017), queer ecopoetics (Winter 2018), and various other strands of ecocritical analysis has not prevented scholars from becoming ever more engaged with the issues of our day. When Nixon (2011) uses the term "writer-activist" (15), he is also nudging scholars in the environmental humanities to take a cue from environmental literary activists, and from postcolonial and ecofeminist colleagues, to become scholar-activists. The findings offered here, together with the thumbnail sketch of recent ecocritical history, suggest that ecocriticism is becoming both increasingly theoretical and increasingly engaged with serious humanitarian and ecological issues.

When Scott Slovic presented the figures from this study at a well-attended session on empirical ecocriticism at the summer 2019 ASLE biennial conference, he noted the trends toward wordiness, heavy use of jargon, and abstraction, drawing dismayed chuckles from the audience. However, these linguistic markers, from a more positive perspective, seem to indicate a flourishing and diverse scholarly and artistic movement, one devoted to exploring the challenges our species faces as we try—as we struggle—to live justly and sustainably on this planet. The growing tendency toward abstract and technical language, detected by our empirical LIWC analysis of materials published in *ISLE* after 2010, exemplifies the evolution of increasingly mature and trenchant ecocritical voices. While many empirical ecocritical approaches highlight audience responses to literature and film, our approach here demonstrates a very different strategy, one that reveals the trajectory of a discipline by carefully describing how linguistic patterns in ecocritical discourse evolved during a critical period in the history of the field.

Notes

Both authors contributed equally to this chapter and share first authorship.

1. In 1995, *ISLE* began publishing not only scholarly articles ("ecocriticism") and book reviews but also creative writing (chiefly poetry, but also literary essays and occasional pieces of short fiction) and intermittent interviews with leading scholars, writers, and artists. For the next thirteen years, from 1995 to 2008, ASLE self-published *ISLE*, handling all editorial processes in house at the University of Nevada, Reno, where Slovic was a faculty member in the English department, and hiring Data Reproductions Corporation, of Michigan, to print approximately 2,000 paper copies of each issue. In 2008, Oxford University Press (OUP), in cooperation with ASLE, added *ISLE* to its roster of North American journals, and the first Oxford issues began to appear quarterly in winter 2009. Editorial processes, still supervised by Slovic, shifted to OUP's ScholarOne Manuscripts processing platform, while the publisher assumed responsibility for producing an online and print edition of each new issue. All of *ISLE*'s back issues, from 1993 to 2008, were scanned and made available on the Oxford Journals website. As of October 2022 (through *ISLE* 29.3), eighty-seven individual issues of the journal had appeared.

2. Franco Moretti (2013), a pioneer in the application of corpus linguistic methodologies to the study of literature, has demonstrated the value of using computers to analyze large numbers of texts as a means of better understanding literary history by moving beyond the necessarily selective body of literature that a single human scholar can read and interpret; see also Oberhelman (2015). One of the implications of the current study is that LIWC, in enabling the linguistic analysis of large numbers of environmental texts, can strengthen our understanding of certain patterns in the development of ecocriticism by reinforcing readers' intuitive sense that the field has moved distinctly toward theoretical sophistication since 2010.

3. SD = 4,185.01 words, Q1 = 584 words, Mdn = 5,490 words, Q3 = 8,778 words. SD indicates standard deviation, or the amount of variation in a set of data points; Q1 indicates first quartile, or the value that represents at most 25 percent of the data; Mdn indicates median, or second quartile, which is the value that represents at most 50 percent of the data; and Q3 indicates third quartile, or the value that represents at most 75 percent of the data.

References

Adamson, Joni, and Scott Slovic. 2009. "The Shoulders We Stand On: An Introduction to Ethnicity and Ecocriticism." Guest editors' introduction to special issue, "Ethnicity and Ecocriticism," *MELUS* 34 (2): 5–24.

Alaimo, Stacy. 2008. "Trans-corporeal Feminisms and the Ethical Space of Nature." In *Material Feminisms,* edited by Stacy Alaimo and Susan Hekman. Bloomington: Indiana University Press, 237–64.

Bennett, Gena R. 2010. *Using Corpora in the Language Learning Classroom: Corpus Linguistics for Teachers.* Ann Arbor: University of Michigan Press ELT, 1–22.

Boyd, Ryan L. 2017. "Psychological Text Analysis in the Digital Humanities." In *Data Analytics in Digital Humanities,* edited by Shalin Hai-Jew, 161–89. Cham, Switzerland: Springer International.

Boyd, Ryan L., and H. Andrew Schwartz. 2021. "Natural Language Analysis and the Psychology of Verbal Behavior: The Past, Present, and Future States of the Field." *Journal of Language and Social Psychology* 40 (1): 21–41.

Buell, Lawrence. 2005. *The Future of Environmental Criticism.* London: Blackwell.

Carmichael, Deborah A. 2010. "Ken Burns' *The National Parks: America's Best Idea* (2009): Missed Opportunities for Environmental Messages." *Environmental Communication* 4 (4): 469–74.

Chung, Cindy K., and James W. Pennebaker. 2007. "The Psychological Functions of Function Words." In *Social Communication,* edited by Klaus Fiedler, 343–59. London: Psychology Press.

Cohen, Michael P. 2004. "Blues in the Green: Ecocriticism under Critique." *Environmental History* 9 (1): 9–36.

Estok, Simon C. 2009. "Introduction: Theorizing Ecophobia, Ten Years In." *ISLE: Interdisciplinary Studies in Literature and Environment* 26 (2): 379–87.

Flesch, R. 1948. "A New Readability Yardstick." *Journal of Applied Psychology* 32 (3): 221–33.

Glotfelty, Cheryll. 1996. "Introduction: Literary Studies in an Age of Environmental Crisis." In *The Ecocriticism Reader: Landmarks in Literary Ecology,* edited by Cheryll Glotfelty and Harold Fromm. Athens: University of Georgia Press, xv–xxxvii.

Glotfelty, Cheryll, and Harold Fromm, eds. 1996. *The Ecocriticism Reader: Landmarks in Literary Ecology.* Athens: University of Georgia Press.

Gonzales, Amy L., Jeffrey T. Hancock, and James W. Pennebaker. 2010. "Language Style Matching as a Predictor of Social Dynamics in Small Groups." *Communication Research* 37 (1): 3–19.

Iovino, Serenella, and Serpil Oppermann, ed. 2014. *Material Ecocriticism.* Bloomington: Indiana University Press.

Ireland, Molly, and M. Mehl. 2014. "Natural Language Use as a Marker of Personality." In *The Oxford Handbook of Language and Social Psychology,* edited by Thomas M. Holtgraves, 201–18. Oxford: Oxford University Press.

Kacewicz, Ewa, James W. Pennebaker, Matthew Davis, Moongee Jeon, and Arthur C. Graesser. 2014. "Pronoun Use Reflects Standings in Social Hierarchies." *Journal of Language and Social Psychology* 33 (2): 125–43.

Larrimore, Laura, Li Jiang, Jeff Larrimore, David Markowitz, and Scott Gorski. 2011. "Peer to Peer Lending: The Relationship between Language Features, Trustworthiness, and Persuasion Success." *Journal of Applied Communication Research* 39 (1): 19–37.

Leopold, Aldo. 1949. "Thinking Like a Mountain." In *A Sand County Almanac and Sketches from Here and There,* 129–33. Oxford: Oxford University Press.

Mackenzie, Louisa, and Stephanie Posthumus. 2013. "Reading Latour Outside: A Response to the Estok–Robisch Controversy." *ISLE: Interdisciplinary Studies in Literature and Environment* 20 (4): 757–77.

Malecki, Wojciech, Bogusław Pawłowski, and Piotr Sorokowski. 2016. "Literary Fiction Influences Attitudes toward Animal Welfare." *PLoS One* 11 (12): e0168695.

Markowitz, David M. 2019. "What Words Are Worth: National Science Foundation Grant Abstracts Indicate Award Funding." *Journal of Language and Social Psychology* 38 (3): 264–82.

Markowitz, David M., and Jeffrey T. Hancock. 2016. "Linguistic Obfuscation in Fraudulent Science." *Journal of Language and Social Psychology* 35 (4): 435–45.

Marx, Leo. 1964. *The Machine in the Garden: Technology and the Pastoral Ideal.* Oxford: Oxford University Press.

Mazel, David. 2001. *A Century of Early Ecocriticism.* Athens: University of Georgia Press.

Moretti, Franco. 2013. "Conjectures on World Literature." In *Distant Reading,* 43–62. London: Verso.

Murphy, Patrick D. 2003. Foreword to *The ISLE Reader: Ecocriticism, 1993–2003,* edited by Michael P. Branch and Scott Slovic, vii–ix. Athens: University of Georgia Press.

Nixon, Rob. 2011. *Slow Violence and the Environmentalism of the Poor.* Cambridge, Mass.: Harvard University Press.

Oberhelman, David D. 2015. "Distant Reading, Computational Analysis, and Corpus Linguistics: The Critical Theory of Digital Humanities for Literature Subject Librarians." In *Digital Humanities in the Library: Challenges and Opportunities for Subject Specialists,* edited by Arianne Hartsell-Gundy, Laura Braunstein, and Liorah Golomb, 53–66. Chicago: Association of College and Research Libraries.

Pennebaker, James W. 2011. *The Secret Life of Pronouns: What Our Words Say about Us.* London: Bloomsbury.

Pennebaker, James W., Roger J. Booth, Ryan L. Boyd, and Martha E. Francis. 2015a. "Linguistic Inquiry and Word Count: LIWC2015." Software. https://www.liwc.app/.

Pennebaker, James W., Ryan L. Boyd, Kayla N. Jordan, and Kate Blackburn. 2015b. "The Development and Psychometric Properties of LIWC2015." Texas ScholarWorks, University of Texas at Austin. http://hdl.handle.net/2152/31333.

Pennebaker, James W., Cindy K. Chung, Joey Frazee, Gary M. Lavergne, and David I. Beaver. 2014. "When Small Words Foretell Academic Success: The Case of College Admissions Essays." *PLoS One* 9 (12): e115844.

Phillips, Dana. 2003. *The Truth of Ecology: Nature, Culture, and Literature in America.* Oxford: Oxford University Press.

Pollock, Lewis. 2018. "Statistical and Methodological Problems with Concreteness and Other Semantic Variables: A List Memory Experiment Case Study." *Behavior Research Methods* 50 (3): 1198–216.

Reed, T. V. 2002. "Environmental Justice Ecocriticism." In *The Environmental Justice Reader: Politics, Poetics, and Pedagogy,* edited by Joni Adamson, Mei Mei Evans, and Rachel Stein, 145–62. Tucson: University of Arizona Press.

Robisch, S. K. 2009. "The Woodshed: A Response to 'Ecocriticism and Ecophobia.'" *ISLE: Interdisciplinary Studies in Literature and Environment* 16 (4): 697–708.

Rueckert, William. 1978. "Literature and Ecology: An Experiment in Ecocriticism." *Iowa Review* 9 (1): 71–86.

Slovic, Scott. 1992. *Seeking Awareness in American Nature Writing: Henry Thoreau, Annie Dillard, Edward Abbey, Wendell Berry, Barry Lopez.* Salt Lake City: University of Utah Press.

Slovic, Scott. 2000. "Ecocriticism: Containing Multitudes, Practicing Doctrine." In *The Green Studies Reader: From Romanticism to Ecocriticism,* edited by Laurence Coupe, 160–62. New York: Routledge.

Slovic, Scott. 2010a. "Editor's Note." *ISLE: Interdisciplinary Studies in Literature and Environment* 17 (4): 639–41.

Slovic, Scott. 2010b. "Elements of This New Alliance." *ISLE: Interdisciplinary Studies in Literature and Environment* 17 (4): 757–59.

Slovic, Scott. 2012. "Editor's Note." *ISLE: Interdisciplinary Studies in Literature and Environment* 19 (4): 619–21.

Slovic, Scott. 2015. "Ecocriticism 101: A Basic Introduction to Ecocriticism and Environmental Literature." *Pertanika* 23 (supplement): 1–14.

Slovic, Scott. 2016. "Seasick among the Waves of Ecocriticism: An Inquiry into Alternative Historiographic Metaphors." In *Environmental Humanities: Voices from the Anthropocene,* edited by Serpil Oppermann and Serenella Iovino, 99–111. Lanham, Md.: Rowman & Littlefield.

Slovic, Scott. 2019. "Editor's Note." *ISLE: Interdisciplinary Studies in Literature and Environment* 26 (3): 513–17.

Tausczik, Yia R., and James W. Pennebaker. 2010. "The Psychological Meaning of Words: LIWC and Computerized Text Analysis Methods." *Journal of Language and Social Psychology* 29 (1): 24–54.

Tuckerman, Henry T. 1864. *America and Her Commentators: With a Critical Sketch of Travel in the United States.* New York: Charles Scribner.

PART III

Reflections

Chapter 10

Empirical Ecocriticism and the Future of (Eco)Narratology

URSULA K. HEISE

Reading the essays in this collection from my vantage point as a narratologist has been both exciting and sobering. On the one hand, empirical ecocriticism builds and elaborates on literary-critical approaches that focus on the distribution and reception of certain kinds of texts—from studies of theater and theater audiences in Shakespearean London to Japanese book-lending libraries in the nineteenth century (Zwicker 2006) and readers of American romance novels in the second half of the twentieth (Radway 1984). On the other hand, it forms part of more recent quantitative approaches that have gained in importance through computational criticism and the digital humanities over the last twenty years, including the groundbreaking work of Stanford's Lit Lab and Archer and Jockers' *The Bestseller Code* (2016). Clearly, both of these are enormously important research directions that deserve more emphasis and widespread attention among environmental humanists.

Putting the study of readership on an empirical and sometimes quantitative basis is unquestionably a crucial tool for testing academic assumptions about the popularity, influence, and political shaping power of texts—especially for an area of study such as ecocriticism, whose research agenda is intertwined with political activism. Already in the early 1980s, Jonathan Culler (1981) argued that in the context of comparative literary studies, the discipline needed not more readings of texts but accounts of why, how, and under what circumstances certain readers produce particular readings of texts. Reader-response theory, cognitive studies of texts (including studies of affect or emotion), and historical studies of readership have contributed to this research. For ecocriticism, the contributions of studies in environmental communication and media have in addition proven invaluable. Empirical ecocriticism, to judge from the essays in this collection, brings together these strands of research to generate fresh insights on what readers read environmental texts, how they interpret them, and what conceptual and practical conclusions they draw from them. This kind of research is urgent and necessary to connect academic work with struggles for environmental conservation and environmental justice outside the university.

In the process, empirical research puts long-cherished assumptions in the humanities about the transformative power of the imagination, art, and literature to the test, and as a consequence, some of its findings struck me as sobering. Matthew Schneider-Mayerson discovered in a study of reader responses to Paolo Bacigalupi's novel *The Water Knife* (2015) that not all readers perceived it to be primarily a novel about climate change, as ecocritics tend to do. He also found that empathy with the climate refugee characters and the life-threatening situations they find themselves in did not necessarily lead readers to adopt a more welcoming attitude toward climate migrants in the real world; instead, it reinforced a fear of migrants in some readers (Schneider-Mayerson 2020). In this volume, a study of Alice Walker's essay/short story "Am I Blue?" by Alexa Weik von Mossner, W. P. Malecki, Schneider-Mayerson, Marcus Mayorga, and Paul Slovic finds that the story, which focuses on the narrator's self-identification with the trauma inflicted on a horse, does not substantially change readers' attitudes toward animal welfare. The effects of environmental narratives on readers or viewers, in both of these studies, turn out not to align neatly with the perspectives and attitudes an environmentalist or ecocritic might expect. As Weik von Mossner and colleagues point out, the ambiguity of texts that literary critics value may also be the reason why their effect on different readers is hard to predict, and why readers can and do find meanings and draw conclusions from them that environmentalists and ecocritics may not find desirable. This is in some sense disappointing news, but it does foreground why empirical research on textual effects is necessary as a counterweight to environmentalist researchers' potential confirmation biases.

Even when environmental texts have some of the effects that ecocritics might wish for, they sometimes only prove temporary. As this volume's study of two short stories about climate change shows, the stories did influence readers' attitudes toward climate change, but this influence did not last, with an immediate effect changing to nonsignificance after a month. At first sight, this may seem unsurprising; especially in the current overcrowded media landscape, much of what we read and view does not leave a lasting impression—or at least not one we're aware of. But when environmentalists and ecocritics discuss the influence of narrative on environmental beliefs and attitudes, they often think of or explicitly invoke examples of books such as Rachel Carson's *Silent Spring* (1962) or documentary films like Davis Guggenheim's *An Inconvenient Truth* (2006)—works that are commonly believed to have shaped public debates and perceptions.[1] But works with so large an audience and so lasting an influence are exceptions, especially now that publics in many countries have access to a far broader range of media outlets from which they can select works that may not be widely shared. Many environmentally themed texts and films are therefore likely to have a much more temporary impact of the kind that the study by Schneider-Mayerson and colleagues in this volume points to. This may in turn be a reason why we need more—and more varied—environmental stories that bring the most important issues to the attention of diverse audiences again and again in different formats and from different perspectives. This is a goal to which LENS, the Lab for Environmental Narrative Strategies at UCLA, which I cofounded with colleagues in 2016, is dedicated—as well as, of course, a broad range of environmental writers, filmmakers, and ecocritics.

Reaching diverse audiences with environmental stories includes attention to the processes and organizations by means of which such narratives are selected for publication and distribution. This dimension of textual reception is wide open for empirical inquiry of the kind that is modeled in this volume: quantitative research on what kinds of environmental stories are published, and by what venues and studios, as well as what kinds of environmental stories win literature or film prizes, and on what grounds; qualitative research involving the editors of environmental presses and magazines, film studio executives, and awards committees; and quantitative as well as qualitative research on translators and translations. Some of this research has of course been modeled in literary studies at large: Janice Radway's (1984) pioneering study of romance readers, mentioned earlier, which relies not only on interviews with readers but also the booksellers who acquire and recommend novels in the genre; analyses of the influence of Oprah Winfrey's televised book club on American readers and reading practices (Hall 2003); studies of the role of the Barcelona-based publisher Seix-Barral for the emergence of Latin American literatures for a global public and that of the London-based publisher Heinemann for African literatures (Currey 2008; Espósito 2009; Saval 2002); James F. English's (2005) study of literary prizes; or Franco Moretti's (1999) study of how European novels circulated in translation across the continent in the mid-nineteenth century. While this research in literary history and criticism provides useful models for quantitative and qualitative research on the mechanisms whereby certain stories reach particular audiences and how these stories move or do not move beyond their original target audience, it has not involved environmentally oriented publishing and distribution. Yet the questions of who selects the environmental stories the public gets to see and hear, on what grounds, and for what potential audiences open up a wide field of research for empirical approaches that would usefully complement the study of readers and viewers themselves.

This research, like the essays in this volume, would cut across the study of literary texts and that of documentary and journalistic ones, and would thereby help to solidify the links between ecocriticism and environmental communication, whose importance was foregrounded by Scott Slovic, Swarnalatha Rangarajan, and Vidya Sarveswaran (2019). Such connections contribute to the identification and analysis of narrative templates and patterns of metaphor that appear in fictional or poetic texts as well as nonliterary ones. Even though literary critics sometimes resist the amalgamation of fictional with nonfictional works out of fear that literature and literary study might be instrumentalized (for the twentieth and twenty-first centuries, at least; studies of earlier periods often do not draw similarly sharp distinctions), several studies as well as some of the analyses in this volume highlight that the fiction/nonfiction difference may not substantially affect reader and viewer responses (Appel and Malečkar 2012; Strange and Leung 1999;

Wheeler, Green, and Brock 1999; Schneider-Mayerson, Gustafson, et al., this volume; Weik von Mossner et al., this volume; see also Gallagher 2018; Myrick and Oliver, this volume). In this way, empirical ecocriticism stands to make crucial contributions to a narratology that integrates anthropology, linguistics, literary criticism, and sociology—all with strands of research on narrative that have often pursued separate trajectories over the last few decades.

The purpose of such an integrated econarratology (James 2015; James and Morel 2020) would not be simply to amalgamate fictional with nonfictional stories about the environment. Rather, its goals would be to identify what storytelling patterns cut across the fiction/nonfiction distinction in specific historic and cultural contexts, which ones do not, and in what ways the differences between fiction and nonfiction matter or do not matter to particular audiences. Such research would also be useful for outlining cultural differences in the way environmental changes are narrated. While printed, televised, or digital news accounts of environmental change are available in many languages and countries, genres such as nonfiction nature writing or scientific nonfiction are not as widely published or read in many parts of the world as they are in Britain and the United States. It therefore makes sense to research empirically whether the distinctions between fictional and nonfictional accounts of environmental crisis work differently in different regions and countries.

By the same token, the genre of climate fiction, which has acquired prominence in ecocriticism and the study of climate change communications in the English language, is not nearly as distinct in, for example, Latin American fiction, where it is often much harder to tell apart fictional engagements with climate change as an ecological phenomenon from climate change as an allegory for political conditions—in, for example, Ignacio de Loyola Brandão's *Não Verás País Nenhum: Memorial Descritivo* (1981) or Pedro Mairal's *El año del desierto* (2005). The very idea of climate fiction as a distinct literary genre may not make sense in this context, and empirical ecocriticism is well placed to explore such differences in a comparatist framework. The beginnings of such promising comparatist work are already visible in this volume (as well as in earlier work) through the cross-national study of the impact of animal stories on readers' attitudes toward the welfare of particular species and of animals in general (Malecki et al. 2019; Weik von Mossner et al., this volume).

To sum up, empirical ecocriticism stands to play a crucial role in linking the study of environmental communication with the study of environmental literature. As I've attempted to outline, its importance reaches beyond these two fields in that it also promises to offer new research directions for comparative literary and cultural studies as well as for narrative theory and analysis. For the moment, the use of these methods has to rely on collaborative efforts, as this volume clearly demonstrates—and the editors encourage those researchers who are as yet unfamiliar with such methods to seek out collaborators across disciplines (Schneider-Mayerson, Weik von Mossner et al., this volume). This is a reasonable and encouraging imperative at a time when researchers located in the humanities rather than the social sciences are still not regularly trained in empirical or quantitative methods. But it is my hope that this volume also contributes to foregrounding how desirable the integration of such training into the humanities would be so as to enhance the many types of humanities research that understand academic work to be in dialogue with social activism and the collective search for justice among and beyond humans.

Note

1. Even for works such as these, more empirical research might be useful. When I researched the influence of Carson's *Silent Spring* (1962) on the emergence of environmental movements internationally in the 1960s and 1970s, I found that Meadows and colleagues' *Limits to Growth* (1972) was mentioned far more often as a triggering text for activism and organizing than Carson's book; see Heise (2017).

References

Appel, Markus, and Barbara Malečkar. 2012. "The Influence of Paratext on Narrative Persuasion: Fact, Fiction, or Fake?" *Human Communication Research* 38: 459–84.

Archer, Jodie, and Matthew L. Jockers. 2016. *The Bestseller Code: Anatomy of the Blockbuster Novel.* New York: St. Martin's Press.

Bacigalupi, Paolo. 2015. *The Water Knife.* New York: Knopf.

Carson, Rachel. 1962. *Silent Spring.* Boston: Houghton Mifflin.

Culler, Jonathan. 1981. *The Pursuit of Signs: Semiotics, Literature, Deconstruction.* Ithaca, N.Y.: Cornell University Press.

Currey, James. 2008. *Africa Writes Back: The African Writers Series and the Launch of African Literature.* Oxford: James Currey.

English, James F. 2005. *The Economy of Prestige: Prizes, Awards, and the Circulation of Cultural Value.* Cambridge, Mass.: Harvard University Press.

Espósito, Fabio. 2009. "Seix Barral y el boom de la nueva narrativa hispanoamericana: Las mediaciones culturales de la edición española." *Orbis Tertius* 14 (15). https://www.researchgate.net/publication/42243389_Seix_Barral_y_el_boom_de_la_nueva_narrativa_hispanoamericana_Las_mediaciones_culturales_de_la_edicin_espaola.

Gallagher, Catherine. 2018. *Telling It Like It Wasn't: The Counterfactual Imagination in History and Fiction.* Chicago: University of Chicago Press.

Hall, Mark R. 2003. "The 'Oprahfication' of Literacy: Reading 'Oprah's Book Club.'" *College English* 65: 646–66.

Heise, Ursula K. 2017. "Comparative Literature and the Environmental Humanities." In *Futures of Comparative Literature: ACLA State of the Discipline Report,* edited by Ursula K. Heise, Dudley Andrew, Alexander Beecroft, et al., 293–301. London: Routledge.

An Inconvenient Truth. 2006. Directed by Davis Guggenheim. Performed by Albert Gore. Hollywood: Paramount.

James, Erin. 2015. *The Storywold Accord: Econarratology and Postcolonial Narratives.* Lincoln: University of Nebraska Press.

James, Erin, and Eric Morel, eds. 2020. *Environment and Narrative: New Directions in Econarratology.* Columbus: Ohio State University Press.

Loyola Brandão, Ignacio de. 1981. *Não Verás País Nenhum: Memorial Descritivo.* Rio de Janeiro: Codecri.

Mairal, Pedro. 2005. *El año del desierto.* Buenos Aires: Interzona.

Malecki, Wojciech, Piotr Sorokowski, Bogusław Pawłowski, and Marcin Cieński. 2019. *Human Minds and Animal Stories: How Narratives Make Us Care about Other Species.* London: Routledge.

Meadows, Donella, Dennis L. Meadows, Jørgen Randers, and William W. Behrens III. 1972. *The Limits to Growth: A Report for the Club of Rome's Project on the Predicament of Mankind.* New York: Universe.

Moretti, Franco. 1999. *Atlas of the European Novel, 1800–1900.* London: Verso.

Radway, Janice A. 1984. *Reading the Romance: Women, Patriarchy, and Popular Literature.* Chapel Hill: University of North Carolina Press.

Saval, José-Vicente. 2002. "Carlos Barral's Publishing Adventure: The Cultural Opposition to Francoism and the Creation of the Latin-American Boom." *Bulletin of Hispanic Studies* 79: 205–11.

Schneider-Mayerson, Matthew. 2020. "'Just as in the Book'? The Influence of Literature on Readers' Awareness of Climate Injustice and Perception of Climate Migrants." *ISLE: Interdisciplinary Studies in Literature and Environment* 27: 337–64.

Slovic, Scott, Swarnalatha Rangarajan, and Vidya Sarveswaran, eds. 2019. *Routledge Handbook of Ecocriticism and Environmental Communication.* London: Routledge.
Strange, Jeffrey J., and Cynthia C. Leung. 1999. "How Anecdotal Accounts in News and in Fiction Can Influence Judgments of a Social Problem's Urgency, Causes, and Cures." *Personality and Social Psychology Bulletin* 25: 436–49.
Wheeler, S. Christian, Melanie C. Green, and Timothy C. Brock. 1999. "Fictional Narratives Change Beliefs: Replications of Prentice, Gerrig, and Bailis (1997) with Mixed Corroboration." *Psychonomic Bulletin and Review* 6: 136–41.
Zwicker, Jonathan E. 2006. *Practices of the Sentimental Imagination: Melodrama, the Novel, and the Social Imaginary in Nineteenth-Century Japan.* Cambridge, Mass.: Harvard University Press.

Chapter 11

Two Cheers for Empirical Ecocriticism

GREG GARRARD

My first response, on reviewing this collection, is that I'm excited by this initiative and grateful to be involved. The overflowing seminar room assigned to the panel on empirical ecocriticism at the 2019 conference of the Association for the Study of Literature and the Environment at the University of California, Davis, showed that there is considerable appetite for approaches that decline to be cavalier with—even contemptuous of—evidence of various kinds. Too often, the canard of positivism is wielded so as to disclaim the need for empirical responsibility altogether. Whereas Timothy Clark's *The Value of Ecocriticism* (2019) rigorously interrogates the field's aspirations to fomenting cultural change, this volume provides not only a new answer to the question of value, but a new *kind* of answer.

Empirical ecocriticism seeks to bridge the gap between the environmental humanities and social sciences by encouraging traffic between fields that seem closely related: ecocriticism and environmental communication. I drew extensively on environmental communication articles by major figures such as Riley Dunlap, Matthew Nisbet, and Stephan Lewandowsky for my contributions to *Climate Change Scepticism: A Transnational Ecocritical Analysis* (2019), coauthored with Axel Goodbody, George B. Handley, and Stephanie Posthumus, in part because anti-environmentalism has hitherto been completely ignored by ecocritics. In 2016, I invited myself to a symposium on "Climate Denialism/Scepticism in a Warming World" at Linköping University in Sweden so I could meet an august group of social scientists and introduce them to ecocriticism. They were welcoming, if bemused.

One of the topics that came up repeatedly was the observation that levels of climate skepticism vary a good deal between nations. A widely cited paper by Min Zhou (2015) using a data set of 45,119 individuals from thirty-two countries, for example, finds that "environmental skepticism stems from insufficient education and self-assessed environmental knowledge, religious and conservative values, lack of trust in general society and science, and other concerns competing with environmental concern" (61). Zhou's sophisticated multivariate analysis, though, fails to explain the large differences between nations (Canada was the lowest and the Philippines the highest) in terms of, for example, relative affluence or objective vulnerability to environmental hazards. To me, the answer seemed obvious: differing cultural histories must have disposed populations to distinct patterns of environmental risk assessment. In acknowledging as much, though, Zhou treats "unique cultural and historical [context]" (73) as a black box inaccessible to empirical analysis. Ideally, the social sciences provide a mesoscale-level analysis of social and discursive patterns that complements macroscale accounts of historical change and microscale analyses such as close textual readings. In practice, mutual incomprehension, licensed and exaggerated by the epistemic hierarchy of disciplines from STEM through social sciences to the humanities and arts, is liable to prevail. Properly reciprocal interdisciplinarity takes time and commitment that few academics can spare.

Knee-jerk rejections of positivism notwithstanding, ecocritics may have reasonable grounds for concern about social science methods. Empirical researchers are, of course, well aware of the problem of ecological validity. They wonder, roughly, does my research actually represent the real world? Still, certain types of objections are less visible and therefore less easily addressed. These include research constructs that use biased language; reification of environmental attitudes; uncritical acceptance of the epistemic hierarchy mentioned above; and overconfidence in Likert scales as a means of translating various aspects of subjective experience into quantitative measures. An example of biased research constructs that we discovered when we reviewed social science studies of climate skepticism was that there was plenty of research into the reasons why climate deniers held their beliefs, but none into why others *were* persuaded by the evidence assembled (Garrard et al. 2019). Indeed, there isn't even a name for the latter category, "presumably because this reasonable default position is shared by the scholars themselves" (26). Contrary to the assumption embedded in much social science research that skeptics don't know or understand the science, Dan M. Kahan's (2015) research suggests that belief in climate change reflects—in the United States, at least—respondents' cultural identity, not their scientific literacy. "Warmists"—the name skeptics give to those in the pro-environmentalist camp—endorsed false survey claims about climate change (e.g., that it would increase skin cancer or reduce photosynthesis) (Kahan 2015, 20). Opinion surveys, organized by the assumption that climate skepticism is a deviation from reason that needs to be explained, purported to quantify belief in climate change, but actually measured America's deep partisan divide. In contrast with overtly subjective critique, empirical research cloaks its biases in a rhetoric of impersonality that typically includes use of passive voice, lack of explicit self-reference, and substitution of an inanimate text ("this book," "this thesis") as the source of truth claims in place of the author. Carolyn R. Miller (1979) describes these crisply as a "series of maneuvers for staying out of the way" (613).

Humanists' other objections to social science methods are related to one another, and they express differences in epistemic culture (Cetina and Reichmann 2015) and scholarly discourse between the humanities and social sciences. Literary theorists are deeply invested in the idea that people are divided against themselves by unconscious conflicts or ideological interpellation, such that the term "subject" is used in preference to "self" or "personality" because it encompasses contradictory meanings: a subject is at once active (e.g., in grammar, the subject of a sentence) and passive (e.g., in law, the subject of a sovereign). Many psychologists, by contrast, consider a given personality to be a unique, stable entity that is accurately, if not exhaustively, described by scores on five major traits: openness to experience, conscientiousness, extraversion, agreeableness, and neuroticism (or OCEAN). This psychometric construct of subjectivity, developed originally by statistical analysis of descriptive terms applied to people, has spawned a vast global industry of personality measurement and management. Whereas personality psychologists are reassured of the validity of the five-factor construct by its consistent replication across multiple studies, a literary theorist might suspect that those research subjects are interpellated, or subjectivated, as possessing certain traits by the studies themselves. Interpellation, as articulated most influentially by Louis Althusser, is best understood as a relationship of hailing and confirmation: "Hey you, are you *this* subject?" "Ah, yes, that's me!"

The theoretical critique of the subject is arguably overdrawn, not to mention founded on Marxist and psychoanalytical constructs still more dubious than the "big five" traits, but it does at least bring needed attention to our capacity for deep ambivalence, as well as context-dependent responses. The desire and seeming capacity to measure personality, moreover, reflects an epistemic hierarchy in which quantification is assumed to be superior (because considered more objective) to qualitative empirical research, which is in turn preferable to the hermeneutic or interpretive methods typical of much humanistic enquiry. The Likert scale is the primary means by which subjective experience is at once elicited and transmuted into quantifiable form. Whether the "facts" thereby generated concern teaching quality, attitudes toward animals, or belief in climate change, they reductively frame the cognitive and affective complexity of, say, living as an omnivore who loves their companion animals as a simple matter of "attitudes toward animal welfare" (W. P. Malecki, this volume). The editors seek to reassure ecocritics that the contributors use "the same methods that scholars, citizens, and policymakers generally trust to provide us with information about climate change, declining biodiversity, and environmental injustice" (Schneider-Mayerson, Weik von Mossner, et al., this volume), but their statement conflates academic disciplines (climatology, sociology, empirical ecocriticism) that produce very different kinds of knowledge. As Ian Hacking (1999) points out, "The fundamental idea is almost too simple-minded. People are self-conscious. They are capable of self-knowledge. They are potential moral agents for whom autonomy has been, since the days of Rousseau and Kant, a central Western value. Quarks and tripeptides are not moral agents and there is no looping [i.e., reflexive] effect for quarks" (59). Climate change is fantastically complex, but climatologists measure and model natural kinds of things, like solar irradiance, levels of greenhouse gases in the atmosphere, alterations of albedo due to land-use changes, and global mean surface temperatures. Such things are both indifferent to their characterization and unable to intervene in it. Sociologists and empirical ecocritics, in contrast, seek to measure what Hacking calls interactive kinds—that is, people who can and do comprehend, internalize, or dispute their classification. Whilst claims of environmental injustice do encompass empirical complexities, they also, centrally, embody notions of justice that rely on inherently reflexive human identities (Walker 2012). It is impossible to subject interactive kinds to quantitative study without some degree of reification.

Malecki's chapter on quantitative methods, for example, betrays such reification. He explains how psychologists use implicit association tests to circumvent subjects' abilities to work out the researchers' hypothesis and intentionally skew the results, and goes on to suggest that these tests could also enable researchers to reveal whether "self-identifying environmentalists may harbor racist attitudes toward climate refugees" (W. P. Malecki, this volume). Note that cognition and reflexivity are represented in this construct as barriers to true knowledge of the human person—hence the need to sneak past them with the help of implicit association tests—and that affective and ethical ambivalence is understood in terms of expressed environmentalist belief and secretly underlying "racist" attitudes. By adopting the ontology of empirical psychology, Malecki necessarily reconstructs or reifies an interactive kind as, functionally, a natural kind for the purposes of measurement.

As it happens, literature too is interested in both the stability and lability of human persons, although its construction of characters—people made of words—is refracted through layers of intentional plotting, generic expectation, historical development of narrative techniques, and, to a degree, changing scientific constructs of the psyche. Take Joseph Conrad's *Lord Jim* (1900), which narrates, primarily through Captain Marlow's retelling of Jim's story to an anonymous first-person narrator (i.e., homodiegetic narration), the life and death of the romantic young man of the title in the colonial Far East. Jim's life and his story both pivot around the moment when, fearing his ship, the *Patna*, was about to sink, he broke the seaman's code by abandoning it without regard for its passengers. Yet he cannot explain why he leapt into a lifeboat from the deck of the *Patna*:

> "I had jumped . . ." He checked himself, averted his gaze. . . . "It seems," he added.
>
> His clear blue eyes turned to me with a piteous stare, and looking at him standing before me, dumfounded *[sic]* and hurt, I was oppressed by a sad

> sense of resigned wisdom, mingled with the amused and profound pity of an old man helpless before a childish disaster.
>
> "Looks like it," I muttered.
>
> "I knew nothing about it till I looked up," he explained hastily. "And that's possible, too."

The rest of Jim's tale consists of his futile, ultimately fatal, efforts to redeem what seems to him, and to those around him, a terrible moral failure. Marlow understands too well, though, that the tragedy of Jim's leap is its elusiveness, not its immorality. Philosopher John Gray (2002) treats *Lord Jim* as an allegory that exposes free will as a cruel illusion: "Stuck in an incessant oscillation between the perspective of an actor and that of a spectator, Lord Jim is unable to decide what it is he has done. He hopes to dredge from consciousness something that will end his uncertainty. He is in search of his own character. It is a vain search. . . . The knowing I cannot find the acting self for which it seeks" (67–68). Again, empirical psychologists and social scientists in general are well aware that knowledge, beliefs, and actions are loosely articulated at best, but the notion of an abyss between "I" and "self" would seem to render moot their efforts to quantify, let alone rectify, their relationship.

Conrad, like his contemporary Thomas Hardy, was fascinated by the contest between the frail, struggling human will and the indifferent, implacable forces that frequently overwhelm it. At the same time, the nested narrators of *Lord Jim* mediate Jim's story, never giving him direct access to the reader. Just over a century later, Ian McEwan's *Saturday* (2005) is a more traditional realist novel with an authorial narrator and a privileged (in both social and narrative terms) focalizer: neurosurgeon Henry Perowne. The narrator's use of free indirect discourse creates, in this case, an intimacy between them that conveys robust Enlightenment optimism in the power of science and medicine to comprehend (and, increasingly, fix) human flaws—flaws that now reside in our genes and our brains, rather than flowing from our place in the cosmos. Perowne is consequently dismissive of literature, especially the "irksome confections" (67) of the magic realists who fascinate his daughter Daisy: "A man who attempts to ease the miseries of failing minds by repairing brains is bound to respect the material world, its limits, and what it can sustain—consciousness, no less. It isn't an article of faith with him, he knows it for a quotidian fact, the mind is what the brain, mere matter, performs. If that's worthy of awe, it also deserves curiosity; the actual, not the magical, should be the challenge" (67).

The "actual" is the challenge that McEwan has set himself in this novel, especially in the figure of Baxter, the plot's antagonist, a working-class crook in the early stages of neurodegenerative Huntington disease. Unlike the case of Jim, there is no mystery about Baxter's behavior: "Anyone with significantly more than forty CAG repeats in the middle of an obscure gene on chromosome four is obliged to share this fate in their own particular way. *It is written.* No amount of love, drugs, Bible classes or prison sentencing can cure Baxter or shift him from his course. It's spelled out in fragile proteins, but it could be carved in stone, or tempered steel" (210). Enraged that Perowne embarrassed him by identifying his condition, Baxter invades the family home with some henchmen. The climax of the novel occurs when Baxter's plans to rape and murder the doctor's family are thrown off course by Daisy's recitation of Matthew Arnold's poem "Dover Beach." Such erratic behavior could just exemplify the horrible determinism of Baxter's deterioration, but McEwan's choice of a poem that pleads for human kindness to mitigate the bleakness of Darwinian disenchantment seems to imply a recuperation of humanism that encompasses scientific knowledge of the self, rather than being antithetical to it (Garrard 2010). Despite the fatality of Baxter's condition and the astonishing capabilities Perowne displays on behalf of medical science, there remains an irreducible singularity in their interaction, as in McEwan's narration of it.

The point is this: just as Likert scales interpellate research subjects as stable, knowable and quantifiable beings, so literary texts interpellate readers as, in Conrad and McEwan, utterly (if differently) inscrutable creatures. The authors collected here argue persuasively that social science methods can complement those of literary studies (and one might also consult co-editor Schneider-Mayerson's superb *Peak Oil* [2015] for an extended example), but they have not yet confronted the question of how empirical and literary interpellations interact or conflict. *Saturday,* for example, hails readers as, ironically, empiricist skeptics of literature, which may account for literature students' near-universal dislike of the novel. One could imagine an empirical study of readers of *Saturday* who initiated an inescapable loop of reflexivity. Readers can obviously be recruited as research subjects, but can such subjects remain *readers,* in a humanistic sense?

Traditional literary scholars would be bothered as much by empiricists' instrumentalizing conception of literature as by their construction of the reading subject, but many ecocritics have conceded that point in advance by touting their work as a form of cultural activism. If the purpose of ecocriticism is to achieve environmentalists' ends, then the best literature just is the most effective, irrespective of supposed literary quality—and why not ensure it is measurably effective? Here too the discourse of empiricism jars with the received assumptions of literary critics. Weik von Mossner and colleagues' fascinating treatment of Alice Walker's "Am I Blue?" defines it as a "radical text" (Weik von Mossner et al., this volume) because it appears to challenge meat eating, and perhaps also because it is hard to allocate to a genre (e.g., autobiography, essay, short story). They discover that the text's explicit analogy between enslavement of humans and our treatment of livestock actually seems to block increased concern with animal welfare—not at all what enthusiasts for interspecies intersectional analysis would anticipate. Compared to *Lord Jim,* though, "Am I Blue?" is anything but radical. It features a single autodiegetic narrator who can be comfortably identified with the esteemed author, experiencing a change of heart with an unambiguous conclusion: she spits out her steak. The implicit constraints on this conception of radicalism would seem to rule modernist texts entirely out of contention.

I once attended a summer matinee performance of Samuel Beckett's *Happy Days* (1961) at the Theatre Royal in Bath. The lead role of Winnie (indeed, the only character, apart from her husband, Willie, who occasionally shouts out newspaper quotations) was played by Felicity Kendal, best known for her charming performance in the British sitcom *The Good Life* (1975–78). The theater was packed with people, mainly pensioners, who evidently had no idea what was about to happen. They were audibly perturbed when Winnie appeared from the outset buried to her waist in a sandy mound. Quite a few people walked out. During the intermission, a couple of elderly women said to me, "You're young! Maybe you can tell us what's going on." I hope my reply persuaded them that the accidental audience was exactly the right one, if they were open to it, because the play is about aging and loss. Kendal's stunning performance, comical and heartbreaking, hangs in my memory on Winnie's oxymoronic query: "What is that unforgettable line?"

Though not in any sense environmental, *Happy Days* is likely to fail on every score of empirical criticism. It seems expressly designed to deflate "self-efficacy" (Schneider-Mayerson, Gustafson, et al., this volume), deter "identification" (my students would say "relatability"), and defer "transportation"—to the point that some audience members simply left. This and other Beckett plays, though, have nevertheless proven influential and, yes, unforgettable. Sara Warner and Jeremy Jimenez's essay in this volume on their participatory climate theater project acknowledges the tension between artistic polysemy and empirical validity; it recites a defense of avant-garde art first proposed by Russian Formalists over a century ago: "It is the difference within repetition that interrupts our habitual ways of experiencing and perceiving—making the familiar strange and the strange familiar—that allows us to consider how things might be otherwise." The problem is, epiphanies like those in *Happy Days* are rare and unpredictable, and hence thoroughly unsuited for systematic, controlled study. While empiricists are measuring modest, transient effects in statistically significant populations they may miss, like the speaker of R. S. Thomas's (1993) "The Bright Field," "the / pearl of great price, the one field that had / treasure in it" (302). This is a question of singularity (Attridge 2004), which has nothing to do with big novels (Schneider-Mayerson, Gustafson, et al., this volume) or "sleeper effects" (Weik von Mossner et al., this volume), but which is something closer to the nonlinear artistic impact broached by Nicolai Skiveren in this volume.

All this carping might seem to suggest I am hostile to empirical ecocriticism, but that is far from the truth. The wide variety of methods encompassed by this collection, including qualitative, phenomenological, and computational ones, anticipates and defuses these objections to a degree—in some cases quite directly. This very diversity implies that empiricism signals a welcome commitment to facts, procedural rigor, and productive interdisciplinarity rather than denoting a predetermined method or cluster of methods. I applaud its challenge to untested assumptions—including my own (Schneider-Mayerson, Gustafson, et al., this volume)—and its openness to interpretive as well as quantitative methods. I am surprised and delighted to discover that other people's brains, too, process fiction and nonfiction just the same (W. P. Malecki, this volume), and my atheistic philosopher's soul thrills to the very idea that an ecocritical proposition might be subjected to "null hypothesis significance testing"—that is, a determination of whether the pattern perceived by the critic has any meaning at all. I am optimistic that in time, we will find that there is more of a continuum between environmental communication and ecocriticism than the binary divide I have rhetorically evoked here.

The missing piece is pedagogy. Despite its activist rhetoric, ecocriticism has always had a better claim to sway students than to impact publics more generally; indeed, Jonathan Bate's *The Song of the Earth* (2000) was uniquely ambitious in this regard. In common with unscholarly fiction readers, students start out preferring relatability (Schneider-Mayerson, Gustafson, et al., this volume), but some, at least, come to appreciate and even expect estrangement over the dozen weeks of a semester. In my own pedagogical work, I have called for more objective evidence that instructors of ecocriticism have the impact on students that we hope for (Garrard 2007); contrarily perhaps, I have also argued that ecocritical pedagogy should respect the autonomy and preexisting knowledge of students (Garrard 2017). There is no need to assume, though, that what John Parham (2006) calls a transactional pedagogy is resistant to empirical study, even though he is critical of both the vocationalizing of higher education and the instrumentalizing of environmental literature. A large-scale cross-cultural mixed-methods study in which students played a role in defining what should count as impact, for example, would satisfy the demands of both empiricism and humanism. I'm game if you are.

References

Attridge, Derek. 2004. *The Singularity of Literature.* London: Psychology Press.

Bate, Jonathan. 2000. *The Song of the Earth.* Cambridge, Mass.: Harvard University Press.

Beckett, Samuel. 1961. *Happy Days.* New York: Random House.

Cetina, Katrin Knorr, and Werner Reichmann. 2015. "Epistemic Cultures." *International Encyclopedia of the Social and Behavioral Sciences* 7: 873–80.

Clark, Timothy. 2019. *The Value of Ecocriticism.* Cambridge: Cambridge University Press.

Conrad, Joseph. 1900. *Lord Jim.* https://gutenberg.org/ebooks/5658.

Garrard, Greg. 2007. "Ecocriticism and Education for Sustainability." *Pedagogy* 7: 359–83.

Garrard, Greg. 2010. "Reading as an Animal: Ecocriticism and Darwinism in Margaret Atwood and Ian McEwan." In *Local Natures, Global Responsibilities: Ecocritical Perspectives on the New English Literatures,* edited by Laurenz Volkmann, 223–42. New York: Rodopi.

Garrard, Greg. 2017. "Towards an Unprecedented Ecocritical Pedagogy." In *Teaching the New English,* edited by Ben Knights, 189–207. London: Palgrave Macmillan.

Garrard, Greg, Axel Goodbody, George B. Handley, and Stephanie Posthumus. 2019. *Climate Change Scepticism: A Transnational Ecocritical Analysis.* London: Bloomsbury Academic.

Gray, John. 2002. *Straw Dogs: Thoughts on Humans and Other Animals.* London: Granta.

Hacking, Ian. 1999. *The Social Construction of What?* Cambridge, Mass.: Harvard University Press.

Kahan, Dan M. 2015. "Climate-Science Communication and the Measurement Problem." *Political Psychology* 36: 1–43.

McEwan, Ian. 2005. *Saturday.* London: Jonathan Cape.

Miller, Carolyn R. 1979. "A Humanistic Rationale for Technical Writing." *College English* 40: 610–17.

Parham, John. 2006. "The Deficiency of 'Environmental Capital': Why Environmentalism Needs a Reflexive Pedagogy." In *Ecodidactic Perspectives on English Language, Literatures, and Cultures,* edited by Sylvia Mayer and Graham Wilson. Trier, Germany: Wissenschaftlicher Verlag.

Schneider-Mayerson, Matthew. 2015. *Peak Oil: Apocalyptic Environmentalism and Libertarian Political Culture.* Chicago: University of Chicago Press.

Thomas, R. S. 1993. *Collected Poems, 1945–1990.* London: Phoenix.

Walker, Gordon P. 2012. *Environmental Justice.* London: Routledge.

Zhou, Min. 2015. "Public Environmental Skepticism: A Cross-National and Multilevel Analysis." *International Sociology* 30: 61–85.

Chapter 12

Empirical Ecocriticism and Modes of Persuasion

DAVID I. HANAUER

As stated in the Introduction to this book and reiterated in different forms in the subsequent chapters, the aim of empirical ecocriticism is to facilitate an interdisciplinary research initiative and discussion designed to enhance, through empirical research, the ways in which the science, causes, and outcomes of a range of socioecological issues (such as plastic pollution, humans' treatment of and relationship to animals, and environmental toxicity) are communicated to the broader public. Put another way, the aim is to leverage scientific methodology on reception in order to facilitate a database from which better modes of persuasion can be enacted so that there is far broader acceptance of the realities of the effects of human activity on the biosphere and climate. From my perspective as an applied linguist, trained to look at the ways in which language works in the world, and as an empirical researcher of literature, trained to explore the psychosocial effects of reading and writing literary texts, I would define the aim of this initiative as one of applied literariness (Hanauer 2018). Applied literariness research utilizes what we know about the reading (and writing) of literary texts (broadly defined across many modalities, including visual media) in order to address real-world social issues. In the case of empirical ecocriticism, applied literariness is being applied to explicating and communicating the various issues associated with a range of socioecological issues.

A careful reading of the chapters in this book makes it clear that the broad agenda in raising awareness and the core underpinning hypothesis in relation to interacting with the human effects in relation to socioecological issues is that once awareness has been raised in enough people, corrective action will emerge. The majority of the empirical chapters presented here have the degree of position change in relation to socioecological issues (whether measured using qualitative or quantitative methods) as the central outcome measure. This seems in line with the underpinning hypothesis that empirical ecocriticism will provide research-informed data on how to persuade others.

The argument that literary forms, and narrative in particular, should be used for the explication of scientific knowledge in order to increase scientific literacy in broad, lay populations is not a new idea. Prior research has both proposed and explored this direction (Avraamidou and Osborne 2009; Klassen 2010). The core argument for narrative in fulfilling this communicative role is situated in the cognitive advantages of this particular form of communication. Empirical research has shown that narrative facilitates ease of comprehension, increased recall, longer-term recall, and intrinsic motivation for continued reading (Graesser, Olde, and Klettke 2002; Glaser, Garsoffky, and Schwan 2009; Graesser and Ottati 1995; Schank and Abelson 1995; Zabrucky and Moore 1999). To an extent, narrative is the most basic form of human comprehension, and as such, it has a privileged position in terms of cognition (Glaser, Garsoffky, and Schwan 2009). It is therefore a natural extension to consider the role of narrative in actually directing a change in the perception of current developments with environmental issues, and climate change in particular.

In many ways, the core argument for narrative, arts-based persuasion in relation to socioecological issues presented in this book parallels the role stated for narrative in the field of medical science. Prior research in this area has looked at the ability of narratives to persuade patients to educate themselves more extensively about breast cancer care and prevention (Wise et al. 2008) and for a wide range of patients to opt in to receiving vaccinations (Betsch et al. 2011; Hopfer 2012; Nan et al. 2015). However, a meta-analysis of these studies did not provide overriding positive support for the role of narrative persuasion in helping patients make appropriate medical decisions (Winterbottom et al. 2008). Two thirds of the studies analyzed did not produce the desired direction of changes in decision making. Anna Winterbottom and colleagues (2008) conclude that within the realm of health care decision-making, the narrative presentation of information should be used cautiously. They found that there was little evidence to suggest that the presentation of narrative information improved patients' decision-making, and there was the option of pushing patients toward a decision detrimental to their health. As reported in a series of recent experiments by Appel (2022), the use of narrative persuasion can result in affective resistance leading to enhanced oppositional positions and behaviors. This is a danger for the implementation of narrative persuasion in relation to medical and ecological issues. The use of narratives in relation to socioecological issues with the aim of changing personal behavior may have mixed outcomes (Schneider-Mayerson 2020).

One avenue for considering the value of narrative is to look at the research on narrative persuasion. Broadly, there is agreement that narratives, whether fictional or factual, have the potential to be persuasive (Green and Brock 2000; Slater and Rouner 2002). However, this broad definition is qualified by a series of conditions. Tied as it is to the depth of emotional and cognitive engagement in the reading of the text, persuasion may be caused by the degree to which a reader is transported into the created world of the text (Green 2006; Green and Brock 2000). As seen in studies of transportation in literary reading, this mode of reading does not happen for every reader under all circumstances, no matter how appealing the text may be. Narratives may be persuasive if there is a degree of emotional involvement and identification with the characters in the story (Green, Brock, and Kaufman 2004; Moyer-Guse 2008). But once again, having an emotional response and producing a sense of identification does not happen uniformly across the population, and it is difficult to systematically engineer on a large scale. While there is no assumption within empirical ecocriticism that a single literary product or setting will make the difference, the problem still stands that even for a collection of specific literary objects, we do not know definitively how they will interact with a wide variety of populations.

There is a connection between factuality and the ability of literary texts to persuade. However, it is unclear what the direction actually is. On the one hand, in some studies, positioning the text as factual increased its persuasive ability; stating the text was fiction made it easier for readers to dismiss the persuasive message (Hanauer 2018). On the other hand, there is evidence that people feel comfortable using knowledge that finds its source in fiction in explaining the world around them, demonstrating the depth to which fiction can produce mental models of the world (Appel and Richter 2007; Busselle and Bilandzic 2008; Marsh, Meade, and Roediger 2003). Research has shown that narratives are persuasive when the intent to persuade is hidden and fail to persuade when the reader has the sense that they are being directly manipulated (Gerrig 1993; Green 2006).

These are just a few of the main conditions that need to be met for a narrative to be persuasive. As known by scholars who have conducted empirical studies on literary and media effects, literary reading and writing often include multivariate interactions in which various variables on the levels of the reader, text, and context play a role in the way the text is received or composed. While we have sophisticated models of literary reading (which are relevant for any media object that utilizes narrative), such as Arthur M. Jacobs's (2015a, 2015b) model of neurocognitive processing, these tend to be large frameworks under which a range of different responses could emerge. Accordingly, the simple movement from narrative to persuasion is not simple at all. There is documented evidence of the failure of narrative to achieve the desired persuasive outcomes (Appel 2022; Slater et al. 2003, 2006; Yoder, Hornik, and Chirwa 1996). The main aim of empirical ecocriticism is to try to produce actionable knowledge with which to convince individuals to change personal behaviors as a result of a narrative exchange. One needs to be realistic in terms of what can be achieved. Indeed, as evidenced by the empirical studies in this book, the results are far from overwhelming. For example, in their essay in this volume, Matthew Schneider-Mayerson and colleagues found no significant long-term effects for the reading of climate fiction; and in their contribution to this volume, Alexa Weik von Mossner and colleagues found that reading the controversial short story "Am I Blue?" did not change the positions of readers who did not already hold a preexisting orientation similar to the story's persuasive intent. As postulated in the Introduction, narrative/literary persuasion may indeed be possible, and it is definitely worth studying. However, as understood by the authors of this book, it is not a simple or straightforward applied research paradigm with clear, actionable results.

The particular problems of using narrative to persuade individuals to change their behavior are compounded in relation to the particular aspects of socioecological issues. Stephen M. Gardiner (2006) clearly explicates the depth of the problem faced by anyone trying to address ecological issues, and climate change in particular, through persuading the public. The problem partly resides in relation to the acute temporal and spatial dispersion of the cause–effect relationships inherent in climate science. Basically, human action (or inaction) taken today will have serious effects in the (distant) future; the human actions at one site may have very serious effects in a different and spatially distant part of the world. The outcomes of human climate behavior today involve a lengthy lag time before their true nature becomes blatantly apparent. Furthermore, human actions taken at one site may have serious outcomes at a future date in a completely different and distanced location with people who one knows nothing about. As explicated by Gardiner (2006), in conceptual and ethical situations like this, immediate individual well-being comes into conflict with broader global aspects of ethical behavior. To what degree should I give up my comfort for some hypothesized future effect which might be better for someone else? It is easy to hedge and delay one's response when the outcomes seem so distant. It is easy to become complacent and believe that others will deal with the issue at the correct time. This is true even when people are convinced that climate change is happening and is the result of human actions. The core question is, to what degree am I willing to change my behavior and to live in a society without fossil fuels, excessive consumerism, meat consumption, airplane travel, private car usage, and so on, in order to address a future and distant negative outcome? Awareness is necessary, but it is not sufficient, even in relation to taking the required moral positions and actively advocating for the required political and policy decisions.

However, beyond these conceptual difficulties with cause and effect and individual responsibility, the central hurdle pinpointed by Gardiner (2006) in relation to reversing or even slowing climate change is that the only truly effective form of action is a global governing response. Unfortunately, governments are not acting quickly enough. According to independent research group Climate Action Tracker (2023), none of the thirty-eight countries they tracked in March 2023 was found to be Paris Agreement compatible, while three quarters were found to be significantly behind in terms of climate action they had already committed to. Under Trumpian and nationalist politics in the European Union and United States, funds for climate change initiatives have been limited (Appelt and Dejgaard 2018; Carrington 2017; Meade 2018), and global leadership roles have diminished (Bäckstrand and Elgström 2013).

In his early philosophical analysis, Gardiner (2006) points out that the biggest hurdle to taking action on climate change is the fragmentation of the world into self-interested nation-states. While every state has an interest in addressing climate change, few are willing to commit to it as aggressively as they should, and all prefer that other states take on a similar or greater burden. At the heart of this reluctance to act is the national perception of the importance of short-term economic growth (Martine and Diniz Alves 2019), the political power of fossil fuel companies (Grasso 2019), and popular support for continual consumerism (Wilk 2017). If at the heart of global agreements is the clause that "economic growth will be sustained and sustainable" (Martine and Diniz Alves 2019, 8), then meaningful reductions in harmful human action in these countries is highly improbable.

Accords that have been created, such as the Paris Agreement, seem to lack enforcement sanctions or sufficient initiatives for serious current action, as is clear when considering the outcomes monitored by Climate Action Tracker. They are a facsimile of global action on climate issues negotiated by nation-states, each more interested in protecting their own interests than solving global climate change. This situation, which Gardiner (2006) terms "institutional inadequacy," makes significant movement on climate change exceedingly difficult. Beyond these challenges, the task of mobilizing global governance has further receded after the Covid-19 pandemic (Levy 2020). In the face of a global health crisis, national governments closed borders, strengthened their own internal controls over their populations, bid against other countries for relevant resources, and either refused or severely limited any international cooperation (Woods et al. 2020). While global cooperation and shared responsibility were required for the health of all, as the pandemic surged, governments across the world were incapable of working together. This does not bode well for future cooperation on socioecological issues.

The Covid-19 pandemic accelerated political processes that were already underway in democratic countries against many forms of global action (Levy 2020). Populist nationalism combined with economic protectionism and xenophobic isolationism, orientations that oppose globalism, were solidified by the Covid-19 pandemic (Norris and Inglehart 2019). Beyond global cooperation, on a national level, the current political climate in the majority of democratic countries exposes deep rifts between more progressive and more conservative factions (Goodhart 2017). Facilitating global cooperation on socioecological issues seems a Herculean task against this backdrop, and Covid-19 has already laid bare the world's lack of effective governance when faced with a global challenge.

An additional complicating factor in using narrative persuasion in relation to socioecological issues results from both the rise of populist nationalism and the Covid-19 pandemic. Covid-19 was not just a pandemic but also an "infodemic" in which misinformation, predominantly in the form of narrative-based conspiracy theories, countered more valid scientific forms of knowledge (Pian, Chi, and Ma 2021; Zarocostas 2020). Conspiracy narratives were spread widely through social media (Naeem, Bhatti, and Khan 2021) and underpinned significant sections of the public's objections to wearing masks, keeping socially distant, and getting vaccinated (Ali 2022; Bertin, Nera, and Delouvee 2020; Biddlestone, Green, and Douglas 2020; Romer and Jamieson 2020). The underpinning structure of a conspiracy theory is a narrative in which social events (such as a pandemic and economic suffering) are presented as malicious acts enacted by secret and powerful groups (Douglas, Sutton, and Cichocka 2017; Douglas et al. 2019). Those who are susceptible to these narratives may be looking for meaning, control, and power in the face of adverse events (Douglas et al. 2016; Kruglanski, Molinario, and Lemay 2021; van Prooijen, Douglas, and de Inocencio 2018). Often beliefs in particular conspiracy theories are deeply rooted in an individual's political and group identity, making it difficult to facilitate a change of positions (Uscinski, Klofstad, and Atkinson 2016). For people with these particular psychological positionings, the use of narrative is powerful and completely overrides other cognitive faculties. In one interesting study, Daniel Romer and Kathleen Hall Jamieson (2020) report that belief in conspiracy theories related to Covid-19 in the United States completely undercut the evaluation of the threat posed by the pandemic. We are seeing a similar dynamic with global ecological issues. The work conducted on conspiracy theories and misinformation during the Covid-19 pandemic highlights the difficulties faced in convincing large segments of the population, particularly those with a more conservative political orientation, of the real threat from socioecological issues.

Anyone involved in promoting an increased understanding of the impacts of human behavior on our planet knows that addressing socioecological issues is complex; it is not something that can be easily reversed with a few narratives, no matter how well they are constructed or how empirically sound they are. As explicated above, understanding the coming (and current) ecological disasters and acting in relation to these global ecological threats involves multiple difficulties: people are not easily persuaded by narratives that are not within their ideological identity positioning; governments have integral, short-term conflicts of interest in conducting meaningful environmental action; and the outcomes of the Covid-19 pandemic have left the world deeply divided. This set of circumstances makes this book's endeavor, however worthy, far more difficult to achieve.

My aim here is not to make readers despair but rather to offer a reminder about the size of the undertaking. I do not assume that empirical ecocriticism plays no role here or that it can do no good. This is not a message of hopelessness. It is, however, a reality check on the limits of the empirical ecocritical approach—limits that result from the size and complexity of the problem of persuading governments across the world of the need for serious policy change and significant action.

What, then, do I perceive to be the role of empirical ecocriticism? Let me start by stating the obvious: awareness and understanding of the science of socioecological issues is necessary. Awareness is a necessary, if insufficient, condition for any sort of political change to occur. Increasing the number of people who understand climate science and the quality of that understanding is a positive thing, even if this is limited to more progressive sections of the population. I do not assume that this knowledge would translate into policy or meaningful large-scale action. That is, I do not assume that this awareness will lead to political action unless additional steps are taken. Within empirical ecocriticism, the focus needs to be on how to facilitate political persuasion and mobilization. How does awareness become political pressure for change, and what might this look like?

One approach I would consider is integrating the method of persuasion with training on collective political and scientific action; in such an approach, the modes of convincing and teaching about socioecological issues are at the same time teaching people how to act collectively and politically. Perhaps this would offer a bridge from persuasion to political, or at least collective, action. One of the approaches outlined in the chapters of this book resonates more directly with my own understanding of how collective action can be influenced by research and literary practices. The chapter in this volume by Sara Warner and Jeremy Jimenez develops the applied drama approach by working collectively with dramatists, students, and scientists in a local setting.

Within science education, there is an educational model that is directly relevant to the approach I advocate. A course-based research experience (Auchincloss et al. 2014; Hanauer et al. 2017) positions first-year undergraduate students in a context in which they conduct real research that contributes to their communities, to science, and to society. Importantly, this is not a facsimile of doing research; it involves actually contributing something new to science. To understand what this shift in educational orientation means and its relevance to the concerns of this book, consider a course designed so that students study local water supplies measuring pollution levels and then construct arts-based ways of communicating this information to a variety of actors, such as local communities and the municipality. What we know about this type of educational approach is that students become engaged and feel extensive ownership over their work (Hanauer et al. 2017). Furthermore, they tell their friends and family about what they are doing, including details of the science (Hanauer et al. 2017).

In recent work, this type of approach was applied to the issue of countering manipulative and coercive language within the sphere of democratic discourse (Hanauer 2022). Using the core design of the research experience, which involves students directly in applied research, stylistic analyses of language are used by students in a variety of courses to analyze and then report on topical political issues in the public sphere. These analyses are then reported back to the communities within which they were generated, creating an informed response on how the manipulation is conducted, with the aim of lessening its effect and facilitating more informed discussion. An approach like this example of an arts-based course research experience may be useful for the aims of empirical ecocriticism.

These types of courses can be extended into broad networks. The Integrated Research and Education Community (iREC) is a design that involves large-scale collaboration across multiple types of institutions, all involved in exploring a shared research question, such as exploring microbiological diversity (Hanauer et al. 2017). The underpinning educational design has three parallel outcomes: it creates a community of practice, it advances scientific understanding, and it makes its results relevant for broader communities of actors. The value of the iREC is that instead of just having a local effect, suddenly there is the potential for thousands of students to work on a similar set of issues across multiple settings.

For this discussion about the role of empirical ecocriticism, let me suggest a particular form of iREC. As with the theater program from Cornell University described by Warner and Jimenez in this volume, the type of persuasion I am interested in involves homegrown and immediate climate-related issues of relevance to local publics, uses large-scale participation of local people, is scientifically sound, enacts collective action, and produces participatory artistic outcomes. This type of approach addresses many of the problems I specified earlier. First, it is easier to envision what is happening in your own backyard as opposed to across the expanse of time and space. Second, rather than just being persuaded, you are actively involved in researching and disseminating your findings in the community where they are happening. Third, it is difficult to affect global or national politics, but it is much easier to influence local politics. A group of activists, basing themselves on real data from a local setting and having the ability to artistically/literarily express their findings in a local community, has the ability to change local policies about specific environmental issues. If an empirical ecocriticism iREC is constructed, multiple groups could act on the same issue across a variety of different sites. This could create the type of political pressure needed to facilitate enactment of the needed environmental policy.

Empirical ecocriticism could also help with the formulation and assessment of programs like this, which could be replicated across the United States and the world. My recommendation for empirical ecocriticism would be to think carefully about setting up networks of individuals working on a wide range of ecological issues at specific locations. Initiatives within science education have gone far in developing models of this for undergraduate education. It would seem not too far a leap to add components to these models that address narrative, artistic expression, and political action.

In this response to the initiation of empirical ecocriticism, my main aim has been to explore the limitations and potential of this approach. Broadly, I see this as an extension of applied literariness approaches in relation to socioecological issues. My hope is that this field will borrow some of the educational approaches now used in the sciences in creating networked, engaged student communities working on real issues in their local and national (even international) communities. The particular role that empirical ecocriticism would play is in harnessing the power of arts-based research practices to create special iREC structures that can collectively have a significant effect. Narrative lends itself to these types of approaches, and if used properly, it could lead to positive outcomes on local and national environmental policies. The task before empirical ecocriticism, similar to that of researchers in environmental communication, is daunting, but that does not mean action should not be taken. Like other grassroots practices, empirical ecocriticism might be able to contribute to other groups with similar agendas. Empirical ecocriticism should provide a platform through which a wide range of individuals working together locally and in a networked fashion can exert political pressure. Although this will not single-handedly resolve the socioecological problems we face, it can be effective, it can contribute at the local level, and it is attainable.

References

Ali, Sana. 2022. "Combatting against Covid-19 and Misinformation: A Systematic Review." *Human Arenas* 5 (2): 337–52.

Appel, Markus. 2022. "Affective Resistance to Narrative Persuasion." *Journal of Business Research* 149: 850–59.

Appel, Marcus, and Tobias Richter. 2007. "Persuasive Effects of Fictional Narratives Increase Over Time." *Media Psychology* 10 (1): 113–34.

Appelt, J., and H. P. Dejgaard. 2018. "An Analysis of the Climate Finance Reporting of the European Union." ACT Alliance E.U., April 2018. https://actalliance.eu/wp-content/uploads/2018/04/Analysis-of-the-climate-finance-reporting-of-the-EU.pdf.

Auchincloss, Laura Corwin, Sandra L. Laursen, Janet L. Branchaw, et al. 2014. "Assessment of Course-Based Undergraduate Research Experiences: A Meeting Report." *CBE Life Science Education* 13: 29–40.

Avraamidou, Lucy, and Jonathan Osborne. 2009. "The Role of Narrative in Communicating Science." *International Journal of Science Education* 31 (12): 1683–707.

Bäckstrand, Karin, and Ole Elgström. 2013. "The E.U.'s Role in Climate Change Negotiations: From Leader to 'Leadiator.'" *Journal of European Public Policy* 20 (10): 1369–86.

Bertin, Paul, Kenzo Nera, and Sylvain Delouvee. 2020. "Conspiracy Beliefs, Rejection of Vaccination, and Support for Hydroxychloroquine: A Conceptual Replication-Extension in the Covid-19 Pandemic Context." *Frontiers in Psychology* 11: 2471.

Betsch, Cornelia, Corina Ulshöfer, Frank Renkewitz, and Tillman Betsch. 2011. "The Influence of Narrative v. Statistical Information on Perceiving Vaccination Risks." *Medical Decision Making* 31 (5): 742–53.

Biddlestone, Mikey, Ricky Green, and Karen M. Douglas. 2020. "Cultural Orientation, Power, Belief in Conspiracy Theories, and Intentions to Reduce the Spread of Covid-19." *British Journal of Social Psychology* 59 (3): 663–73.

Busselle, Rick, and Helena Bilandzic. 2008. "Fictionality and Perceived Realism in Experiencing Stories: A Model of Narrative Comprehension and Engagement." *Communication Theory* 18: 255–80.

Carrington, Damian 2017. "Fossil Fuels Win Billions in Public Money after Paris Climate Deal, Angry Campaigners Claim." *Guardian,* October 12, 2017.

Climate Action Tracker. 2023. "Climate Action Tracker: Overview." Accessed March 17, 2023. https://climateactiontracker.org/countries/.

Douglas, Karen M., Robbie M. Sutton, Mitchell J. Callan, Rael J. Dawtry, and Annelie J. Harvey. 2016. "Someone Is Pulling the Strings: Hypersensitive Agency Detection and Belief in Conspiracy Theories." *Thinking and Reasoning* 22: 57–77.

Douglas, Karen M., Robbie M. Sutton, and Aleksandra Cichocka. 2017. "The Psychology of Conspiracy Theories." *Current Directions in Psychological Science* 26: 538–42.

Douglas, Karen M., Joseph E. Uscinski, Robbie M. Sutton, et al. 2019. "Understanding Conspiracy Theories." *Political Psychology* 40: 3–35.

Gardiner, Stephen M. 2006. "A Perfect Storm: Climate Change, Intergenerational Ethics and the Problem of Moral Corruption." *Environmental Studies* 15 (3): 397–413.

Gerrig, Richard J. 1993. *Experiencing Narrative Worlds: On the Psychological Activities of Reading.* Boulder, Colo.: Westview.

Glaser, M., B. Garsoffky, and S. Schwan. 2009. "Narrative-Based Learning: Possible Benefits and Problems." *Communications—European Journal of Communication Research* 34 (4): 429–47.

Goodhart, David. 2017. *The Road to Somewhere: The Populist Revolt and the Future of Politics.* London: Hurst.

Graesser, Arthur C., Brent Olde, and Bianca Klettke. 2002. "How Does the Mind Construct and Represent Stories?" In *Narrative Impact: Social and Cognitive Foundations,* edited by Melanie C. Green, Jeffrey J. Strange, and Timothy C. Brock, 229–62. Mahwah, N.J.: Lawrence Erlbaum.

Graesser, Arthur C., and Victor Ottati. 1995. "Why Stories? Some Evidence, Questions, and Challenges." In *Knowledge and Memory: The Real Story,* edited by Robert S. Wyer, 121–31. Hilldale, N.J.: Lawrence Erlbaum.

Grasso, Marco. 2019. "Oily Politics: A Critical Assessment of the Oil and Gas Industry's Contribution to Climate Change." *Energy Research and Social Science* 50: 106–15.

Green, Melanie C. 2006. "Narratives and Cancer Communication." *Journal of Communication* 56: 163–83.

Green, Melanie C., and Timothy C. Brock. 2000. "The Role of Transportation in the Persuasiveness of Public Narratives." *Journal of Personal Social Psychology* 79 (5): 701–21.

Green, Melanie C., Timothy C. Brock, and G. E. Kaufman. 2004. "Understanding Media Enjoyment: The Role of Transportation into Narrative Worlds." *Communication Theory* 14: 311–27.

Hanauer, David I. 2018. "Intermediate States of Literariness: Poetic Lining, Sociological Positioning, and the Activation of Literariness." *Scientific Study of Literature* 8 (1): 114–34.

Hanauer, David I. 2022. "Pedagogical Stylistics in the Service of Democracy." In *Pedagogical Stylistics in the 21st Century,* edited by S. Zyngier and G. Watson, 55–74. London: Palgrave-Macmillan.

Hanauer, David I., Mark J. Graham, SEA-PHAGES, et al. 2017. "An Inclusive Research Education Community (iREC): Impact of the SEA PHAGES Program on Research Outcomes and Student Learning." *Proceedings of the National Academy of Sciences of the United States of America* 114: 13531–36.

Hopfer, Suellen 2012. "Effects of a Narrative HPV Vaccination Intervention Aimed at Reaching College Women: A Randomized Controlled Trial." *Prevention Science* 13 (2): 173–82.

Jacobs, Arthur M. 2015a. "Neurocognitive Poetics: Methods and Modals for Investigating the Neuronal and Cognitive-Affective Bases of Literature Reception." *Frontiers in Human Neuroscience* 9: 1–22.

Jacobs, Arthur M. 2015b. "Towards a Neurocognitive Poetics Model of Literary Reading." In *Towards a Cognitive Neuroscience of Natural Language Use,* edited by R. Willems, 135–95. Cambridge: Cambridge University Press.

Klassen, Stephen. 2010. "The Relation of Story Structure to a Model of Conceptual Change in Science Learning." *Science Education* 19 (3): 305–17.

Kruglanski, Arie W., Erica Molinario, and Edward P. Lemay. 2021. "Coping with Covid-19-Induced Threats to Self." *Group Processes and Intergroup Relations* 24: 284–89.

Levy, David L. 2020. "Covid-19 and Global Governance." *Journal of Management Studies* 58 (2): 562–66.

Marsh, Elizabeth J., Michelle L. Meade, and Henry L. Roediger. 2003. "Learning Facts from Fiction." *Journal of Memory and Language* 49 (4): 519–36.

Martine, George, and José Eustáquio Diniz Alves. 2019. "Disarray in Global Governance and Climate Change Chaos." *Revista Brasileira de Estudos de População* 36: 1–30.

Meade, Natalie 2018. "Trump's Cuts in Climate-Change Research Spark a Global Scramble for Funds." *New Yorker*, July 7, 2018.

Moyer-Guse, Emily. 2008. "Toward a Theory of Entertainment Persuasion: Explaining the Persuasive Effects of Entertainment-Education Messages." *Communication Theory* 18: 407–25.

Naeem, Saman Bin, Rubina Bhatti, and Aqsa Khan. 2021. "An Exploration of How Fake News Is Taking Over Social Media and Putting Public Health at Risk." *Health Information Library Journal* 38: 143–49.

Nan, Xiaoli, Michael F. Dahlstrom, Adam Richards, and Sarani Rangarajan. 2015. "Influence of Evidence Type and Narrative Type on HPV Risk Perception and Intention to Obtain the HPV Vaccine." *Health Communication* 30 (3): 301–8.

Norris, Pippa, and Ronald Inglehart. 2019. *Cultural Backlash: Trump, Brexit, and Authoritarian Populism.* Cambridge: Cambridge University Press.

Pian, Wenjing, Jianxing Chi, and Feicheng Ma. 2021. "The Causes, Impacts and Countermeasures of Covid-19 'Infodemic': A Systematic Review Using Narrative Synthesis." *Information Processing and Management* 58: 102713.

Romer, Daniel, and Kathleen Hall Jamieson. 2020. "Conspiracy Theories as Barriers to Controlling the Spread of Covid-19 in the U.S." *Social Science and Medicine* 263: article 11335.

Schank, Roger C., and Robert Abelson. 1995. "Knowledge and Memory: The Real Story." In *Knowledge and Memory: The Real Story*, edited by Robert S. Wyer, 1–86. Hilldale, N.J.: Lawrence Erlbaum.

Schneider-Mayerson, Matthew. 2020. "'Just as in the Book'? The Influence of Literature on Readers' Awareness of Climate Injustice and Perception of Climate Migrants." *ISLE: Interdisciplinary Studies in Literature and Environment* 27 (2): 337–64.

Slater, Michael D., David B. Buller, Emily Waters, Margarita Archibeque, and Michelle LeBlanc. 2003. "A Test of Conversational and Testimonial Messages versus Didactic Presentations of Nutrition Information." *Journal of Nutrition Education and Behavior* 35: 255–59.

Slater, Michael D., and Donna Rouner. 2002. "Entertainment–Education and Elaboration Likelihood: Understanding the Processing of Narrative Persuasion." *Communication Theory* 12 (2): 173–91.

Slater, Michael D., David Rouner, and Marilee Long. 2006. "Television Dramas and Support for Controversial Public Policies: Effects and Mechanisms." *Journal of Communication* 56: 235–52.

Uscinski, Joseph E., Casey Klofstad, and Matthew D. Atkinson. 2016. "What Drives Conspiratorial Beliefs? The Role of Informational Cues and Predispositions." *Political Research Quarterly* 69: 57–71.

Van Prooijen, Jan-Willem, Karen M. Douglas, and Clara de Inocencio. 2018. "Connecting the Dots: Pattern Perception Predicts Belief in Conspiracies and the Supernatural." *European Journal of Social Psychology* 48: 320–35.

Wilk, Richard. 2017. "Without Consumer Culture, There Is No Environmental Crisis." Panel contribution to Population-Environment Research Network Cyberseminar, "Culture, Beliefs, and the Environment," May 15–19, 2017. https://www.populationenvironmentresearch.org/pern_files/statements/PERN_Cyberseminar2017_Wilk.pdf.

Winterbottom, Anna, Hilary L. Bekker, Mark Conner, and Andrew Mooney. 2008. "Does Narrative Information Bias Individual's Decision Making? A Systematic Review." *Social Science and Medicine* 67 (12): 2079–88.

Wise, Meg, Jeong Y. Han, Bret Shaw, Fiona McTavish, and David H. Gustafson. 2008. "Effects of Using Online Narrative and Didactic Information on Healthcare Participation for Breast Cancer Patients." *Patient Education and Counseling* 70 (3): 348–56.

Woods, Eeric Taylor, Robert Schertzer, Liah Greenfeld, Chris Hughes, and Cynthia Miller-Idriss. 2020. "Covid-19, Nationalism, and the Politics of Crisis: A Scholarly Exchange." *Nations and Nationalism* 26: 807–25.

Yoder, P. Stanley, Robert Hornik, and Ben C. Chirwa. 1996. "Evaluating the Program Effects of a Radio Drama about AIDS in Zambia." *Studies in Family Planning* 27: 188–203.

Zabrucky, Karen M., and DeWayne Moore. 1999. "Influence of Text Genre on Adults' Monitoring of Understanding and Recall." *Educational Gerontology* 25 (8): 691–710.

Zarocostas, John. 2020. "How to Fight an Infodemic." *Lancet* 395: 676.

Chapter 13

Stories about the Environment for Diverse Audiences

Insights from Environmental Communication

HELENA BILANDZIC

Audiences form their views about the environment through a number of sources, some involving direct contact with nature and the environment, but many requiring a facilitation by different types of media, be they traditional news articles, social media, fictional books, or films. This indirect information and experience through media is necessary and relevant, as many ecological issues evolve in a way undetected by normal human senses—for example, chemical pollutants in water, radioactivity, or the rise of global temperature over a long time span. Environmental issues also concern remote locations, which can only be experienced through extensive travel—or through media. Apart from facts, media may also convey deeper understanding of the causes and consequences of environmental issues. For example, the popular Netflix nature documentary series *Our Planet* (2019) shows the interconnectedness of oceans, forests, deserts, species diversity, and climate change; it particularly focuses on human impact. Fictional examples include the novel *Gray Mountain* by John Grisham (2014), which deals with the environmental and health dangers of strip-mining coal, or *Heat and Light* by Jennifer Haigh (2016), which revolves around the serious environmental and health issues caused by fracking. Apart from providing facts and background information, such media can stimulate a sense of relevance and urgency as well as establish emotional ties to the issue—for example, by showing how characters in a novel are affected by environmental issues in their lives and livelihood.

Environmental communication is concerned with such media presentations of environmental topics and the effects they have on audiences. It represents a branch of communication studies that "includes research and practices regarding how different actors (e.g., institutions, states, people) interact with regard to topics related to the environment and how cultural products influence society toward environmental issues" (Antonopoulos and Karyotakis 2020, 551).

A notable point of convergence between communication research and empirical ecocriticism is the exploration of the role of narratives. In communication research, the field of narrative persuasion has gained traction in the past two decades, researching the effects of narratives on beliefs and attitudes, and, most important, the mechanisms and conditions of these effects (Bilandzic and Busselle 2013; Braddock and Dillard 2016; Green et al. 2019). In environmental communication, narratives have also garnered considerable attention as a communication mode that creates relevance and urgency for the environment in a wide audience (Seelig 2019).

Fictional narratives (as opposed to news stories, documentaries, and the like) are relatively less researched, but they nonetheless constitute an important part of the audience's sources for knowledge of the environment (Bilandzic and Kalch 2021a). Notably, environmental topics have become widespread in written and filmic narrative (Christensen et al. 2018; Seelig 2019). Indeed, this book is concerned with narratives that report about, analyze, and reflect on the environment—for example, the chapters in this volume by Schneider-Mayerson and colleagues, Myrick and Oliver, and Weik von Mossner and colleagues.

Human knowledge is a patchwork derived from sources as different as news, documentaries, novels, and films. At face value, we should expect audiences to disregard sources that they deem inappropriate for the belief or judgment at hand. Normally we should not use information from a fictional film about fracking to make real-world judgments. However, people do this all the time; they use information from fictional, unreliable, and untrustworthy sources for real-world beliefs and attitudes. One way to explain this phenomenon is the sleeper effect (Kumkale and Albarracin 2004), which assumes that the actual information and the source of the information do not remain equally long or equally accessible in memory.

Instead, the information itself is more persistent than the source tag. As a result, while at first people do discount unreliable and inappropriate sources from their judgments, with time, the source information decays or becomes less accessible, and the initially unreliable and inappropriate information gets incorporated into real-world judgments (Kumkale and Albarracin 2004). This is the reason why long-term exposure to fiction generates effects on real-world beliefs (as argued in cultivation research; Shrum 1995, 1997). For narratives, there is an additional phenomenon not related to long-term processes. Research has found that effects of narratives do not differ if a story is labeled as fact or fiction (Green and Brock 2000; Green et al. 2006; Hartung et al. 2017). Malecki (this volume) confirms this finding using an environmental topic. In a re-view of narrative effects, Green and colleagues (2019) conclude that when creating persuasive narratives, the quality of the narrative—in terms of its realism as well as its ability to transport and create strong emotions—is more important than presenting the frame as fact or fiction. Therefore, although it does not seem to matter whether a narrative claims that the events have actually happened (factual frame), it does matter whether the narrative events could happen or are likely to happen in actual life (realism).

From an ecocritical view, this has a good side and a bad side for the role of narratives. On the one hand, it means that environmental fiction has a solid chance to improve societal awareness of and ability to respond to contemporary environmental challenges. On the other hand, it depends on the authenticity of a fictional text, whether it heightens the understanding and recognition of environmental issues or creates skepticism. Fiction does not pretend to assert truth—nor is it expected to (Busselle and Bilandzic 2008). Depending on the plot and (implicit) message, climate fiction can increase environmental beliefs and attitudes, as Schneider-Mayerson, Gustafson, and colleagues (this volume) have found using two different climate fiction stories. Of course, should the story line be a skeptical one, narrative persuasion predicts effects in the skeptical direction.

In the following sections, I will reinforce three aspects implied in empirical ecocriticism and use them to provide insights from an environmental communication perspective: exploring the aesthetic and the formal, dealing with persuasive intent, and making good use of the fictional frame.

The Exploration of Aesthetic and Formal Features

Schneider-Mayerson, Weik von Mossner, Malecki, and Hakemulder (this volume) point out a limitation in environmental communication: that scholars in this field "have rarely been concerned with the formal and aesthetic features of the texts they have studied." Indeed, the focus is almost always on the content of environmental texts—for example, frames (Bilandzic, Kalch, and Soentgen 2017), geographic distance (Huang and Ells 2021), or kinship appeals between humans and nonhumans (Morris and Qirko 2020). The reason for this focus is presumably that most of these studies use texts from mass media and news, social media, or campaigns that often are formally straightforward and make less use of stylistic devices compared to fiction. Accordingly, mass-produced media fare is not consumed in a reflective and appreciative way by the audience. The necessity of considering formal features therefore does not seem to be urgent, at least most of the time. However, when we are considering fiction, the matter becomes more interesting—and perhaps urgent after all. A good example is the German film *Hell* (2011), which has been analyzed by at least one ecocritic (Weik von Mossner 2012) and tested by communication scholars for its ability to change audience attitudes on climate change (Bilandzic and Sukalla 2019). In a detailed analysis of narrative engagement, Freya Sukalla and I found that it was narrative presence—the feeling of being inside in the narrative world of the film rather than the actual world—that was the most important dimension of narrative engagement to statistically explain whether respondents felt guilt after viewing the film and behavioral intentions to take concrete actions to reduce their carbon footprint (see supplemental material in Bilandzic and Sukalla 2019). Although other studies have shown that emotional engagement was more important (Busselle and Bilandzic 2009; de Graaf et al. 2009), this film was effective through the experience of being in the film. We noted that this result makes sense "considering the film's extensive emphasis of visuals of the glistening and blinding sun" (Bilandzic and Sukalla 2019, supplemental material, 2). This echoes a central proposition of Alexa Weik von Mossner's (2012) analysis, which starts from the title, *Hell:* "At first glance, even German viewers tend to attribute the rather gruesome English meaning to the title, but director Fehlberg has explained that the German meaning of the word hell—'bright'—is what he originally had in mind. Indeed, the two meanings turn out to be closely related, since the unbearable brightness of the sun in the film makes being outside a hellish experience. The sunlight is now so strong that it burns everything on the surface of the planet" (44). The formal features of the film (bright colors and overexposure) have tilted the narrative experience toward a strong visual impression, which then shaped the effects.

This case is a lucky coincidence of two groups of scholars working on the same material independently. In most cases, communication scholars are not that lucky to benefit from coincidence. However, let us say that we learned from this lucky coincidence and decided to better coordinate our efforts. There still would be a major obstacle to overcome. In the example of *Hell,* an atypical result could be explained by a film analysis. However, causal evidence does not prove that the film's visuals were responsible for the effects. Perhaps it was the close-ups, which increased narrative presence; perhaps it was the stark coherence of the narrative world, or the familiarity of a German audience with a German setting. To provide firm evidence, we would need to conduct an experiment that compares the original version of the film to a version with different visuals—that is, no bright colors and overexposure (on experimental designs for environmental research, see Malecki, this volume). Needless to say, this is not an ordinary experimental manipulation but the production of a new film—something out of the scope for regular academic folks, and almost certainly out of the scope for a research budget, especially if we want to use whole films rather than clips. The usual tricks of the trade—cut a bit of film here and there—does not work with aesthetic and formal features.

The situation is a bit different for written narratives, where rewriting is not impossible; on the contrary, it is quite common in communication research. A good example is the study presented by Jessica Myrick and Mary Beth Oliver in this volume on the use of exemplar voice in news stories—that is, direct quotes from people concerned by unsafe drinking water. They compared one version of a news story, which only used an exemplar voice once, at the beginning (a typical news structure), to another news story that used exemplar voice as the dominant sources in the news item. The factual content was held constant across conditions. This is a clean, valid way to use the same content while changing formal features. Any differences between the experimental groups can be attributed to the formal feature, absent any a priori differences between the groups.

Experimentally researching formal and aesthetic features of (especially) filmic environmental narratives remains a challenge. In some cases, we can certainly modify formal features in a film (e.g., change the music); in other cases, it is possible to modify by prior instruction to participants (e.g., information that a film is entirely fictional or is based on facts). In yet other cases, the formal features permeate the film and cannot be extricated (e.g., the visuals, the genre).

Dealing with Persuasive Intent in an Environmental Narrative

The environment is a politically charged issue. One meta-analysis found that the more respondents leaned toward the liberal end of the political spectrum, the more environmental concern they reported (Cruz 2017). The divide is particularly strong for climate change, where left/liberal citizens are more concerned about climate change and support action to a greater extent than right/conservative citizens, a result found consistently across different countries (McCright and Dunlap 2011; McCright, Dunlap, and Marquart-Pyatt 2016). In this situation, openly partisan or persuasive texts may evoke resistance to the message (Chinn and Hart 2021; Ma, Dixon, and Hmielowski 2019). Fictional narratives offer ample opportunities to communicate with audiences that are indifferent to environmental issues or even opposed to environmental protection. Audiences expect to be entertained, not educated, by fiction, which may make a difference for reaching resistant audiences. In addition, the pace of fictional influence may be slow and gradual, possibly through repeated exposures to similar narratives, which can be more acceptable for potentially resistant audiences. Climate change skeptics will not choose to watch a film on climate change, but they will choose to watch an action thriller about a father and son separated by an ice storm freezing New York, with lots of fascinating special effects and a suspenseful plot. As we know, *The Day After Tomorrow* (2004) was an immense success. To be successful with a "difficult" audience, a narrative should not foreground the environment; rather, it should use the environment as a backdrop for an intriguing plot. Spreading environmental issues across various, everyday entertaining narratives may be a mainstreaming approach to environmental communication: a bit about recycling on the Netflix comedy series *Modern Family* (2009–20), a bit about aggressive urbanization in the animated film *Over the Hedge* (2006), a bit about industrial livestock farming in the horror film *The Bay* (2012). The *Chicago Review of Books,* for example, lists nine "Novels To Give Your Favorite Climate Change Skeptic This Holiday Season," claiming the books "might also help to broach the subject of climate change in a non-threatening way" (Brady 2019). Such narratives can be effective with resistant audiences.

Conversely, audiences that are open to environmental issues but need a final push to action can tolerate (or may need) stronger stimuli, like radical or activist texts (Weik von Mossner 2021; Weik von Mossner et al., this volume). For each subaudience, addressed at its appropriate level (strategic communication would speak of "tailoring" or "targeting" messages), these narratives can stimulate thoughts and discussions about the environment and create an openness for future exposures to other environmental information. For example, Matthew Schneider-Mayerson (2018) reports that readers of climate fiction often discuss the books with their friends and family. Although an overtly environmentally activist book may not serve this purpose with a skeptical audience, a plot that is not primarily concerned with an environmental issue, but that is set against the backdrop of one, may be a promising strategy to reach skeptical audiences.

Making Good Use of the Fictional Frame

W. P. Malecki (this volume) argues that a fictional frame facilitates thinking about difficult topics, as audiences can allow more and more extreme feelings which would be too "gruesome or too morally demanding to watch." He argues that such unpleasant negative reactions may inhibit exposure and engagement with a narrative, ultimately preventing people from taking action, and "a fictional frame might neutralize such reactions." Indeed, entertainment media can provide safe spaces to try out emotions and moral stances, simulating social situations that would be otherwise unavailable to the viewer or reader. In fiction, people do not run the risks that are usually associated with these feelings and thoughts in actual life, such as assuming responsibility to act or alienate one's peers (who may be anti-vegan, anti–climate protection, anti-environment). Fictional books and films have become both more available and popular in the past years, reaching audiences that rarely seek out environmental topics (Bilandzic and Kalch 2021a).

Media entertainment research has demonstrated that emotionally moving media content and experiences can stimulate reflection in audiences (Bartsch 2012) and provide meaningful experiences (Oliver et al. 2018). However, these processes do not automatically lead to a change in attitudes and behaviors. Rather, in search of effects, we need to keep in mind the whole transitional process that people go through when they change a belief or worldview that is deeply ingrained in their lives. The focus shifts from looking for short-term effects of environmental media toward gradual, slow, long-term processes that happen in almost imperceptible steps, possibly with long periods with no change. For example, consider a viewer of *The Colony* (2013), a science fiction film set in a future in which humans have tried to control global warming but instead caused a new ice age. This viewer gets interested in the topic of climate change and seeks to answer pressing questions like, is it really that bad? What can I do? Next, the viewer watches the documentary *Tomorrow* (2015), which is concerned with strategies to manage and create a sustainable future. After watching the documentary, the viewer may choose to ride a bicycle rather than drive a car on one day. Then, becoming more invested in the topic of sustainable consumption, the viewer turns to *Cowspiracy: The Sustainability Secret* (2014), a documentary on the environmental impacts of animal agriculture—something the viewer never would have watched without the somewhat gentler and more positive preparation by *Tomorrow.* This exposure may stimulate reflection about options of sustainable behavior and about implementing those in one's life. For example, this individual may—from time to time—incorporate a vegan meal into their diet. Friends may notice the changes in this person's behavior, and actually enjoy the vegan meal they were served at dinner. In this circle of friends, more people may become inspired to change, and the norms within this peer group may gradually shift. People within this group may post their new habits on social media, reaching a wider audience.

Fiction may not evoke sudden, dramatic change, but it may well initiate a series of smaller changes that include other, factual media texts as well as personal contacts.

Conclusions

Fictional media can facilitate pro-environmental beliefs and thoughts by providing a nonthreatening, fictional frame without persuasive intent; aesthetic features may carry and support a narrative's message. All three message properties—fictionality, no persuasive intent, and aesthetic features—can build up mental images of the environment and specific issues, emotions, and a sense of urgency. However, what ultimately helps the environment is *action*. The transition from pro-environmental beliefs, attitudes, and feelings to actions is difficult. In concrete situations, there are obstacles to realizing a pro-environmental behavior, such as time pressures, convenience, and resource restrictions. This does not necessarily feel at odds with a general pro-environmental worldview, but it is perceived as small exceptions in specific circumstances. The problem is that the small exceptions in specific circumstances add up to an overall behavior that is not sustainable. Models of environmental behavior include factors such as self-efficacy, or people's belief that they are capable of performing actions and implementing them in their lives (Bilandzic and Kalch 2021b). Self-efficacy is low, for example, if there is no infrastructure for recycling or public transport. Here, individual action is limited. Other domains of self-efficacy can be addressed on an individual level—for example, when dealing with situational boundaries such as motivation, time pressure, and convenience. Finally, what narratives can do is inform people about options to act in an environmentally friendly way. Alarmist texts may frighten and discourage people because they get the impression that their contribution does not matter. However, narratives can convey that individual contributions do matter, along with feelings of hope and the motivation to act.

Many studies focus on single media texts. However, in their daily lives, people use different media in different combinations. The media mix may contain a daily online newspaper, a fictional film, some social media content, and an environmental documentary. These sources may interact and reinforce or weaken each other; another option is that one exposure may create interest and motivate exposure to another text. Such differentiated processes are hard to track, but they may explain why individual texts are sometimes limited in their capacity to affect attitudes or behaviors. Qualitative approaches can reconstruct individual trajectories to sustainable life on all levels, including changing individual behaviors to reduce one's carbon footprint, supporting environmentally beneficial policies, or even engaging in political engagement and activism. Such studies could help inform us about the role of and interaction between different media to build up awareness and motivate environmental action.

Ultimately, scholarly insight into the relationship between narratives and the environment, of symbolic representations in a collective discourse and mental representations in people's minds, requires a multifaceted research approach. Regular citizens do not make strict distinctions between factual and fictional texts; they patchwork-source their knowledge on the environment from different sources. Accordingly, research from angles as different as environmental communication and empirical ecocriticism, and more specifically synergies between the research traditions, is needed to understand the full complexity of the relationship between humans and the environment.

References

Andersen, Kip, and Keegan Kuhn, dir. 2014. *Cowspiracy: The Sustainability Secret.* Santa Rosa, Calif.: AUM Films.

Antonopoulos, Nikos, and Minos-Athanasios Karyotakis. 2020. "Environmental Communication." In *The Sage International Encyclopedia of Mass Media and Society,* edited by Debra L. Merskin, 551–52. Thousand Oaks, Calif.: Sage.

Bartsch, Anne. 2012. "Emotional Gratification in Entertainment Experience: Why Viewers of Movies and Television Series Find It Rewarding to Experience Emotions." *Media Psychology* 15 (3): 267–302.

Bilandzic, Helena, and Rick W. Busselle. 2013. "Narrative Persuasion." In *The Sage Handbook of Persuasion: Developments in Theory and Practice,* edited by James Price Dillard and Lijiang Shen, 200–219. Los Angeles: Sage.

Bilandzic, Helena, and Anja Kalch. 2021a. "Fictional Narratives for Environmental Sustainability Communication." In *The Sustainability Communication Reader: A Reflective Compendium,* edited by Franzisca Weder, Larissa Krainer, and Matthias Karmasin, 123–42. New York: Springer.

Bilandzic, Helena, and Anja Kalch. 2021b. "Models of Attitudes, Intentions and Behaviors in Environmental Communication." In *Handbook of International Trends in Environmental Communication,* edited by Bruno Takahashi, Julia Metag, Jagadish Thaker, and Susan Evans Comfort. New York: Taylor & Francis/Routledge.

Bilandzic, Helena, Anja Kalch, and Jens Soentgen. 2017. "Effects of Goal Framing and Emotions on Perceived Threat and Willingness to Sacrifice for Climate Change." *Science Communication* 39 (4): 466–91.

Bilandzic, Helena, and Freya Sukalla. 2019. "The Role of Fictional Film Exposure and Narrative Engagement for Personal Norms, Guilt and Intentions to Protect the Climate." *Environmental Communication* (8): 1069–86.

Braddock, Kurt, and James Price Dillard. 2016. "Meta-analytic Evidence for the Persuasive Effect of Narratives on Beliefs, Attitudes, Intentions, and Behaviors." *Communication Monographs* 83 (4): 446–67.

Brady, Amy. 2019. "Novels to Give Your Favorite Climate Change Skeptic This Holiday Season." *Chicago Review of Books,* November 12, 2019. https://chireviewofbooks.com/2019/12/12/novels-to-give-your-favorite-climate-change-skeptic-this-holiday-season/.

Busselle, Rick W., and Helena Bilandzic. 2008. "Fictionality and Perceived Realism in Experiencing Stories: A Model of Narrative Comprehension and Engagement." *Communication Theory* 18: 255–80.

Busselle, Rick W., and Helena Bilandzic. 2009. "Measuring Narrative Engagement." *Media Psychology* 12: 321–47.

Chinn, Sedona, and P. Sol Hart. 2021. "Climate Change Consensus Messages Cause Reactance." *Environmental Communication.* https://doi.org/10.1080/17524032.2021.1910530.

Christensen, Miyase, Anna Åberg, Susanna Lidström, and Katarina Larsen. 2018. "Environmental Themes in Popular Narratives." *Environmental Communication* 12 (1): 1–6.

The Colony. 2013. Film. Directed by Jeff Renfroe. Toronto: Alcina Pictures.

Cruz, S. M. 2017. "The Relationships of Political Ideology and Party Affiliation with Environmental Concern: A Meta-analysis." *Journal of Environmental Psychology* 53: 81–91.

The Day After Tomorrow. 2004. Film. Directed by Roland Emmerich. Los Angeles: 20th Century Fox.

de Graaf, Anneke, Hans Hoeken, José Sanders, and Hans Beentjes. 2009. "The Role of Dimensions of Narrative Engagement in Narrative Persuasion." *Communications—European Journal of Communication Research* 34 (4): 385–405.

Green, Melanie C., Helena Bilandzic, Kaitlin Fitzgerald, and Elaine Paravati. 2019. "Narrative Effects." In *Media Effects: Advances in Theory and Research,* edited by Mary Beth Oliver, Arthur A. Raney, and Jennings Bryant, 130–45. New York: Routledge.

Green, Melanie C., and Timothy C. Brock. 2000. "The Role of Transportation in the Persuasiveness of Public Narratives." *Journal of Personality and Social Psychology* 79 (5): 701–21.

Green, Melanie C., Jennifer Garst, Timothy C. Brock, and Sungeun Chung. 2006. "Fact versus Fiction Labeling: Persuasion Parity Despite Heightened Scrutiny of Fact." *Media Psychology* 8 (3): 267–85.

Grisham, John. 2014. *Gray Mountain.* New York: Doubleday.

Haigh, Jennifer. 2016. *Heat and Light.* New York: Ecco.

Hartung, Franziska, Peter Withers, Peter Hagoort, and Roel M. Willems. 2017. "When Fiction Is Just as Real as Fact: No Differences in Reading Behavior between Stories Believed to Be Based on True or Fictional Events." *Frontiers in Psychology* 8: 1618.

Hell. 2011. Film. Directed by Tim Fehlbaum. Munich: Caligari Film- und Fernsehproduktions.

Huang, Jialing, and Kevin Ells. 2021. "Risk Here vs. Risk There: Intention to Seek Information about Gulf Coastal Erosion." *Environmental Communication* 15 (3): 386–400.

Johnson, Tim, and Karey Kirkpatrick, dirs. 2006. *Over the Hedge*. Glendale, Calif.: DreamWorks Animation.

Kumkale, G. Tarcan, and Dolores Albarracin. 2004. "The Sleeper Effect in Persuasion: A Meta-analytic Review." *Psychological Bulletin* 130 (1): 143–72.

Levinson, Barry, dir. 2012. *The Bay*. Culver City, Calif.: Automatic Entertainment.

Ma, Yanni N., Graham Dixon, and Jay D. Hmielowski. 2019. "Psychological Reactance from Reading Basic Facts on Climate Change: The Role of Prior Views and Political Identification." *Environmental Communication* 13 (1): 71–86.

McCright, Aaron M., and Riley E. Dunlap. 2011. "The Politicization of Climate Change and Polarization in the American Public's Views of Global Warming, 2001–2010." *Sociological Quarterly* 52 (2): 155–94.

McCright, Aaron M., Riley E. Dunlap, and Sandra T. Marquart-Pyatt. 2016. "Political Ideology and Views about Climate Change in the European Union." *Environmental Politics* 25 (2): 338–58.

Modern Family. 2009–20. Created by Steven Levitan and Christopher Lloyd. Los Angeles: 20th Century Fox Television.

Morris, David S., and Hector N. Qirko. 2020. "Saving 'Little Sister': A Test of the Effectiveness of Kinship Appeals in Conservation Marketing." *Environmental Communication* 14 (4): 481–91.

Oliver, Mary Beth, Arthur A. Raney, Michael D. Slater, et al. 2018. "Self-Transcendent Media Experiences: Taking Meaningful Media to a Higher Level." *Journal of Communication* 68 (2): 380–89.

Our Planet. 2019. Film. Produced by Alastair Fothergill and Keith Scholey. Bristol, U.K.: Silverback Films.

Schneider-Mayerson, Matthew. 2018. "The Influence of Climate Fiction: An Empirical Survey of Readers." *Environmental Humanities* 10 (2): 473–500.

Seelig, Michelle. 2019. "Popularizing the Environment in Modern Media." *Communication Review* 22 (1): 45–83.

Shrum, L. J. 1995. "Assessing the Social Influence of Television: A Social Cognition Perspective on Cultivation Effects." *Communication Research* 22 (4): 402–29.

Shrum, L. J. 1997. "The Role of Source Confusion in Cultivation Effects May Depend on Processing Strategy—A Comment on Mares (1996)." *Human Communication Research* 24 (2): 349–58.

Tomorrow. 2015. Documentary film. Directed by Cyril Dion and Mélanie Laurent. Paris: Move Movie.

Weik von Mossner, Alexa. 2012. "Visceralizing Ecocide in Science Fiction Films: The Road and *Hell.*" *Ecozon@* 3 (2): 42–56.

Weik von Mossner, Alexa. 2021. "From Mindbombs to Firebombs: The Narrative Strategies of Radical Environmental Activism Documentaries." In "The Rhetoric of Ecology in Visual Culture," edited by Katarzyna Paszkiewicz and Anna Bendrat, special issue, *Res Rhetorica* 8 (2): 21–37.

Acknowledgments

Our great thanks go out to everyone who contributed to making this book possible.

To all our contributors, thank you for your enthusiasm, patience, and dedication. Working with you has been a pleasure, and we deeply appreciate the diligence of revisions, the promptness of submissions, and above all the quality of scholarship.

We would like to thank the people who served as peer reviewers for the case studies or read other parts of this book: Stephan Dickert, Karin Fikkers, Berenike Herrmann, Tom Idema, Matthias Klestil, Jennifer Ladino, Cameron Leader-Picone, Eggo Müller, Philip Taylor, and Loris Vezzali.

We are indebted to Douglas Armato, director of the University of Minnesota Press, who recognized the timeliness and value of this project and supported it at an early stage of development, and to the entire team at the University of Minnesota Press. Two anonymous readers provided insightful comments and suggestions for revisions.

We are especially grateful to the two directors of the Rachel Carson Center for Environment and Society, Christof Mauch and Helmuth Trischler, for their generous support of the "Empirical Ecocriticism" workshop, which we were able to organize at the RCC in Munich in December 2018, and which generated many of the ideas that continue to inform the field. In this context, we would also like to thank the people who attended the workshop—among them Jan Alber, Pat Brereton, Marco Caracciolo, Judith Eckenhoff,

Greg Garrard, Harriet Hawkins, Salma Monani, Meryl Shriver-Rice, and Eline Tabak—for helping us think through and critically interrogate the project of empirical ecocriticism, heightening our awareness of its potential and challenges.

Special thanks go to people who have helped us build empirical ecocriticism in other ways. Above all that is Scott Slovic, who also attended the workshop in Munich and who has done so much to help empirical ecocriticism develop and flourish, from inviting us to organize a special thematic cluster in *ISLE: Interdisciplinary Studies in Literature and Environment* to helping us find partners for interdisciplinary experiments. In this vein, we also thank Don Kuiken and Moniek Kuijpers, presidents of the Society for the Empirical Study of Literature, for their encouragement to form a research coalition within that organization.

We thank the institutions and venues that have invited us to give talks about empirical ecocriticism, along with participants in the ensuing conversations. These include the Association for the Study of Literature Conference at the University of California, Davis, USA; the Yale Environmental Humanities Initiative at Yale University, USA; the Environmental Humanities Initiative at the University of Minnesota, USA; Hebei Normal University, China; Penn State University, USA; the Intermedial Ecocriticism Conference at the University of Babeş-Bolyai, Romania; the Greenhouse Environmental Humanities Series at the University of Stavanger, Norway; the PhD seminar on "Ecocritical Theory: Literature, Culture and the Environment" at the University of Agder, Kristiansand, Norway; the Environmental Science Center at the University of Augsburg, Germany; the NARMESH Project at Ghent University, Belgium; two European Association for the Study of Literature, Culture, and Environment webinars; the annual conference of the Association for Media Studies (GfM) in Innsbruck, Austria; the Transatlantic Conversations series on "Science, Technology, and Innovation for a Sustainable Future" organized by the Austrian Studies Program at the University of California, Berkeley, USA; the Environmental Humanities Lecture Series at the University of Konstanz, Germany; the "Understanding Risk Asia" conference in Singapore; the biennial conference of the European Association for the Study of Literature, Culture, and Environment in Granada, Spain; RWTH Aachen University, Germany; and Utrecht University, Netherlands.

Special thanks go to the individuals who organized these talks and conversations: Paul Sabin, Daniel J. Philippon, Yaqin Wang, Eduardo Mendieta, Jørgen Bruhn, Doru Pop, Dolly Jørgensen, Reinhard Hennig, Hubert Zapf, Marco Caracciolo, Nikoleta Zampaki, Julia Ditter, Petra Massomelius, Timo Müller, Margarita Carretero González, Jan Alber, Jos van Berkum, and Liesbeth van de Grift.

Empirical research often depends on funding availability. We would therefore like to thank the institutions that provided grants to do the empirical ecocritical research that led to this book. In particular, we received support from the Austrian Science Fund (P31189-G30), the National Science Foundation of Poland (2012/07/B/HS2/02278 and 43/B/HS2/03268), the Research Council of the University of Klagenfurt, and Yale–NUS College (IG17-LR104).

We acknowledge that some of the editing of this book was carried out on the traditional territory of the Orang Laut, and the unceded territory of the Lenape, Wappinger, Paugusset, Schaghticoke, and Wabanaki peoples.

Finally, in addition to human beings, communities, and institutions, we are grateful for the nonhuman beings that make our work possible. Thanks to the nematodes for recycling soil nutrients to enable agriculture; to the bees, butterflies, bats, and birds that pollinate so many of the plants that we eat; to the trees and the phytoplankton that produce the oxygen we breathe every day; to the bacteria and other microbes that comprise over half of the cells in our bodies; and to all the other nonhuman beings that constitute the endless network of life from which this book, and all other human endeavors, developed.

Contributors

MATTHEW BALLEW is research affiliate at the Yale Program on Climate Change Community of Yale University.

HELENA BILANDZIC is professor of communication research at the University of Augsburg, Germany.

REBECCA DIRKSEN is Laura Boulton professor of ethnomusicology and associate professor in the department of folklore and ethnomusicology at Indiana University. She is author of *After the Dance, the Drums Are Heavy: Carnival, Politics, and Musical Engagement in Haiti* and coeditor of *Performing Environmentalisms: Expressive Culture and Ecological Change.*

GREG GARRARD is professor of environmental humanities and associate dean of research in the faculty of creative and critical studies, University of British Columbia Okanagan. He is author of *Ecocriticism,* coauthor of *Climate Change Skepticism: A Transnational Ecocritical Analysis,* and editor of *The Oxford Handbook of Ecocriticism.*

MATTHEW H. GOLDBERG is associate research scientist at the Yale Program on Climate Change Communication at Yale University.

ABEL GUSTAFSON is assistant professor of public relations and environmental communication in the department of communication at the University of Cincinnati.

FRANK HAKEMULDER is associate professor of liberal arts and sciences at Utrecht University and affiliated professor at the Reading Center at the University of Stavanger. He is author of *The Moral Laboratory.*

DAVID I. HANAUER is professor of applied linguistics at Indiana University of Pennsylvania and the lead assessment coordinator of the SEA-PHAGES program in the department of biological sciences at the University of Pittsburgh. He is author of eight books, including *Poetry as Research* and *Scientific Writing in a Second Language.*

URSULA K. HEISE teaches in the department of English and at the Institute of the Environment and Sustainability at UCLA. Her books include *Chronoschisms: Time, Narrative, and Postmodernism; Sense of Place and Sense of Planet: The Environmental Imagination of the Global;* and *Imagining Extinction: The Cultural Meanings of Endangered Species.* Heise is editor of *Futures of Comparative Literature: The ACLA Report on the State of the Discipline* and coeditor of *The Routledge Companion to the Environmental Humanities.*

JEREMY JIMENEZ is associate professor in the foundations and social advocacy department at State University of New York Cortland.

ANTHONY LEISEROWITZ is founder and director of the Yale Program on Climate Change Communication and senior research scientist at the Yale School of the Environment.

W. P. MALECKI is university professor of literary theory at the University of Wrocław, Poland. His most recent books are *Human Minds and Animal Stories: How Narratives Make Us Care about Other Species* and the coedited collection *What Can We Hope For?*

DAVID M. MARKOWITZ is assistant professor in the school of journalism and communication at the University of Oregon.

MARCUS MAYORGA works as a behavioral scientist in industry.

JESSICA GALL MYRICK is associate professor of media studies in the Donald P. Bellisario College of Communications at Penn State University.

MARY BETH OLIVER is Donald P. Bellisario Professor of Media Studies at Penn State Univeristy in the department of film/video and media studies. She is coeditor of several books, including *Media Effects: Advances in Theory and Research.*

YAN PANG is visiting assistant professor of music at Point Park University. She is coauthor of *Cool Math for Hot Music, All About Music, Basic Music Technology,* and *The Future of Music.*

MARK PEDELTY is professor of communication studies and anthropology at the University of Minnesota and fellow at the Institute on the Environment. He is author of several books, including *Ecomusicology: Rock, Folk, and the Environment* and *A Song to Save the Salish Sea: Musical Performance as Environmental Activism.*

SETH A. ROSENTHAL is project director at the Yale Program on Climate Change Communication.

ELJA ROY is assistant professor in the department of communication and film at the University of Memphis.

MATTHEW SCHNEIDER-MAYERSON is associate professor of English and environmental studies at Colby College. He is author of *Peak Oil: Apocalyptic Environmentalism and Libertarian Political Culture,* coeditor of *An Ecotopian Lexicon* (Minnesota, 2019), and editor of *Eating Chilli Crab in the Anthropocene: Environmental Perspectives on Life in Singapore.*

NICOLAI SKIVEREN is principal investigator of the research project "Bridging the Gap: Qualitative Empirical Ecocriticism and the Impact of Environmental Narrative," funded by the Carlsberg Foundation.

PAUL SLOVIC is president of Decision Research and professor of psychology at the University of Oregon.

SCOTT SLOVIC is University Distinguished Professor of Environmental Humanities at the University of Idaho. He is coeditor of *Nature and Literary Studies* and *The Bloomsbury Handbook to Medical-Environmental Humanities.*

NICOLETTE SOPCAK is the qualitative research lead on the BETTER WISE project in the department of family medicine at the University of Alberta. She is a certified member of the Canadian Counselling and Psychotherapy Association (CCPA) and works as psychotherapist in private practice.

PAUL SOPCAK teaches philosophy and comparative literature at MacEwan University, Canada.

SARA WARNER is associate professor in the department of performing and media arts and director of LGBT studies at Cornell University. She is author of *Acts of Gaiety: LGBT Performance and the Politics of Pleasure.*

ALEXA WEIK VON MOSSNER is associate professor of American studies at the University of Klagenfurt, Austria. She is author of *Cosmopolitan Minds: Literature, Emotion, and the Transnational Imagination* and *Affective Ecologies: Empathy, Emotion, and Environmental Narrative;* editor of *Moving Environments: Affect, Emotion, Ecology, and Film;* and coeditor of *The Anticipation of Catastrophe: Environmental Risk in North American Literature and Culture* and *Ethnic American Literatures and Critical Race Narratology.*